Electron Crystallography of Organic Molecules

NATO ASI Series

Advanced Science Institutes Series

A Series presenting the results of activities sponsored by the NATO Science Committee, which aims at the dissemination of advanced scientific and technological knowledge, with a view to strengthening links between scientific communities.

The Series is published by an international board of publishers in conjunction with the NATO Scientific Affairs Division

A Life Sciences	Plenum Publishing Corporation
B Physics	London and New York
C Mathematical	Kluwer Academic Publishers
and Physical Sciences	Dordrecht, Boston and London
D Behavioural and Social Sciences	
E Applied Sciences	
F Computer and Systems Sciences	Springer-Verlag
G Ecological Sciences	Berlin, Heidelberg, New York, London,
H Cell Biology	Paris and Tokyo

Series C: Mathematical and Physical Sciences - Vol. 328

Electron Crystallography of Organic Molecules

edited by

John R. Fryer
Electron Microscope Centre,
University of Glasgow,
Glasgow, U.K.

and

Douglas L. Dorset
Electron Diffraction Department,
Medical Foundation of Buffalo,
Buffalo, NY, U.S.A

SPRINGER-SCIENCE+BUSINESS MEDIA, B.V.

Proceedings of the NATO Advanced Research Workshop on
'Electron Crystallography: The Application of Electron Microscopy to the Structural
Determination of Small Organic Molecules' and the 17th International School of
Crystallography
Erice, Sicily, Italy
April 22–28, 1990

Library of Congress Cataloging-in-Publication Data

```
Electron crystallography of organic molecules / editors, John R. Fryer
and Douglas L. Dorset.
      p.   cm. -- (NATO ASI series. Series C, Mathematical and
physical sciences ; vol. 328)
   "Proceedings of the NATO advanced research workshop held at Erice,
Sicily, 22nd-28th April 1990. This meeting was the 17th of the
International School of Crystallography."
   Includes index.
   ISBN 978-94-010-5447-8     ISBN 978-94-011-3278-7 (eBook)
   DOI 10.1007/978-94-011-3278-7
   1. Crystallography--Congresses.  2. Electron microscopy-
-Congresses.  3. Organic compounds--Congresses.   I. Fryer, J. R.
II. Dorset, Douglas L., 1942-   . III. International School of
Crystallography.  IV. Series: NATO ASI series.  Series C,
Mathematical and physical sciences ; no. 328.
QD906.7.E37E42   1990
548--dc20                                               90-20708
```

ISBN 978-94-010-5447-8

Printed on acid-free paper

CONTENTS

These proceedings are dedicated to the memory of Wolfgang Kunath

WOLFGANG KUNATH   1930-1990

OBITUARY

On 19th January 1990 our colleague and friend Wolfgang Kunath passed away at the age of 59, after patiently enduring a serious illness over the last three years. It was impressive and touching to see him continue his work up to the last days, fully aware of his fate yet positive towards life. Wolfgang was a hard worker. He pursued his scientific problems with stamina. More than forty publications are the result of his activity.Some of his works I want to point out:

First, his diploma thesis(under Ernst Ruska) in which he worked out both theoretically and experimentally an effective method for measuring the spherical aberration of electron microscopical objective lenses. It was typical of him not to be content merely with the measurement of spherical aberration but also to seek its correction. In his doctoral thesis(under Ernst Ruska and Friedrich Lenz), therefore, he investigated theoretically a three-lens system in which the spherical aberration is partially corrected for a hollow cone beam. Also our paper on the coma-free alignment of high resolution electron microscopes was a step forward in improving the optics of the electron microscope. Again, logically taking this paper as a base, Wolfgang studied the particularities of hollow-cone imaging. As a splendid result he suceeded in imaging heavy-atom clusters with atomic resolution. The experiments confirmed his theoretical findings.

With an electron microscope operating in a very low dose mode, the impacts of single electrons can be observed on the monitor of a TV camera. Wolfgang studied the statistics of these low-dose images. These studies paved the way for a method to reconstruct an image of the original molecule from a series of images of increasingly damaged molecules. This "play-back" method found practical application in the elucidation of the fine structure of glutamine synthetase.

Even in the last year of his life in which he increasingly suffered from his illness,he published another four papers, and from these the one on circular harmonic averaging should be particularly mentioned.

Wolfgang Kunath's clear and critical thinking was combined with a sense of modesty.Besides physics, he was interested in philosophy, as a compensation for these mental activities he went in for active sports-high seas sailing, wind-surfing, skiing and jogging. All these kinds of sports he enjoyed with inherent enthusiasm and intensity. Wolfgang spent thiry years of his life in the Department of Electron Microscopy of the Max-Planck-Gesellschaft in Berlin-Dahlem, first under Ernst Ruska and from 1977 on under Elmar Zeitler. His colleagues all liked him, and his early death had a great and sad impact. We lost a friend.

In   memorium   Fritz   Zemlin
                           Fritz-Haber-Institut
                           der Max-Planck-Gesellschaft

ELECTRON CRYSTALLOGRAPHY OF ORGANIC MOLECULES

There are many problems in structural research for which
samples cannot be prepared to adequate dimensions for single-crystal
x-ray diffraction measurements.   Electron scattering measurements,
to obtain either electron diffraction patterns or high resolution
images, have been used as a means to overcome the crystallite size
restriction, given the relative scattering cross-sections of matter
for electrons and x-rays.

Organic crystals are notoriously sensitive to beam damage by
the same electron source used to obtain the elastically scattered
signal; hence structural information is often only partially
retrieved before the specimen itself is destroyed.   That scientists
should concern themselves with such a masochistic activity reflects
the importance of gleaning any details about the molecular structure
from difficultly-crystallized materials, especially when details
about local defects are needed for the understanding of their
properties.

Because inorganic crystals are less susceptible to radiation
damage than are the organics, such materials have been employed to
furnish a sound theoretical basis for understanding the electron
scattering from thin crystal-line foils over the last few decades.
This basis includes the dynamical scattering process as well as the
effects of other perturbations to either electron diffraction
intensity data or the image itself, including distortions imposed by
the instrument.   More recently, despite the problems imposed by
inelastic scattering, the interpretation of electron scattering data
from organic crystals has become more quantitative, to the point where
one can now speak of "electron crystallography" in the sense that an
x-ray crystallographer would describe his art.   Practitioners of
electron crystallography interested in organic molecules have been
spread not only around the world but also across a variety of
disciplines, including polymer physics, membrane biophysics, the
physical chemistry of pigments and surfactants and even protein
crystallography.   The latter has the advantages of possessing large
crystallographic details, or large units of substructure so that the
function of the protein can be described without high resolution
structural information.   Hence the application of electron crystallography to protein has become well established, whilst scientists
studying small molecules, with high resolution examination, have been
isolated in their diverse applications.

It was thus suggested that these scientists might have a
meeting with fellow practitioners of the art in order to compare
their different approaches to quantitative analysis of their data.
We therefore initiated a forum for such discussion in form of a
workshop to establish guidelines by which the molecular structures
of organic compounds could be determined using the electron microscope.
The International School of Crystallography, directed by Professor

Blundell, agreed that the Ettore Majorana Centre for Scientific
Culture in Erice, Sicily would be an appropriate venue for this
workshop.   The prospect of interactions with a concurrent meeting
at the site on the uses of direct phasing methods in crystallography
was thought to be an added attraction, given recent interest in
applying such techniques to electron scattering data.   The
collaboration between the two meetings was arranged by Professor
Riva di Sanseverino and support for the electron crystallographic
workshop provided by NATO whose financial sponsorship was both
generous and flexible.

The workshop was planned to ensure the maximum involvement of
all participants.   Hence invitations were confined to experienced
scientists, all of whom presented papers, and it was requested that
these papers be concerned more with methods than specific structural
results.   The papers engendered quite lively discussion both during
and after their presentations, to the point where an observer from
the direct phasing meeting characterized the group as being somewhat
unruly.   His comment on this behaviour during the joint session with
the direct phasing school was explained as being characteristic of
scientists working in a difficult discipline who have to be critical
of all data and procedures, especially their own.   The final
session of the workshop was a discussion of each facet of the subject,
led by acknowledged experts in each area.   An attempt to reproduce
this discussion in the final chapter will hopefully serve as a guide-
line to those hoping to utilize this structural probe, giving insights
into the expectations of experienced users.

The objective of the workshop was to establish parameters,
within which electron crystallography could be applied with confidence
to organic structural problems.   This goal was achieved in most
convivial, stimulating atmosphere, marked by a remarkable spirit of
cooperativity by the participants.   The organizers are most
grateful to Dr. Paola Spadon for ensuring that the use of the site was
so well arranged that the cooperative spirit was well-advanced from
the very start.   We are also most grateful to Professor Lodovico
Riva di Sanseverino for his tireless efforts in ensuring that the
meeting would take place at all.

       John R.Fryer   Douglas L.Dorset

PARTICIPANTS AT ERICE 1990

Back row (L to R):J.R.Fryer,M.Sherman,J.Turner,D.Vesely,I.Peterson,
        A.Gavezzotti,G.Duckett,R.Wade,B.Lotz,W.Baumeister,J.Gimzewski,
        R.Scaringe,G.Miller,J.Lando,D.Martin,J.Wittmann,S.Hovmoller,
        J-F.Revol.

Front Row:M.Ashida,B.Jap,E.Orlova,F.Zemlin,F.H.Li,E.Zeitler,F.Brisse
        S.Perez,D.Dorset,S.Hui,I.Voigt-Martin,N.Uyeda

PARTICIPANTS

Professor M.Ashida,
Faculty of Engineering,
Kobe University,
Rokko,
Kobe 657,
Japan

Professor W.Baumeister,
Max-Planck-Institut fur Biochemie,
D-8033 Matinsried bei Munchen,
Federal Republic of Germany.

Dr F.Brisse,
Department of Chemistry,
Universite de Montreal,
CP6128 Succ.A,
Montreal,
Quebec H3C 3J7
Canada.

Dr D.L.Dorset
Medical Foundation of Buffalo,
Research Laboratories,
73 High St.,
Buffalo,
New York 14203,
USA.

Dr G.Duckett,
Research Laboratories,
British Petroleum Ltd.,
Sunbury-on-Thames,
Middlesex,
England.

Dr J.R.Fryer,
Electron Microscope Centre,
Chemistry Building,
University of Glasgow,
Glasgow G12 8QQ,
Scotland,U.K.

Professor A.Gavezzotti,
Department of Physical Chemistry,
Via Golgi 19,
20133 Milan,
Italy.

Dr J.K.Gimzewski,
IBM Zurich Research Laboratory,
CH-8803 Ruschlikon,
Switzerland.

Dr S.Hovmuller,
Department of Structural Chemistry,
Arrhenius Laboratory,
University of Stockholm,
S10691 Stockholm,

Dr S.W.Hui
Electron Optic Lab.,
Biophysics Dept.,
Roswell Park Memorial Institute,
666 Elm St.,
Buffalo N.Y.14263
USA

Dr Bing Jap,
Lawrence Berkeley Laboratory,
1 Cyclotron Road,
Berkeley,
California 94720,
USA.

Professor J.B.Lando,
Department of Macromolecular Science,
Case Western Reserve University,
Cleveland,Ohio 44106
USA.

Professor Li Fang Hua
Institute of Physics,
Academia Sinica,
Beijing 100080
China.

Dr B.Lotz,
CNRS Centre de Recherches sur les Macromolecules
6,rue Boussingault,
F67083 Strasbourg,
France.

Dr D.C.Martin
Central Research and Development,
E.I.du Pont de Nemours and Co.,
Experimental Station,
PO Box 80356,
Wilmington,DE 19880-0356,
USA.

Professor D.P.Miller,
Department of Physics and Astronomy,
College of Sciences,
Clemson University,
120 Kinard Laboratory of Physics,
Clemson,
South Carolina 29634-1901,
USA.

Mr G.Miller,
Electron Microscopy Centre,
Chemistry Building,
University of Glasgow,
Glasgow G12 8QQ,
Scotland UK.

Dr E.Orlova,
Institute of Crystallography,
Academy of Sciences of the USSR,
Leninsky pr.59,
Moscow 117333,
USSR.

Dr S.Perez,
Laboratoire de Physicochemie des Macromolecules,
INRA Nantes,
France

Dr I.Peterson,
Institut f.Physikalische Chemie,
Universitat Mainz,
Jacob Welder-Weg 11,
D-6500 Mainz,
W.Germany

Dr J-F.Revol,
Department of Chemistry,
Pulp and Paper Research Centre,
McGill University,
3420 University St.,
Montreal,Quebec H3A 2A7
Canada

Dr R.P.Scaringe
Chemistry Division
Eastman Kodak Research laboratories
Rochester,N.Y.14650
USA.

Dr M.Sherman,
Institute of Crystallography,
Academy of Sciences of the USSR,
Leninsky pr.59,
Moscow 117333,
USSR.

Dr J.Turner,
New York Department of Health,
Wandsworth Center,
Empire State Plaza,
PO Box 509,
Albany,
New York 12201-0509,
USA.

Professor N.Uyeda,
Institute for Chemical Research,
Kyoto University,
Uji,Kyoto-Fu 611
Japan

Dr D.Vesely
Department of Non-metallic Materials,
Brunel University,
Uxbridge,
Middlesex,
England.

Dr I.G.Voigt-Martin,
Institut f.Physikalische Chemie,
Universitat Mainz,
Jacob Welder-Weg 11,
D-6500 Mainz,
W.Germany

Dr R.H.Wade
Laboratoire de Biologie Structurale,
LBio/DRF-G,
Ceng 85X,
38041 Grenoble Cedex,
France

Dr J.Wittmann,
CNRS Centre de Recherches sur les Macromolecules
6,rue Boussingault,
F67083 Strasbourg,
France

Dr F.Zemlin
Fritz-Haber-Institut der Max Planck gesellschaft,
Faradayweg 4-6
D-1000
Berlin 33
W.Germany

Professor Dr E.Zeitler,
Fritz-Habe-Institut der Max-Planck-Gesellschaft,
4-6 Faradayweg,
D-1000 Berlin,
Germany.

ELECTRON DIFFRACTION STRUCTURE ANALYSIS OF ORGANIC CRYSTALS

Douglas L. Dorset

Medical Foundation of Buffalo, Inc.
73 High Street
Buffalo, New York 14203 U.S.A.

I.   ELECTRON DIFFRACTION STRUCTURE ANALYSIS

A.   Why Electron Diffraction?

Because matter scatters electrons far more efficiently than either x-rays or neutrons, electron diffraction techniques in some instances are preferable to more conventional crystallographic probes for elucidation of crystal structure. Vainshtein's[1] monograph describing the geometrical aspects of transmission electron diffraction is still a useful guide to the technique. More recent reviews[2,3] of the method have updated several points in light of current diffraction theory[4], including the treatment of intensity data for *ab initio* structure analysis.

Electron diffraction experiments are usually carried out in an electron microscope, which emphasizes the relationship between diffraction and imaging in any optical system. Nearly the same sample handling capabilities exist for electron diffraction as for x-ray or neutron diffraction. For example, eucentric goniometer stages which allow $\pm 60°$ tilt and 360° rotation of the specimen are standard accessories for electron microscopes. Heating and cooling stages are also available for the study of phase transitions, e.g. between $-170°C$. to 150°C. In special cases, environmental chambers exist for the study of fully solvated specimens[5].

B.   Advantages of Electron Diffraction

1. **Small wavelength.** Because the wavelength of e.g. a 100kV electron is some 40-fold smaller than that of a CuKα x-ray, the Ewald sphere can be approximated by a plane. This means that a virtually undistorted reciprocal net from a zonal projection can be photographed at one orientation of the specimen[6]. With sensitive photographic emulsions an electron diffraction pattern can be photographed within several seconds when a reasonably low radiation dose rate is used (e.g. 10^{-6} A/cm^2 or 6.2×10^{-4} e/Å^2-sec).

2. **High coherence.** For nearly parallel illumination with very low beam currents, the coherence of the electron beam is far greater than

1

J. R. Fryer and D. L. Dorset (eds.), Electron Crystallography of Organic Molecules, 1–10.
© 1990 *Kluwer Academic Publishers.*

for normal laboratory x-ray sources. For example, selected area techniques[7] can easily produce a coherence width of 10^4Å compared to 700 to 800Å expected from a rotating anode x-ray generator. This will aid in the visualization of superlattice structures which would not be detected by normal x-ray methods.

3. Small sample sizes. Normal selected area techniques utilize the electron microscope as an optical bench. Hence, an aperture can be inserted in the image plane of any lens to define the area of the specimen contributing to an electron diffraction pattern. Typical selected area diameters are in the range 1–10µm. Alternate methods for area selection exist which employ the illumination system to give a defocussed image of a condensor aperture, and allow even smaller specimen diameters to be defined. With the large scattering cross section of matter for electrons, a very thin (<100Å) microcrystal or a portion of it can be used as a specimen such that a single crystal diffraction pattern will be obtained. This can be indexed and evaluated for zonal symmetry and measured to determine unit cell constants. Selection of small areas is important for the identification of twins and/or defects in terms of "shape effects" or continuous diffuse scattering.

4. Use of conjugate low magnification images. Diffraction contrast bright- or dark-field images obtained by aperture selection of direct or diffracted beams can be of great assistance for the evaluation of crystal texture (e.g. *via* "bend contours") and defects (e.g. *via* moiré imaging)[8]. These images can be obtained at relatively low magnification (e.g. 5,000–10,000x) so that few demands need be made on the specimen in terms of radiation damage.

C.   Difficulties with Electron Diffraction

1. n-beam dynamical scattering. Since a whole zone of diffracted beams is excited for any given crystal orientation for a radiation which is strongly scattered by matter, the single scattering (or kinematical) approximation necessary for *ab initio* crystal structure analysis is valid only under certain conditions[4]. Otherwise, the multiple interactions of diffraction beams with each other and the incident beam are far more complicated than two-beam *Pendellösung* diffraction ("primary extinction") so that the recorded diffracted intensities may have no simple relationship to the crystal structure. Three major determinants exist which effect the magnitude of this effect:

a.   atomic number of the molecular components. If the component atoms have a low Z value, as is the case with organic structures, the contribution due to dynamical scattering is minimized;

b.   sample thickness. The path of the electron beam through the sample is another important factor – hence, the thinner the crystal, the better;

c.   electron wavelength. With increasing accelerating voltage (or lowered wavelength) the scattering cross section is diminished.

These considerations have been described in detail by Jap and Glaeser[9].

Although n-beam dynamical scattering has been demonstrated experimentally for organic crystals[2] (thus showing that the original two beam arguments based on a mosaic crystal model[1] are in error), they can be minimized appropriately to allow *ab initio* structure analysis.

2. Crystal deformation. Insistence that only thin crystals be used for electron diffraction means that these are prone to elastic bend deformations, e.g. imposed by the uneven support film on the electron microscope grid. With a coherent source, bend deformations of several degrees over a selected area can result in an apparent loss of diffraction coherence along the incident beam, particularly if the unit cell dimension in this direction is large (e.g. >20Å). This was originally demonstrated by Cowley[10] for silicate structures and later in our laboratory for linear chain molecules[2]. If flatter specimens or smaller selected areas are not feasible, epitaxial crystallization techniques can be employed to give a more optimal projection of the unit cell[11,12].

3. Incoherent multiple scattering. If the growth of thin crystals favors a layered stack of thin sheets, then strongly scattered beams from upper layers can behave as primary beams for the lower layers. Cowley et al.[13] described the consequences of such incoherent multiple scattering for intensity data from solution crystallized paraffin. As modeled by the self-convolution of diffracted intensities for an excited zone of reflections, this scattering leads to violations of space group forbidden reflections as well as affecting the observed diffraction intensity of other reflections. It can also cause a spurious increase of resolution in the diffraction pattern. The consequences of such scattering for analysis of the polyethylene crystal structure with electron diffraction data have been discussed both for projections along the chain[14] or onto the chain[15]. Previously uncharacterized extra reflections from epitaxially crystallized n-paraffins have been attributed to this scattering phenomenon[16] and, most recently, it has been found necessary to correct for this scattering in the crystal structure analysis of phospholipids with lamellar electron diffraction data.

4. Radiation damage. Perhaps the most mentioned fault with electron microscopy (but consider also synchrotron x-ray diffraction experiments!), radiation damage due to inelastic interaction of the electron beam with the specimen can be suitably minimized by the simultaneous use of low beam currents (see above) and suitably sensitive photographic film (e.g. an x-ray film). Much effort has been spent investigating this aspect of electron microscopy, almost to the point of necrophilia. Although it is very difficult to characterize the changes in crystal structure during beam damage separately in terms of molecular and textural events, various studies have been made in linear chain systems which appear to make chemical and physical sense[17-19]. Nevertheless, the best approach is to avoid detectable damage with low beam doses and, if necessary, with the additional use of low temperature specimen stages[20].

5. <u>Solvent loss, specimen sublimation</u>. Occasionally a crystal with included solvent molecules is unstable in the electron microscope vacuum (typically, 10^{-5} to 10^{-7} torr). Either a differentially-pumped environmental stage can be used to keep a continuous stream of solvent over the sample[5] or the samples can be stabilized at low temperature[21]. The latter method can also be employed when the crystals themselves sublime *in vacuo* (e.g. the case of many aromatics).

D.   Crystal Structure Analysis

The feasibility of using electron diffraction intensity data for *ab initio* quantitative crystal structure analysis relies, of course, on the minimization of limiting factors discussed above. This implies a more extensive knowledge of the crystalline specimen than is usually realized in x-ray crystallography where corrections for dynamical scattering and crystal texture are relatively simple and made *a priori* rather than *a posteriori*. If all conditions are satisfied then electron diffraction intensities from a thin organic microcrystal can be used much as they are in x-ray crystallography - i.e. one assumes that the kinematical approximation is valid. Various aspects of structure analysis are reviewed:

1. <u>Phasing techniques</u>. Use of single crystal diffraction data for quantitative structure analysis often has been limited to a particular zone, either due to the characteristics of the unit cell (e.g. the effect of crystal bending for structures with one long cell dimension) or the tilt limitations of the goniometer stage. More recently three-dimensional data from polymers[15,22,23] and epitaxially crystallized alkanes[24] have been used for successful structure analyses. Phasing techniques are similar to the ones used in x-ray crystallography and include the following:

a.   <u>Trial and error</u>. Rather than being a totally random process, trial and error methods usually begin with a good guess at the molecular conformation. For example, translational searches have been used to analyze phospholipid structures[25-27] and, minimization of non-bonded potential functions have been used to provide a good estimate of a linear polymer conformation[28,29] and are often based on complete x-ray structures of short oligomeric segments.

b.   <u>Patterson function</u>. The autocorrelation function has often been employed for electron diffraction structure analysis and is effective for finding the orientation of regular structural segments (e.g. rings, alkane chains)[2,24]. On the other hand, due to the rather narrow range of scattering factor values in comparison to x-ray diffraction, it may not be very effective for determining the position of heavy atoms in an organic structure[2].

c.   <u>Direct methods</u>. The first actual use of direct methods for electron diffraction structure analysis was for the determination of two polymethylene chain packings, respectively, for an alkane and a phospholipid[30], and these utilized estimates for three- and four-phase invariants comprised from zonal data. More recently these techniques have been used for data from epitaxially-crystallized polymethylene chain compounds, including several phospholipids, a paraffin, a paraffin solid solution, and polyethylene. It is particularly useful to have high resolution image data to provide additional phase information. Model studies assessing the effect of n-beam dynamical scattering[31] and crystal bending[32] on the successful phase determination have also been carried out. Recently[33], Hoppe-Gassmann type phase refinement[34] has been proposed as a method of utilizing low angle phase information obtained from direct images and extending the determination to the resolution limits of the electron diffraction pattern. As also discussed in an accompanying paper, such direct image data have already been used to assist direct phasing with three phase structure invariants.

2. <u>Structure refinement</u>. Of the two techniques used for structure refinement, Fourier refinement seems to be preferable to least-squares techniques, since the latter can easily lead to geometrically unreasonable structures. Two problems are inherent here. One is that the use of only zonal diffraction data can easily lead to false minima in the crystallographic residual. More important perhaps is that any inaccuracies in the data due to dynamical scattering and/or crystal texture must be realistically accounted for as part of the refinement procedure. For limited data sets, the use of the crystallographic residual as the only figure of merit is very imprecise[35], as was also found in the early days of x-ray crystallography[36]. The structure refinement must be physically reasonable so that anticipated scattering behavior for a particular crystalline texture is accounted for as well as the crystal structure having a reasonable geometry and packing energy. Electron diffraction structure determinations are not anticipated to be as accurate as the best x-ray determinations but, nevertheless, will give a good representation of molecular conformation, if not the most precise determination of valence and thermal parameters.

E.   <u>Examples of Structure Determinations</u>

1. <u>"Perfect crystals"</u>. A fairly complete catalog of crystal structures determined from electron diffraction data is given in recent reviews[2,3]. For crystals grown from solution it is well known[37] that the longest unit cell axis is nearly perpendicular to the best developed crystal face; hence the problem of elastic crystal bending can be rather severe for some materials. Fortunately, in linear polymer crystals, the fiber repeat along the molecular axis is generally short enough to allow *ab initio* structure determination, and thus it is no accident that this class of materials represent the greatest number of quantitative structure analyses based on electron diffraction data from thin lamellae. Biological lipids which contain alkane chains have also been extensively studied[2]. In this case, the pseudocell of the polymethylene repeat is often the only major contributor to the electron diffraction intensity[38] so that features of the head group region are lost. Thus, for solution grown lamellae, only the direction and packing of the acyl chains can be determined. Nevertheless, new alkyl chain packings have been determined[2]. (Such data are useful for the study of lipid polymorphism). Electron diffraction data from more complex materials with no regular structure will not be useful for structure analysis if the projected cell length is too long[32].

One way of overcoming this constraint is to use an epitaxial crystallization technique to give a different orientation of the unit cell. For example, aromatics can often be crystallized on various salt substrates[39] to give a view down a very short projection axis. The utility of electron diffraction data from n-alkanes[24,40], linear polymers[15,41], and lipids[25-27,42-44] which had been epitaxially-crystallized on organic substrates[11] for determination of complete unit cell contents has also been demonstrated and also includes *ab initio* determinations of previously unknown crystal structures.

2. <u>Imperfect crystals</u>. Electron diffraction often has been particularly effective for studying crystalline defects in samples which would not be easily examined by single crystal x-ray techniques. Two cases can be discussed:

a. <u>Thermotropic transitions</u>. Thermotropic behavior of linear chain compounds studied by electron diffraction includes projections along and onto the chain axes. In the former case, the thermally-induced lattice expansion of linear polyethylene was studied by Charlesby[45] and similarly for n-alkanes by us[46] using solution crystallized samples. Phase transitions in single hydrated phospholipid bilayers have been studied by Hui and co-workers[47]. A study of epitaxially-crystallized samples has been especially effective for describing the structure of the "rotator" phase in paraffins[46]. In this determination, the qualitative use of diffuse continuous scatter is as useful as the observed change in Bragg intensities. A more recent, quantitative analysis[48] indicates that much of the disorder is due to longitudinal chain translations. Similarly, epitaxially crystallized samples of cholesteryl esters[42] and phospholipids[49] have been useful for visualization of crystalline-to-smectic transitions as well as subambient crystal-crystal transitions.

b. <u>Binary solids</u>. By nature, linear chain materials are rarely monodisperse and, hence, it is useful to determine rules for formation of stable continuous solid solutions. Electron diffraction and diffraction contrast images have been used to study the phase separation of phospholipid mixtures in a view onto a single bilayer[50]. Again, epitaxially prepared samples have been most useful for studying comiscibility in terms of Vegard's law. Symmetry and molecular size effects were evaluated for binary paraffin mixtures[51-53] and, in one case, the crystal structure of a solid solution was determined from the zonal diffraction data. The process of fractionation was also considered in an examination of two binary systems, one which forms a metastable solid solution which slowly fractionates and another which only forms a eutectic[54]. In both cases the binary solid was found to be a superlattice. Further applications of this technique have been made in a study of cholesteryl esters[43] and phospholipids[49]. In some cases, the rules for solid solution formation are not in accord with those formulated by Kitaigorodsky[55].

F. <u>Conclusions</u>

The structure determinations described here are based mainly on the use of single crystal diffraction intensities. While several earlier determinations used data from polycrystalline or fiber specimens, the reader is cautioned that the treatment of n-beam dynamical scattering from such samples is by no means easy[56]. Any new structure analysis requires a critical evaluation of its potential success based on the nature of the material involved (i.e. atomic composition) and the characteristics of the crystals obtained from the compound being investigated. The texture of thin crystals can often be assessed by direct high resolution images. However, with this caveat in mind, there is no doubt that electron diffraction data can effectively be used to determine new crystal structures, including disordered packing arrays.

REFERENCES

1. B.K. Vainshtein, 1964, "Structure Analysis by Electron Diffraction," Pergamon, Oxford.
2. D.L. Dorset, 1985, Electron crystal structure analysis of small organic molecules, J. Electron Miscrosc. Techn., 2:89-128.
3. D.L. Dorset, 1989, Electron diffraction from crystalline polymers, in: "Comprehensive Polymer Science, Vol. 1," Sir. G. Allen, ed., Pergamon, Oxford, p. 651-668.
4. J.M. Cowley, 1981, "Diffraction Physics," 2nd Ed., North-Holland, Amsterdam.
5. S.W. Hui, G.G. Hausner and D.F. Parsons, 1976, A temperature-controlled hydration or environmental stage for the Siemens Elmiskop IA, J. Phys. E. Sci.-Instr., 9:69-72.
6. R.D. Heidenreich, 1964, "Fundamentals of Transmission Electron Microscopy", Interscience, New York.
7. R.P. Ferrier, 1969, Small angle electron diffraction in the electron microscope, Adv. Optical Electron Microsc., 3:155-216.
8. P.B. Hirsch, A. Howie, R.B. Nicholson, D.W. Pashley, and M.J. Whelan, 1965, "Electron Microscopy of Thin Crystals," Butterworths, London.
9. B.K. Jap and R.M. Glaeser, 1980, The scattering of high energy electrons. II. Quantitative validity domains of the single-scattering approximations for organic crystals, Acta Cryst., A36:57-61.
10. J.M. Cowley, 1961, Diffraction intensities from bent crystals, Acta Cryst., 14:920-927.
11. J.C. Wittmann and B. Lotz, 1986, Epitaxial crystallization of long chain molecules: Morphological and structural investigations and their applications, Inst. Phys. Conf. Ser. Chapter 11, 78:417-422. (Proceedings of EMAG85 Conference, Newcastle Upon Tyne, England).
12. D.L. Dorset, W.A. Pangborn, and A.J. Hancock, 1983, Epitaxial crystallization of alkane chain lipids for electron diffraction analysis, J. Biochem. Biophys. Methods, 8:29-40.
13. J.M. Cowley, A.L.G. Rees, and J.A. Spink, 1951, Secondary scattering in electron diffraction, Proc. Phys. Soc. (London), A64:609-619
14. D.L. Dorset and B. Moss, 1983, Crystal structure analysis of polyethylene with electron diffraction intensity data. Deconvolution of multiple scattering effects, Polymer, 24:291-294.

15. H. Hu and D.L. Dorset, 1989, Three-dimensional electron diffraction structure analysis of polyethylene, Acta Cryst., B45:283-290.
16. H. Hu, D.L. Dorset and B. Moss, 1989, Space group symmetry and the location of forbidden reflections due to incoherent multiple scattering from epitaxially oriented paraffins, Ultramicroscopy, 27:161-170.
17. D.L. Dorset, F.M. Holland, and J.R. Fryer, 1984, The "quasi-thermal" mechanism for electron beam damage of paraffins, Ultramicroscopy, 13:305-310.
18. D.L. Dorset and F. Zemlin, 1985, Structural changes in electron-irradiated paraffin crystals at <15K and their relevance to lattice imaging experiments, Ultramicroscopy, 17:229-236.
19. D.L. Dorset and F. Zemlin, 1987, Specimen movement in electron-irradiated paraffin crystals-a model for initial beam damage, Ultramicroscopy, 21:263-270.
20. G. Lefranc, E. Knapek, and I. Dietrich, 1982, Superconducting lens design, Ultramicroscopy, 10:111-123.
21. H. Chanzy, C. Guizard, and R. Vuong, 1977, Electron diffraction of frozen hydrated polysaccharides, J. Microsc., 111:143-149.
22. F. Brisse, B. Remillard, and H. Chanzy, 1984, Poly (1,4-*trans*-cyclohexanediyl dimethylene succinate): A structural determination using x-ray and electron diffraction, Macromolecules, 17:1980-1987.
23. H. Chanzy, S. Perez, D.P. Miller, G. Paradosi, and W.T. Winter, 1987, An electron diffraction study of mannan I. Crystal and molecular structure, Macromolecules, 20:2407-2413.
24. B. Moss, D.L. Dorset, J.C. Wittmann, and B. Lotz, 1984, Electron crystallography of epitaxially grown paraffin, J. Polym. Sci.-Polym. Phys. Ed., 22:1919-1929.
25. D.L. Dorset, A.K. Massalski, and J.R. Fryer, 1987, Interpretation of lamellar electron diffraction data from phospholipids, Z. Naturforsch, 42a:381-391.
26. D.L. Dorset, 1987, Molecular packing of a crystalline ether-linked phosphatidylcholine: an electron diffraction study, Biochim. Biophys. Acta, 898:121-128.
27. D.L. Dorset, 1988, Two untilted lamellar packings for an ether-linked phosphatidyl-N-methylethanolamine. An electron diffraction study, Biochim. Biophys. Acta, 938:279-292.
28. F. Brisse, 1989, Electron diffraction of synthetic polymers: The model compound approach to polymer structure, J. Electron Microsc. Techn., 11:272-279.
29. S. Perez and H. Chanzy, 1989, Electron crystallography of linear polysaccharides, J. Electron Microsc. Techn., 11:280-285.
30. D.L. Dorset and H.A. Hauptman, 1976, Direct phase determination for quasi-kinematical electron diffraction intensity data from organic microcrystals, Ultramicroscopy, 1:195-201.
31. D.L. Dorset, B.K. Jap, M.-S. Ho, and R.M. Glaeser. 1979, Phasing of electron diffraction data from organic crystals. The effect of n-beam dynamical scattering, Acta Cryst., A35:1001-1009.
32. B. Moss and D.L. Dorset, 1982, Effect of crystal bending on direct phasing of electron diffraction data from cytosine, Acta Cryst., A38:207-211.

33. K. Ishizuka, M. Miyazaki, and N. Uyeda, 1982, Improvement of electron microscope images by the direct phasing method, Acta Cryst. A38:408–413.
34. W. Hoppe and J. Gassmann, 1968, Phase correction, a new method to solve partially known structures, Acta Cryst., B24:97–107.
35. W.C. Hamilton, 1964, "Statistics in Physical Science," Ronald, New York, p. 157–162.
36. P.R. Pinnock, C.A. Taylor, and H. Lipson, 1956, A redetermination of the structure of triphenylene, Acta Cryst., 9:175–178.
37. L.H. Jensen, 1970, Molecular packing in organic crystals, J. Polym. Sci., C28:47–63.
38. D.L. Dorset, 1983, Electron crystallography of alkane chain lipids; identification of long chain packing, Ultramicroscopy, 12:19–28.
39. J.R. Fryer, 1989, High-resolution imaging of organic crystals, J. Electron Microsc. Techn., 11:310–325.
40. D.L. Dorset, 1986, Electron diffraction structure analysis of epitaxially crystallized n-paraffins, J. Polym. Sci.-Polym. Phys. Ed., 24:79–87.
41. B. Moss, D.L. Dorset, J.C. Wittmann, and B. Lotz, 1985, Quantitative analysis of electron diffraction data from epitaxially grown crystals, J. Macromol. Sci.-Phys., B24:99–118.
42. D.L. Dorset, 1985, Thermotropic mesomorphism of cholesteryl myristate. An electron diffraction study, J. Lipid Res., 26:1142–1150.
43. D.L. Dorset, 1987, Cholesteryl esters of saturated fatty acids: co-solubility and fractionation of binary mixtures, J. Lipid Res., 28:993–1005.
44. D.L. Dorset and W.A. Pangborn, 1988, Polymorphic forms of 1,2-dipalmitoyl-sn-glycerol. A combined x-ray and electron diffraction study, Chem. Phys. Lipids, 48:19–28.
45. A. Charlesby, 1945, Effect of temperature on the structure of highly polymerized hydrocarbons, Proc. Phys. Soc. (London), 57:510–518.
46. D.L. Dorset, B. Moss, J.C. Wittmann, and B. Lotz, 1984, The pre-melt phase of n-alkanes: Crystallographic evidence for a kinked chain structure, Proc. Natl. Acad. Sci., U.S.A., 81:1913–1917.
47. S.W. Hui, M. Cowden, D. Papahadjopoulos, and D.F. Parsons, 1975, Electron diffraction study of hydrated phospholipid single bilayers. Efects of temperature, hydration and surface pressure of the "precursor" monolayer, Biochim. Biophys. Acta, 382:265–276.
48. H. Hu and D.L. Dorset, unpublished data.
49. D.L. Dorset and A.K. Massalski, 1987, Co-solubility in binary phospholipid crystals, Biochim. Biophys. Acta, 903:319–332.
50. S.W. Hui and D.F. Parsons, 1975, Direct observation of domains in wet lipid bilayers, Science, 190:383–384.
51. D.L. Dorset, 1985, Crystal structure of n-paraffin solid solutions: an electron diffraction study, Macromolecules, 18:2158–2163.
52. D.L. Dorset, 1987, Role of symmetry in the formation of n-paraffin solid solutions, Macromolecules, 20:2782–2788.
53. D.L. Dorset, 1990, Chain length and the co-solubility of n-paraffins in the solid state, Macromolecules, 23:623–633.

10

54. D.L. Dorset, 1986, Crystal structure of lamellar paraffin eutectics, Macromolecules, 19:2965-2973.
55. A.I. Kitaigorodsky, 1984, "Mixed Crystals," Springer, Berlin.
56. P.S. Turner and J.M. Cowley, 1969, The effects of n-beam dynamical diffraction on electron diffraction intensities from polycrystalline materials, Acta Cryst. A25:475-481.

CRYSTAL STRUCTURES BY ELECTRON DIFFRACTION: DIACETYLENE MONOMERS AND POLYMERS IN LANGMUIR-BLODGETT FILMS AND SINGLE CRYSTALS

JEROME B. LANDO
Department of Macromolecular Science
Case Western Reserve University
Cleveland, Ohio 44106, USA

ABSTRACT: We have investigated the crystal structures of a number of diacetylene monomers and polydiacetylenes in which single crystal electron diffraction relative intensity data is utilized for structure refinement. A number of conditions allow the kinematic approximation to be made, the thinness of polymer single crystals and Langmuir-Blodgett films, the small number of unit cells in the beam direction (large crystallographic spacing in the beam direction) and generally low atomic number. Examples are the polyfunctional diacetylene poly(1,11-dodecadiyne) and its crosspolymerized product, a monolayer of the polymerized lithium salt of 10,12-nonacosadiynoic acid and multilayers of the tubule forming amphiphile 1,2-bis(10,12-tricosadiynoyl)-sn-glycerol-3-phosphocholine.

INTRODUCTION

Approximately eighteen years ago an investigation of the possible use of electron diffraction intensity data from polymer single crystals in crystal structure determination was explored and led to the publication of a paper (1) on the structure of Penton, a chlorine containing polyether in which the electron diffraction based structure determination improved the previous x-ray structure and established that dynamical structure factors calculated from the determined crystal structure did not differ within experimental error from kinematic structure factors.

A major factor that allows the kinematic approximation is the thinness of polymer single crystals (~100Å) (and Langmuir-Blodgett) (LB) monolayers and multilayers (~30Å/monolayer). A factor that is not usually recognized involves the unit cell repeat in the chain direction in many polymers and the side chain repeat in LB films. Since the packing in the plane of an LB film or perpendicular to a polymer chain is similar to low molecular weight materials and the other unit cell dimension tends to be much longer, the unit cell volume tends to be higher than usual. This lessens dynamical effects on the relative intensities for a given thickness. Finally these organic molecules tend to contain atoms of relatively low atomic number. It should be noted that in the paper mentioned above , the presence of chlorine atoms in the polymer repeat did not preclude the use of the kinematic approximation.

This paper contains a brief review of efforts to use electron diffraction intensity data for crystal structure determination. In each case the kinematic approximation was made.

11

J. R. Fryer and D. L. Dorset (eds.), Electron Crystallography of Organic Molecules, 11–18.
© 1990 *Kluwer Academic Publishers.*

Poly(1,11-dodecadiyne and It's Cross-Polymerized Product (2,3)

The cross-polymerization of the macromonomer poly(1,11-dodecadiyne) using UV, x-ray or ^{60}Co γ radiation has recently been reported (2). Upon cross-polymerization the sample changes from colorless to dark blue. The term macromonomer is used to describe the original polymer, which has a chemical repeat unit $[(CH_2)_8-C{\equiv}C-C{\equiv}C-]_x$. The term cross-polymerization is utilized to distinguish systematic polymerization of the diacetylene units to a crystalline structure composed of sheets from the more familiar random cross-linking that many polymers undergo when exposed to radiation. The crystal structure of the macromonomer and cross-polymerized material has been reported earlier (3) and is briefly described here. The crystal structures (macromonomer and cross-polymerized) were refined by using electron diffraction data because of the limited information obtained from x-ray fiber patterns (2). Moreover, electron diffraction analysis provides more accurate information about the location of the hydrogen atoms in comparison to what we get from x-ray analysis (4).

The macromonomer was prepared by oxidative coupling of $HC{\equiv}C-(CH_2)_8.C{\equiv}CH$ using copper-pyridine catalyst. After purification (1) the polymer was dissolved in chloroform to make a dilute solution. A drop of this solution at 4 degrees C on a carbon-coated copper grid was evaporated to obtain single crystals. The diffraction pattern was obtained by a JEOL JEM 100B electron microscope at 100 kV under very low beam intensity using high-speed x-ray films.

The cross-polymerization was effected by γ irradiation (100 Mrd) of the macromonomer crystals. A second orientation of the crystals was obtained by casting a macromonomer film at room temperature. The intensity data were collected with a high-precision photodensitometer. The diffraction maxima were quite sharp, so the peak heights were taken as the relative intensities.

Electron diffraction patterns for the macromonomer and the crosspolymerized material are shown in Figure 1. The x-ray fiber pattern indicated a two fold screw axis along b, whereas the a*c* lattice net contains the systematic absences h + l = odd for both materials. The space groups for both is $P2_1/n$. The lattice constants are shown in Table I.

Figure 1. Selected-area diffraction patterns of (a)
the macromonomer and (b) the cross-polymerized material.

Table I.
Comparison of Unit Cell Dimensions

cell parameter	macromonomer	cross-polymerized material
a, Å	13.25	9.17
b, Å	14.15	12.25
	(chain axis)	(hydrocarbon chain axis)
c, Å	7.63	9.92
		(diacetylene chain axis)
β, deg	118.50	123.50

The two structures were refined using the linked atom least square programs of Arnott and co-workers (7). A planar zigzag structure was assumed for the methylene groups so that the refinement involved determination of the orientation of the zigzag and the sense of the chain. In the crosspolymerized material the angle between the plane of the zigzag and the diacetylene chain is zero. In the macromonomer structure the plane of the zigzag makes an angle of 34 degrees with the a axis. In each case the final residual was 0.13 assuming equal weight (w=1) for all reflections. The final structures are shown in Figures 2 and 3. Note that the doubling of the diacetylene repeat is caused only by the hydrogen position, a situtation that would be difficult to determine using x-ray intensity data.

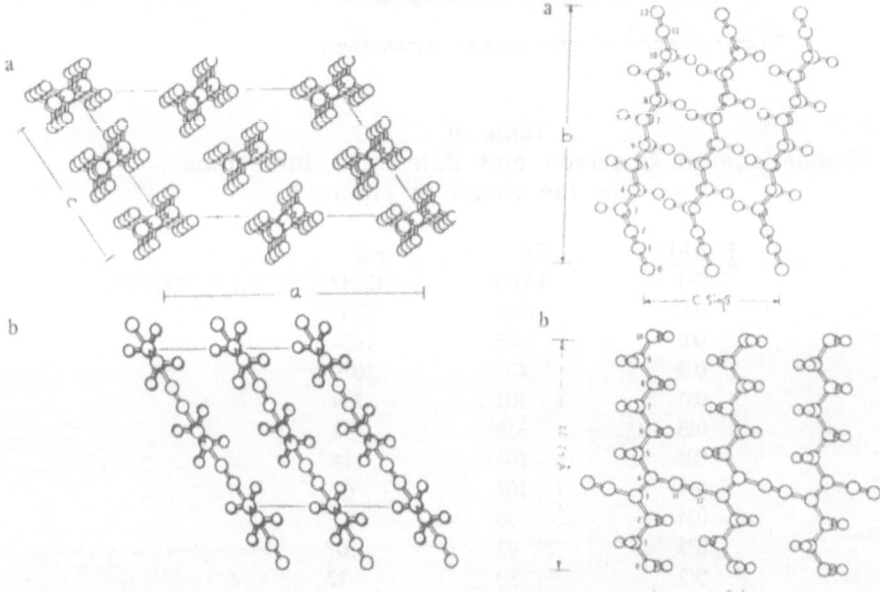

Figure 2. ac projections of (a) the macromonomer and (b) the cross-polymerized material.

Figure 3. bc projections of (a) the macromonomer and (b) the cross-polymerized material.

The Polymerized Lithium Salt of 10,12-Nonacosadiynoic Acid (8)

A monolayer of the monomer was polymerized with ultraviolet radiation as a monolayer at the nitrogen water interface and picked up on an electron diffraction grid by a horizontal dipping mechanism. The electron diffraction pattern yielded an orthogonal b* c* (see Figure 4) reciprical lattice net, with results with spacings b = 8.11Å and c = 4.89 (polydiacetylene chain axis). The axis was 71.0Å, the bilayer repeat, determined from x-ray diffraction data.

The structure was determined from the electron diffraction data where k + l = odd were absent. For a monolayer this indicates a base centering, although in the full unit cell there is an n glide.

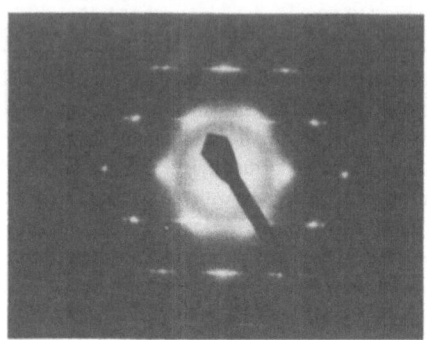

Figure 4. Electron diffraction pattern of a monolayer.

Table II
Comparison of Observed and Calculated Intensities
for the Model in Figure 5.

hkl	Fo	Fc
011	12000	12187
020	7011	6367
002	2406	2824
022	481	1079
031	301	264
013	310	372
004	202	114
040	102	67
024	70	23
033	93	107
042	30	32
051	25	16
060	10	11

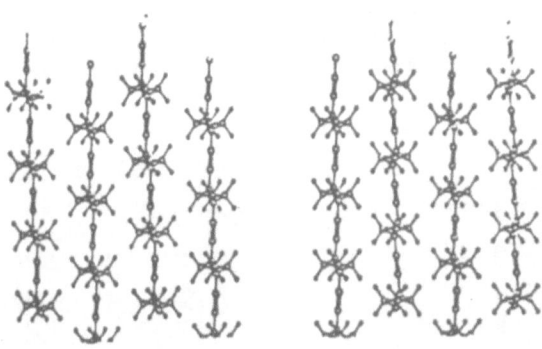

Figure 5. bc projections: (left) base centered; (right) glide related.

Table II gives a comparison of calculated and observed reflections for the base centered projection (left hand projection in Figure 5). A final unweighted residual of 0.09 was obtained and good agreement between calculated and observed intensities occurred over a range of relative intensities from 10 to 12,000. However, the same agreement occurs for the right hand projection in Figure 4, since only two carbon atoms are moved slightly. The k and l odd reflections have no appreciable intensity caused by this slight change. The right hand structure is favored by the arrangement in the monomer structure through application of the least motion principal of Cohen and Schmidt in topochemical reactions.

Finally, a look at Figure 4 shows that there is a progressively greater smearing as the value of l increases indicating a misregister along the polydiacetylene chain. However, there are maxima within the smears indicating a regular behavior. Optical diffraction from models indicate that the misregister is ±0.8 Å every third unit cell. Note that the ideal structure was determined using the total integrated intensity in each of the smeared diffraction maxima.

1,2 bis (10,12-Tricosadiynol)-sn-Glycerol-3-Phosohocholine (DC$_{8,9}$PC) (9)

The phospholipid 1,2 bis (10,12-tricosadiynol)-sn-glycerol-3-phosphocholine (DC$_{8,9}$PC) can form hollow, cylindrical structures known as tubules. In an effort to investigate the mechanisms of tubule formation we have studied the crystal structures of DC$_{8,9}$PC in both tubules and Langmuir-Blodgett (LB) films. We find that in both cases DC$_{8,9}$PC lies within a monoclinic cell of space group P2$_1$ with two molecules per unit cell. The lattice parameters are a = 5.18Å and b = 7.79Å for tubules and a = 5.09Å, b = 7.74Å, c = 78.5Å and β = 117 degrees for LB films. As seen in Figures 6 and 7 the patterns are the same except for the indicated differences in lattice constants and the doubling of the spots caused by the pattern being from two walls of the hollow tubules (850Å wall thickness and outer diameter 3000 to 5000 Å).

We produced the tubules by direct crystallization from water-ethanol mixtures and the multilayers by LB deposition from the gas-water interface.

The structure of the multilayers was determined so that packing calculations might be undertaken to determine the energy difference associated with the difference in lattice constants assuming the same basic structure. This would aid our understanding of why these curved crystals (tubules) form.

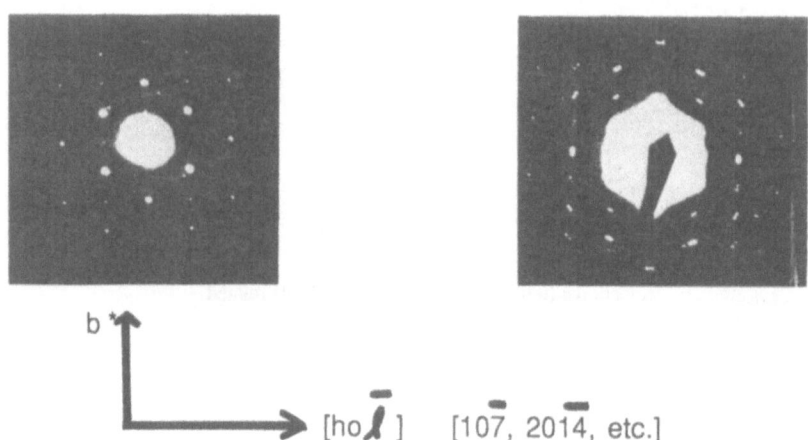

Figure 6. Electron diffraction pattern from 3 bilayers (thickness 236Å)

Figure 7. Electron diffraction pattern from 850Å thick walls)

Since the amount of water contained in the crystal structure of DC$_{8,9}$PC depends on the humidity and the electron diffraction experiments were performed in a vacuum the monoclinic cell and the indexing of Figure 7 was determined as part of the structure refinement using the LALS (7) program. A projection of the structure is shown in Figure 8 and comparison of calculated and observed structure factors is shown in Table III. An unweighted residual of 0.175 was obtained.

CONCLUSIONS

This paper demonstrates the utility of electron diffraction intensity data in crystal structure determination of thin polymer and LB monolayer and multilayer crystals. Unique features such as the enhanced scattering power of hydrogen in electron diffraction as opposed to x-ray diffraction, symmetry differences between monolayer and bulk crystals, defect structures and indexing problems have been pointed out and the practical consequences indicated.

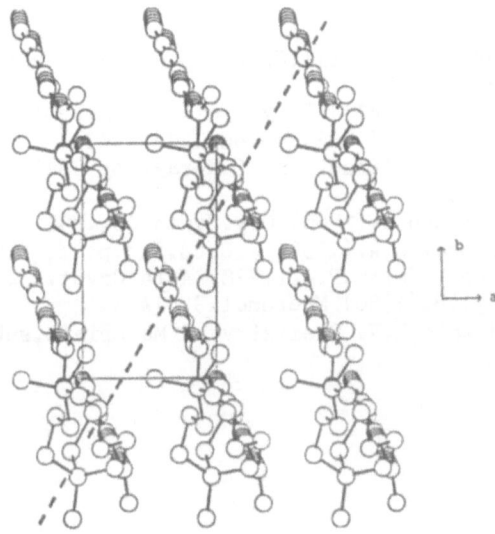

Figure 8. Projection of the structure, dotted line indicates possible polymerization direction.

Reflection	Fo	Fc	Difference
	Table III.		
(1,0,-7)u	1.56	2.80	-1.24
(2,0,-14)	15.24	14.83	- 0.40
(3,0,-21)u	1.56	.64	0.07
(0,1,0)a	0.00	0.00	0.00
(1,1,-7)	20.89	20.25	0.64
(2,1,-14)	5.62	3.97	1.65
(3,1,-21)	7.16	7 .09	0.07
(0,2,0)	18.71	18.22	0.49
(1,2,-7)	6.73	7.31	-0.58
(2,2,-14)	9.83	9.58	0.26
(3,2,-21)	4.64	3.22	1.42
(0,3,0)a	0.00	0.00	0.00
(1,3,-7)	11.27	11.51	-0.24
(2,3,-14)	6.00	5.93	0.07
(3,3,-21)	2.91	3.19	-0.28
(1,4,-7)	3.88	4.06	-0.19
(2,4,-14)	4.20	2.62	1.58
(3,4,-21)	3.50	1.28	2.22
(0,5,0)a	0.00	0.00	0.00
(1,5,-7)	3.87	2.81	1.06
(2,5,-14)	3.18	0.77	2.41
(3,5,-21)	3.18	1.40	1.78

a = systematic absences   u = unobserved

REFERENCES

1.Chaffey,W.,Gardner,K.,Blackwell,J.,Lando,J.B.and Geil,P.H.(1974)
  Phil.Mag.10, 1223.
2.Day,D.R.and Lando,J.B.(1981)J.Polym.Sci.,Polym.Lett.Ed.19, 227.
3.Thakur,M.K.and Lando,J.B.(1983)Macromol.16, 143.
4.Vainshtein,B.K.(1966)'Diffraction of X-rays by Chain Molecules',
  Elsevier,New York,p53.
5.Campbell,I.and Eglinton,G.(1966) Org.Synth.45, 39.
6.Eglinton,G.and Galbraith,A.(1959) J.Chem.Soc.pt I, 889.
7.Arnott,S.and Campbell Smith,P.J.(1978) Acta Cryst.A34, 3.
8.Day,D.R.and Lando,J.B.(1980) Macromol.3, 1483.
9.Lando,J.B.and Sudiwala,R.V. Chemistry of Materials,submitted for
  publication.

OBSERVATION OF HEXATIC BEHAVIOR IN LIQUID CRYSTALS BY *IN SITU* ELECTRON DIFFRACTION

S.W. HUI
Roswell Park Memorial Institute
Buffalo, NY 14263

M. CHENG and J.T. HO
State University of New York at Buffalo
Buffalo, NY 14260

R. PINDAK
AT&T Bell Laboratories
Murray Hill, NJ 07974

ABSTRACT.   The translational and orientational orders of free-standing liquid-crystal films at temperatures between the crystal and liquid phases were measured by the reflection intensity profiles of single-domain electron-diffraction patterns. Both translational and orientational orders fit those described for a hexatic phase. By examining the scaling relation among the $6n$-fold orientational order parameters, it is demonstrated that, as the number of molecular layers increases, its property departs from that predicted for a two-dimensional system.

1.   Two-Dimensional Melting.

1.1   THE HEXATIC STATE OF MATTER

The melting of a three-dimensional solid crystal to a liquid represents a phase transition from a well ordered to a much less ordered state. In a solid crystal, the periodicity of mass density is described by long-range translational and orientational orders. At the transition temperature T_c, both the translational and orientational symmetries are broken simultaneously. Only a short-range order, as described by an exponentially decaying mass density correlation function, remains in the liquid state. If the crystal is reduced to two dimensions, the constraint in the third dimension vanishes, leading to increasing structural fluctuations. In this case, the fluctuations of the phase of the order parameter become dominant over those of its amplitude.   It is shown quantitatively by Halperin and Nelson (1978) that the 2-D crystal melts in two stages. At one temperature T_m, existing pairs of translational defects (dislocations) would uncouple, resulting in the loss of translational order. At a higher temperature T_h,

J. R. Fryer and D. L. Dorset (eds.), Electron Crystallography of Organic Molecules, 19–32.
© *1990 Kluwer Academic Publishers.*

existing pairs of orientational defects (disclinations) would uncouple, resulting in the loss of orientational order, and the material becomes a 2-D liquid. In between the temperatures T_m and T_h, the 2-D material is neither a solid crystal nor a liquid, but in a state of having pseudo-long-range orientational order without long-range translational order. This state is a peculiarity of the 2-D system, and is referred to as the hexatic state. Its existence was predicted to be in between the solid and liquid phase boundary in the 2-D phase diagram (Fig. 1). The orientational correlation function in a hexatic structure decays as a function of the distance r algebraically as $r^{-\eta(T)}$, as compared to an exponential decay in a liquid, and a finite constant in solid crystals.

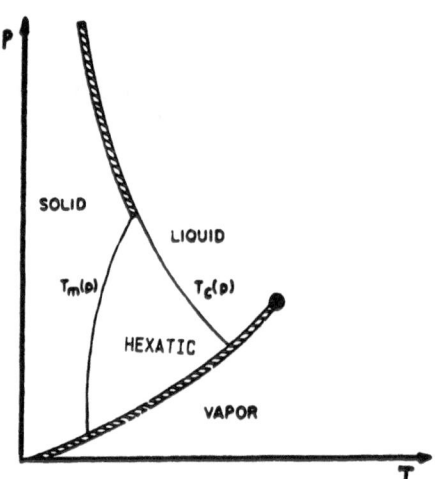

Fig. 1. A schematic 2-D phase diagram showing the place of the expected hexatic phase.

Experimentally, free-standing smectic-liquid-crystal films have proven to be a most attractive system in which to verify the various aspects of the theory of defect-mediated 2-D melting (Strandberg, 1988).

1.2 MEASURING ORDER PARAMETERS

The ordering of a 2-D system can be described in terms of a translational correlation function and an orientational correlation function. Both functions have constant values in a perfect 3-D crystal (i.e., the periodicity does not decay with distance). In a 2-D system, according to the 2-D melting theory, the translational correlation function decays as $r^{-\eta(T)}$ in the solid phase, i.e., a pseudo-long-range order. In the 2-D or 3-D liquid state, the translational correlation decays as $e^{-r/\zeta}$ where ζ is the short-range correlation length. The orientational order is long range in the 2-D solid, pseudo long range in the hexatic, and short-range in the liquid phase.

In reciprocal space, the translational and orientational orders of the system are described by the Fourier transforms, $S(q_{//})$ and $S(\chi)$, where $q_{//}$

and χ are the radial and azimuthal polar coordinates on the diffraction plane. Measuring the intensity broadenings $I(\Delta q_{//})$ and $I(\Delta\chi)$ from the pseudo Bragg diffraction peaks, and comparing them to those expected from $S(q_{//})$ and $S(\chi)$, one may gauge the degree of disorder of a particular system.

The broadening of $I(\chi)$ may be conveniently described by harmonic scaling (Aharony et al., 1986). The $6n$-fold symmetric azimuthal spreading of $I(\chi)$ may be expressed as:

$$I(\chi)=I_0 \left\{ 1/2 + \sum_n C_{6n} \cos 6n(\pi/2 - \chi) \right\} + I_{background}$$

where C_{6n} represents the $6n$-fold orientational order parameters.

As the system becomes more disordered, the sharp, pseudo-Bragg diffraction spots give way to broader, sinusoidal angular "smears" in the χ-plot. In other words, the higher harmonic coefficients C_{6n} diminish. Thus the scaling of C_{6n} gives an excellent way to monitor the orientational order.

It was shown by Aharony et al. (1986) that the scaling relation of C_{6n} obeys:

$$C_{6n} = C_6^{\sigma(n)} = C_6^{n+\lambda n(n-1)},$$

where λ is approximately 0.3 for a 3-D **crystal**, while $\lambda=1$ $[\sigma(n)=n^2]$ for a truly 2-D system (Paczuski and Kardar, 1988). Therefore the scaling of C_{6n} provides a convenient way to measure the dimensionality effect on melting.

X-ray diffraction experiments on free-standing liquid-crystal films, especially those employing a finely collimated synchrotron source for high-resolution measurements of $I(q_{//})$, have confirmed the behavior of the translational order during 2-D melting (Davey et al., 1984; Brock et al., 1986). However, the orientational disorder can be measured by x-ray diffraction only from thick (>>20 molecular layers) films or magnetically orientated films of tilted smectic phases, in which the domains of orientational coherence are large in comparison to the cross section of the x-ray beam. Otherwise, within the sampling area of the film, there could be several differently oriented domains. The superposition of diffraction from these domains may render the interpretation of the azimuthal diffraction results exceedingly difficult, especially in terms of harmonic scaling. The interesting dimensionality effect on orientational disorder cannot easily be obtained from x-ray diffraction measurements.

Electron diffraction provides a way to sample in principle an area as small as 100 Å in diameter, using a micro-diffraction beam. The strong beam-sample interaction enables diffraction data to be collected within seconds, as against many hours of x-ray exposure. Diffraction patterns may be obtained from mono-molecular layers (Hui, 1977). Therefore electron diffraction is undoubtedly the method of choice for probing orientational disorder in 2-D melting.

2.   Electron Diffraction Experiments

2.1. FREE-STANDING THIN FILMS.

The liquid crystals we chose to study are n-(p-butoxybenzylidene)-n-n-octylaniline (40.8), s-n-[4-(3,7-dimethyloctanoxy)-2-hydroxybenzylidene]-4-n-nonylaniline, or a mixture of 25 wt.% of 4-propionyl-*trans*-(4-n-pentyl)cyclohexane carboxylate (PP5CC) in n-hexyl-4'-n-pentyloxybiphenyl-4-carboxylate (650BC). The latter two are known to form hexatic smectic liquid crystals in the bulk. The mixture was chosen for its stability and its hexatic-B to smectic-A phase transition temperature to be within the experimentally feasible temperature range. Free-standing films of 2 to >100 molecular layers thick were formed by drawing a spreader, wetted with the material, across a circular hole on a stainless steel sample holder. The hole is typically 400 μm in diameter. These free standing films were stable at 30 Torr of pressure in nitrogen, but were not stable in vacuum. The film thickness was measured by optical reflectance. The reflected intensities of films of various thicknesses from a He-Ne laser fell in discrete groups. Each step indicated the reflectance due to the addition of one molecular layer. A calibration curve could thus be constructed by plotting the reflected intensity I against n^2 , where n is an integer denoting the number of molecular layers (Fig. 2).

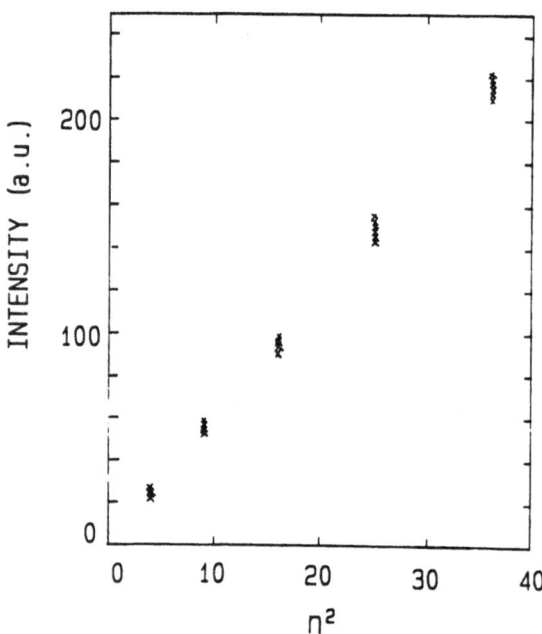

Fig. 2. The light reflectance (I) from films consisting of different numbers of molecular layers. The number of layers n is identified by fitting I against n^2 to a straight line.

2.2. CONTROL OF SPECIMEN ENVIRONMENT.

Since the thin film was not stable in vacuum, an environmental stage was used to protect the specimen from the vacuum of the Siemens Elmiskop 1A electron microscope. The environmental stage was a differentially-pump type (Hui et al., 1976). The inner chamber, where the specimen holder was situated, opened to the outer chamber through two 100 μm apertures. Adjustment of the specimen position was actuated through O-ring sealed X and Y push rods. The chamber pressure was regulated by a needle-valve-controlled connecting tube to a nitrogen tank outside the electron microscope. The temperature of the entire (thermally insulated but electrically grounded) stage was regulated by a thermoelectric wafer, and measured by a thermocouple. Any deviation of the specimen temperature from the stage was monitored by a miniature thermistor placed next to the film on the holder. The outer chamber of the stage opened to the microscope vacuum through another set of apertures, the upper one being 200 μm and the lower one 400 μm in diameter. These 4 apertures and the specimen hole were aligned so to allow the transmission of the electron beam. The apertures were sufficiently small to maintain the pressure difference between each chamber and the vacuum. For a typical experiment, the inner chamber was maintained at 30 Torr, the outer chamber at 0.1 Torr by a rotary pump, and the microscope vacuum at 10^{-4} Torr.

The environmental stage was placed above the objective pole piece, the specimen plane being 2 cm from the center of the pole gap (Fig. 3). This distance is within the remote focusing range of the electron microscope. The optical parameters of the stage was measured and reported elsewhere (Hui et al., 1976). The distance between the inner apertures, and those between the inner and outer apertures were 2 mm. The scattering by nitrogen gas within these gaps was less than 5%.

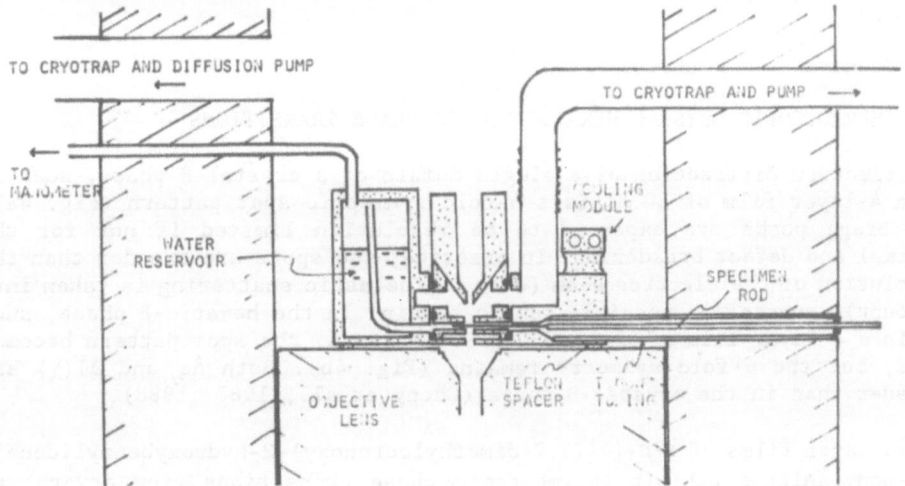

Fig. 3. The structure of the differentially-pumped environmental stage used in the experiment.

2.3. RADIATION DAMAGE MEASUREMENT.

The radiation sensitivity of the material used was monitored by measuring the changes in the electron diffraction patterns after prolonged exposure to the electron beam. The beam current was measured by an electrically insulated screen connected to an electrometer. The collection efficiency of the screen was about 0.6 as measured against a Faraday cage. A change of diffraction intensity was not detected even after a cumulated exposure of 10 e/Å^2. This was considerably more stable than hydrated phospholipid bilayers studied by the same criterion (Hui, 1980).

In order to reduce the possibility of radiation damage to the minimum, high speed Kodak No-screen X-ray film was used to record diffraction patterns. A dose 0.5 e/Å^2 was needed for a typical recording. The electron beam current was set to a minimum, and was further limited by a 5 μm second condenser aperture to about 0.1 e/Å^2/sec. The faint pattern was not visible on the screen even after dark adaptation.

2.4. PATTERN RECORDING AND REDUCTION

An illuminated area of 5 μm in diameter was formed on the specimen by placing the crossover slightly below the specimen. The convergence was sufficient to achieve a resolution of 0.01 Å^{-1} on the diffraction plane. To locate the specimen, an out-of-focus image could be seen by defocusing the intermediate lens by a pre-set value. Diffraction pattern may be obtained by immediately refocusing the intermediate lens.

The photographic recording was scanned by a high resolution video camera and converted into a 720x720x8-bit digital information. Subsequent analysis were made on a VAX-780 computer.

3. Results

3.1 THERMOTROPIC CRYSTAL-HEXATIC-LIQUID PHASE TRANSITIONS

The electron diffraction of a single domain of a crystal-B phase, such as in a 4-layer film of 40.8, is a 6-fold symmetric spot pattern (Fig. 4a). The Bragg peaks are expected to be resolution limited if not for the thermal and defect broadening. In practice, the spots are broader than the resolution of the electron beam (even if inelastic scattering is taken into account), suggesting possible sample strain. In the hexatic-B phase, such as in a 4-layer film of the 650BC-PPSCC mixture, the spot pattern becomes arcs, but the 6-fold symmetry remains (Fig. 4b). Both $\Delta q_{//}$ and $\Delta I(\chi)$ are broader than in the crystal-B phase (Cheng et al., 1987, 1988).

Three-layer films of s-n-[4-(3,7-dimethyloctanoxy)-2-hydroxybenzylidene]-4-n-nonylaniline exhibit thermotropic phase transitions from crystal to hexatic-F at 28.5°, and from hexatic-F to liquid-C at 40.5°C, as observed

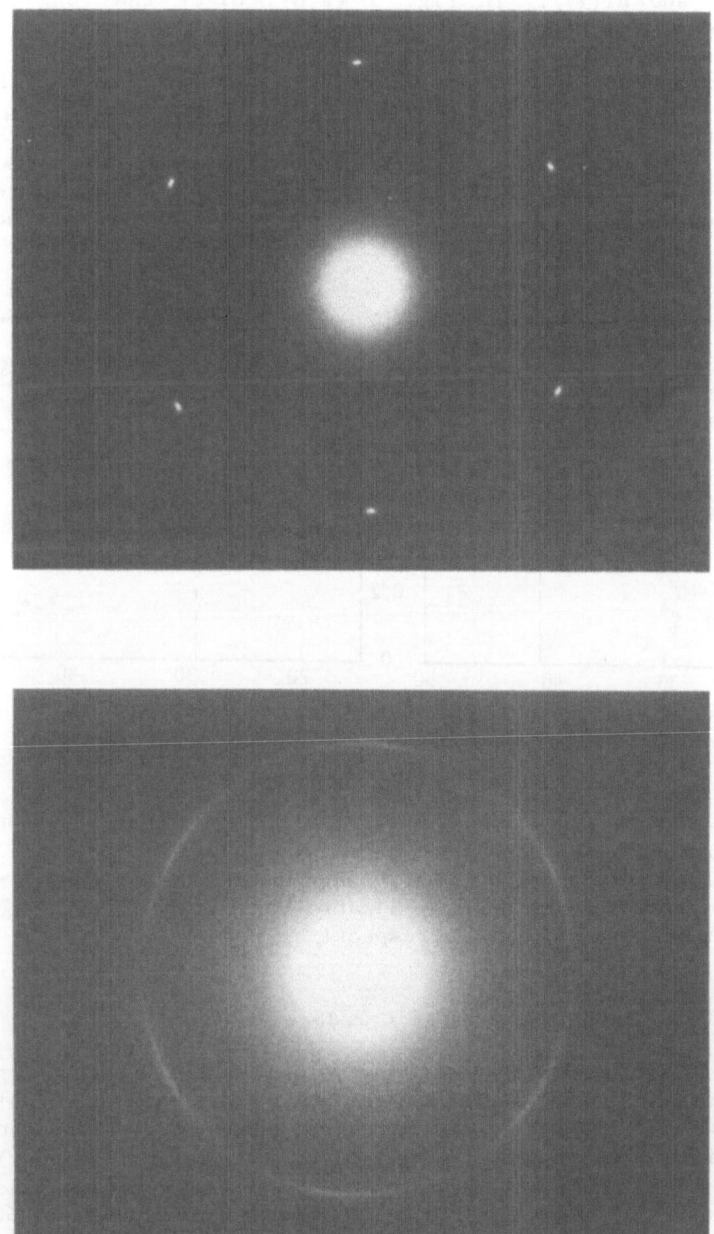

Fig. 4. Electron-diffraction patterns of (a) a 4-layer film of 40.8 in the crystal-B phase at 30⁰C; (b) a 4-layer film of a mixture of 25 wt% of PP5CC in 650BC in the hexatic-B phase at 45°C.

by polarized reflected-light microscopy. The measured $S(q_{//})$ and $S(\chi)$ are plotted as $\Delta q_{//}/q_0$ and $\Delta I(\chi)/I_0$ in Fig. 5 (a,b). Since the long rod molecules are tilted with respect to the layer plane, the pattern does not retain the 6-fold symmetry as expected from a hexatic-B or crystal-B film (Hui, 1989). In fact the $\Delta q_{//}$ and $\Delta I(\chi)$ can only be measured from 4 of the arcs only, since the other 2 are modulated by a component of $q_\perp$. Nonetheless, both the translational order as indicated by $\Delta q_{//}/q_0$, and the orientational order, as indicated by $\Delta I(\chi)/I_0$, show marked decreases upon phase transitions. The orders at the hexatic state are in between the crystal and the liquid states. Similar transitions were observed in other films containing 2 and 4 layers.

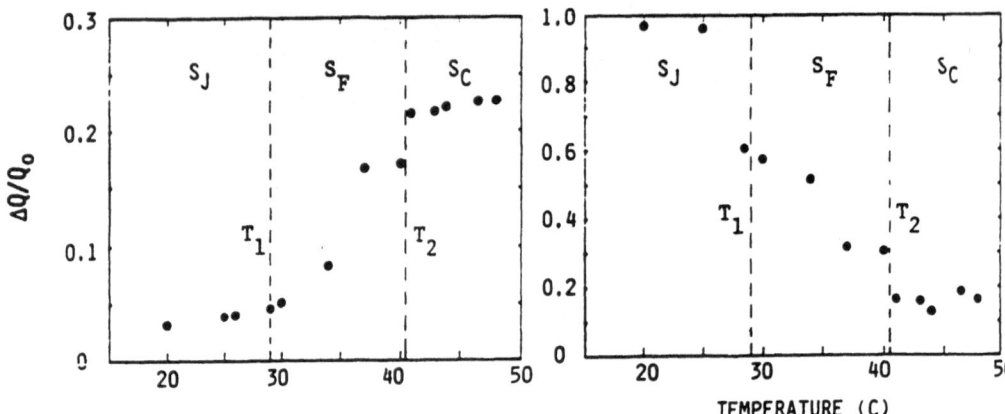

Fig. 5.   The radial and azimuthal spreads of diffraction reflections ($\Delta q_{//}/q_0$ and $\Delta I(\chi)/I_0$ respectively) from a 3-layer film of S-n-[4-(3,7-dimethyloctanoxy)-2-hydroxybenzylidene]-4-n-nonylaniline are plotted as functions of temperature. The crystal-J to hexatic-F to liquid-C transition temperatures are indicated as T_1 and T_2 respectively.

3.2. THE TRANSLATIONAL ORDER IN THE HEXATIC PHASE.

According to the 2-D melting theory, the translational order is short range in the hexatic phase. The radial spread of the reflections should resemble a Laurentzian or similar broadening trends rather than a delta function. The reflection profile of a 3-layer film of S-n-[4-(3,7-dimethyloctanoxy)-2-hydroxybenzylidene]-4-n-nonylaniline at 30° is shown in Figure 7. Each datum point represents the intensity $I(q_{//})$ integrated over the arc of 60° after background subtraction. The profile fits a Lorentzian function $S(q_{//})=[(q_{//}-q_0)^2 + \Delta q_{//}^2]^{-1/2}$, where $\Delta q_{//}$ is the width of the peak. The fit yields a correlation length of about 110 Å.   The correlation length decreases to 20 Å in the liquid state.

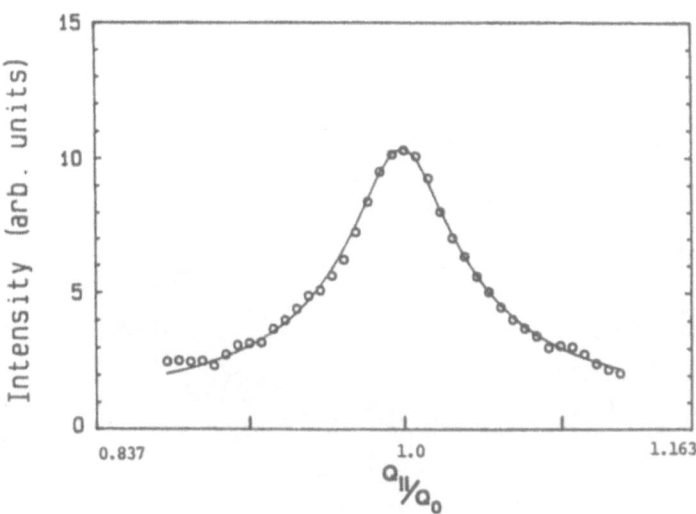

Figure 6.   The radial profile of a 3-layer film of *s*-*n*-[4-(3,7-
dimethyloctanoxy)-2-hydroxybenzylidene]-4-*n*-nonylaniline at 30°. The solid
curve represents a Lorentzian fit.

It is apparent that the translational order in the hexatic phase is a short
range one, although it is still considerably longer than that in the liquid
state. The Lorentzian fit may not be a unique one, but the profile is
definitely not resolution limited. Alternative fit is not attempted because
the resolution along $q_{//}$ is not high enough to produce an unambiguous fit.

3.3. THE ORIENTATIONAL ORDER IN THE HEXATIC PHASE.

Electron diffraction patterns of thin films of 2 and 6 layers of a mixture
of 25 wt% of PP5CC in 650BC in the hexatic-B phase at 45°C are shown in
Fig. 7. Their azimuthal profiles (together with that from a 4-layer film,
Fig. 4b) are shown in Fig. 8. The intensity spread along the arc was
analyzed by the harmonic scaling method, by fitting the Fourier
coefficients C_{6n} against $C_6^{n+\lambda n(n-1)}$. Figure 10 shows the fitting of $\sigma_n =
n+\lambda n(n-1)$ for the first 4 orders of C_{6n}, for several representative film
thicknesses and temperatures. The parameter λ thus obtained would indicate
the dimensionality of the system.

Figure 10 shows the value of λ for the 2-, 4-, and 6- layer films at
different temperatures. Since the value of λ is expected to be 1 for a 2-
D system, and 0.3 for a 3-D system, the proximity of λ to the value of
unity for the 2-layer film indicates the 2-layer film behaves like a true
2-D system. The general decrease of λ for thicker films shows an apparent
departure from a 2-D behavior even for a 4- or 6- layer film. The small
increase of λ with increasing temperature indicates that the interlayer
interaction is progressively weakened with increasing temperature,
resulting in each layer behaving independently as a 2-D system. This trend
is particularly apparent for the 6-layer film.

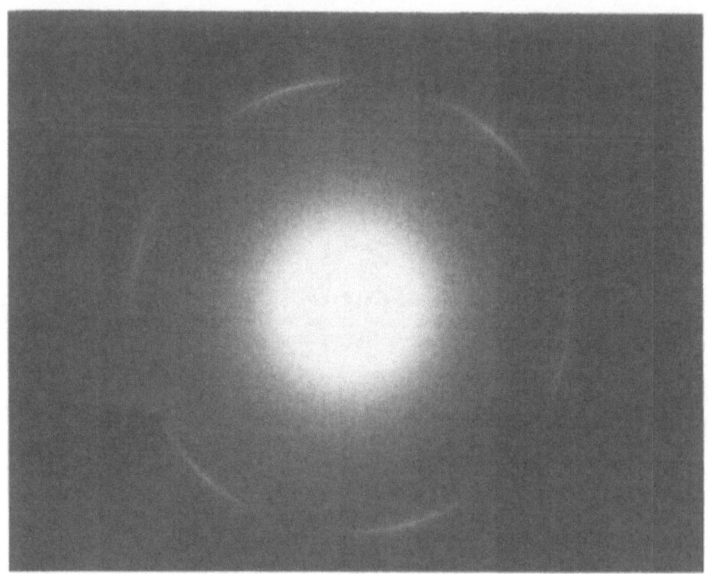

Fig. 7. Electron-diffraction patterns of thin films of (a) 2 and (b) 6 layers of a mixture of 25 wt% of PP5CC in 650BC in the hexatic-B phase at 45°C.

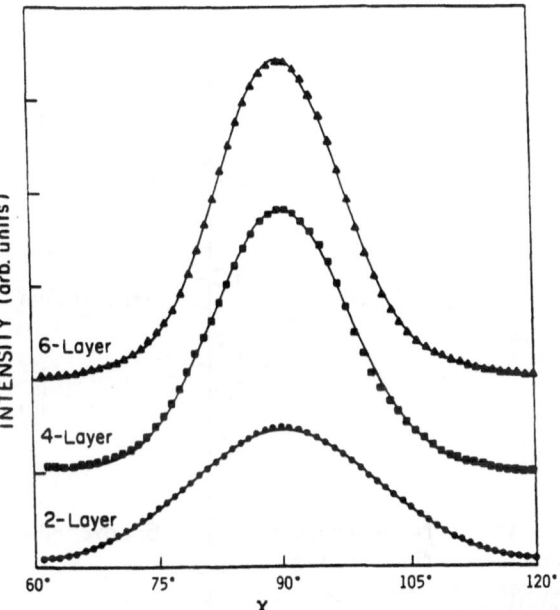

Figure 8. The azimuthal spread (angular χ scan) of diffraction intensities measured from Fig. 8a, 4b and 8b. The points are experimental values; and the solid curves are fits by: $I(\chi)=I_o\{1/2 + \sum_n C_{6n}\cos 6n(\pi/2 - \chi)\} + I_{background}$

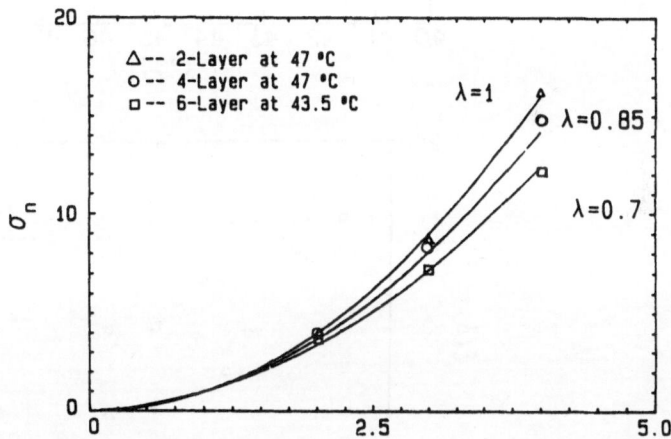

Figure 9. The fitting of $\sigma_n = n + \lambda n(n-1)$ to n for the determination of λ from C_{6n} obtained from Fig. 9.

A corresponding analysis of $\Delta q_{//}$ was also made for these samples. The translational correlation lengths ζ deduced from these measurements are shown in Fig. 11 for comparison with azimuthal measurements. The correlation length decreases with increasing temperature, as expected. Due to the relatively low resolution of electron diffraction, the variation among the different thicknesses should not be taken too seriously.

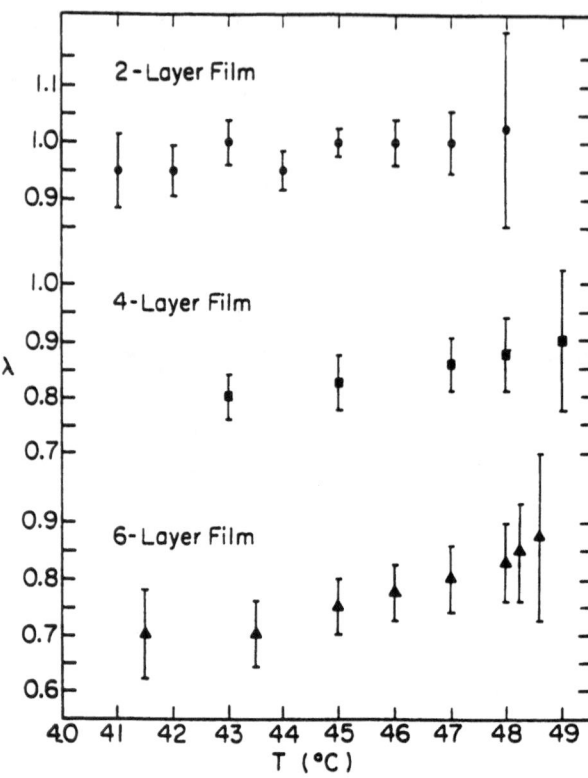

Fig. 10. Temperature dependence of the scaling parameter λ in $C_{6n} = C_6^{n+\lambda n(n-1)}$ for 2, 4 and 6 layer films of 25 wt% of PP5CC in 650BC in the hexatic-B phase.

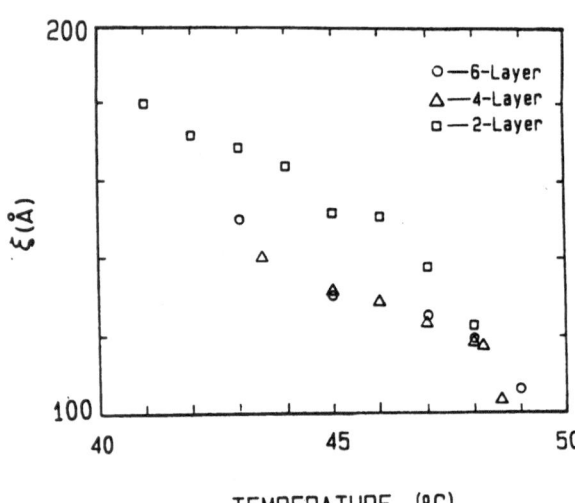

Fig. 11. The correlation length ζ determined from $\Delta q_{//}$ as a function of temperature for 2, 4 and 6 layer films of 25 wt% of PP5CC in 650BC in the hexatic-B phase.

4. Discussion

The problem of phase translations in reduced dimensions has received much attention in recent years. The theory applies not only to the melting of liquid crystals but also to a variety of quasi-2D systems, such as CuO_2 lamellar superconductors, superfluids and spin density waves in rare earth metals. Although the role of defects in mediating 2D melting has long been suspected, a rigorous theory was proposed only recently (Nelson and Halperin, 1978). Since then, a number of experiments were performed to test the quantitative predictions of the theory.

The intensity profile measurement by high-resolution x-ray diffraction studies (Davey et al., 1984) convincingly showed the short-range translational order in hexatic liquid crystals. Azimuthal profiles measured from thick films also suggested the retention of orientational order in these materials. The extrapolation to a true 2-D system from thick films has been hampered by the weak interaction between x-rays and thin films, and the small size of quasi-coherent domains from which azimuthal measurement can be obtained, without the complication of superimposing patterns from differently oriented domains. Therefore the dimensionality criterion of the scaling parameter λ cannot be verified unambiguously.

Electron diffraction offers an obvious solution to the problem. The well focused beam can sample areas as small as 100 Å in diameter. Because of the requirement for a non-convergent beam to improve the resolution of the diffraction, we had to settle for the rather large beam size of 5 μm in diameter. Even with this beam size, single domains can be sampled with ease. A lateral movement of several μm did not affect the diffraction pattern, indicating that the domains were larger than the illumination spot on the specimen. This is a significant improvement from a typical x-ray beam size of mm^2. The strong interaction between electrons and the sample enables the diffraction from a few molecular layers to be recorded in seconds. The measurable radiation damage dose to our specimen was at least an order of magnitude higher than our experimental dose. Thus the technique is a logical choice.

The one remaining problem is the vacuum damage to the specimen. This is circumvented by the use of an environmental stage, which also offers temperature control. The combination of low dose and *in situ* electron diffraction gives us the opportunity to observe and quantitate, for the first time, the orientational order in a true 2-D system. The scaling of these order parameters provides the first definitive evidence for the existence of 2-D hexatic ordering. It has been suggested that our observed departure from 2-D behavior in thicker films may be interpreted in terms of interlayer coupling (Aharong and Kardar, 1988).

Lipid bilayers are special cases of liquid crystals. The phase transitions are well known, and single-domain patterns have been observed from molecular bilayers (Hui et al., 1974). Due to the quasi-isotropic nature of the fatty acyl chains, the transition from a crystal to a less ordered phase relaxes the atomic position but retains the molecular rotomer position(Garoff et al.,1986).Its mechanical fragility and extreme susceptibility to radiation damage render it a difficult system for quantitative measurement.This special case, widely adapted for biological functions,poses a challenge to future theoretical and experimental studies.

Acknowledgement.

We would like to thank John Goodby and David Johnson for our liquid-crystal materials. This work was supported by the National Institutes of Health (Grant No. GM28120 to SWH) and the National Science Foundation (Grant No. DMR-8717674 to JTH).

References

A. Aharony and M. Kardar, Phys. Rev. Lett. 61, 2855 (1988).

A. Aharony, R.J. Birgeneau, J.D. Brock, J.D. Litster, Phys. Rev. Lett. 57, 1012 (1986).

J. D. Brock, A. Aharony, R.J. Birgeneau, K.W. Evans-Lutterodt, J.D. Litster, P.M. Horn, G. B. Stephenson, and A.R. Tajbakhsh, Phys. Rev. Lett. 57, 98 (1986).

M. Cheng, J.T. Ho, S.W. Hui, and R. Pindak, Phys. Rev. Lett. 61, 550-53 (1988).

M. Cheng, J.T. Ho, S.W. Hui, and R. Pindak, Phys. Rev. Lett. 59, 1112-15 (1987).

S.C. Davey, J. Budai, J.W. Goodby, R. Pindak, and D.E. Moncton, Phys. Rev. Lett. 53, 2139 (1984).

S. Garoff, H.W. Deckman, J.H. Dunsmuir, M.S. Alvarez and J.M. Bloch, J. Physique 47, 701-709 (1986)

B.I. Halperin and D.R. Nelson, Phys. Rev. Lett. 41, 121-24 (1978).

S.W. Hui, Biochim. Biophys. Acta, 472, 345-371 (1977).

S.W. Hui, J. Electron Microscopy Technique, 11, 286-97 (1989).

S.W. Hui, Ultramicroscopy, 5, 505-12 (1980).

S.W. Hui, G.G. Hausner, and D.F. Parsons, J. Phys. E:Scientific Instruments 9, 69-72 (1976).

S.W. Hui, D.F. Parsons, and M. Cowden, Proc. Nat. Acad. Sci. USA, 71, 5068-72 (1974).

M. Paczuski and M. Kardar, Phys. Rev. Lett. 60, 861 (1988).

K.J. Strandberg, Rev. Mod. Phys. 60, 161 (1988).

A PRIORI CRYSTAL STRUCTURE MODELING OF POLYMERIC MATERIALS

Serge PEREZ
Laboratoire de Physicochimie des Macromolécules
Institut National de la Recherche Agronomique
BP 527
44026 Nantes
France.

ABSTRACT. Under appropriate conditions micron-sized crystals of linear polymers can be grown. Electron diffraction data collected on such crystals provide highly reliable lattice constants, test for symmetry elements, and base plane structure amplitudes. The joint use of molecular modeling and electron diffraction has been invaluable in the quantitative elucidation of crystals and molecular structures for both synthetic and natural linear polymers. The modeling technique can be combined with the information derived from electron diffraction, i.e. accurate unit-cell parameters, and unambiguous determination of the space group symmetry to help solving three-dimensional structures. A method for predicting the packing relationship of two polymer chains has been developed. Given a rigid model of an isolated chain, its interaction with a second chain is studied at varied helix-axis translations and mutual rotational orientations while keeping the helices in van der Waals contact. The stability of each structure is evaluated by an energy calculated using atom-atom potentials that include compensation for hydrogen bonding. The structures derived from this process are surprisingly good approximations to the experimental ones. These results indicate that isolated chain-pair models can provide useful information about possible modes of interactions between the chains, as well as their three-dimensional modes of association. Moreover, they provide a unique opportunity to explain why the crystal structures are the stable forms.

Introduction

Many recent advances in the theory and application of molecular modeling to structural elucidations of polymers have produced a wide range of useful results. It is fair to state that, in combination with experimental methods, computer modeling has become an integral part of the elucidation scheme of 3-dimensional structures, both in solution and in the condensed phase. In the field of electron crystallography of polysaccharides, a recent review [1] concluded that apart from the knowledge of the primary structure (i.e. nature of the repeating units and type of linkage between these units), the only experimental data required for a crystal structure elucidation are accurate unit-cell parameters and an unambiguous determinaton of the space group. From this information follow all the structural details.

Now the question arises as to whether molecular modeling has any predictive power?

J. R. Fryer and D. L. Dorset (eds.), Electron Crystallography of Organic Molecules, 33–53.
© 1990 *Kluwer Academic Publishers.*

Starting from only the primary structure of a stereoregular polymer, it is quite straightforward to identify all the low energy conformers which are likeley to occur in vacuum. Taking into account solvation, even in a qualitative way, the equilibrium between these conformational states may be assessed, providing some insight into the inherent conformational characteristics of individual macromolecules free of interactions with other like or unlike species. Modeling of solution conformations of polymers provides some insight not only into the overall shape of the random coil, but also into the occurrence of local helical regions that might be appropriate to further ordering. After these regions are identified it is essential to evaluate all the possible stable modes of interactions and/or aggregations between these regions. Methods for investigating the inter-helix structure and energy through non-bonded forces have been suggested by a number of workers [2-7]. Those procedures minimize the inter-helix energy. We have developed a method [8-9] where the helices are moved as close to each other as is possible without causing interpenetration of the van der Waals radii of atoms of the two different helices. After the helices are positioned to the shortest interhelical distance for a given helix-helix translation and rotation, the energy is calculated. The results indicated that our modeling scheme could provide useful information about possible modes of interactions between chains.

The hydroxylated character of biopolymers influences between-chain interactions through networks of hydrogen bonds that occur during crystallization. Frequently, several possible attractive interactions exist that lead to different packing arrangements, and several allomorphic crystalline forms have been observed for biopolymers such as polysaccharides. There was therefore a need to extend our method to compensate for interchain hydrogen bonding [10].

In the present work, we report progress on the method where some aspects of the crystalline arrangements, i.e. position and orientation of the polymer chains, symmetry group and lattice constants can be predicted. We have selected four polymer systems of increasing complexity: poly(1,4-trans-cyclohexanediyl-dimethylene terephthalate), polyethylene terephthalate, chitin and the crystalline moieties of starch. We describe the calculated structures of several polymer chains and make direct comparison with experimental data.

Methodology

Here, we consider two chains interacting, the two chains being parallel or antiparallel. In what follows, the denomination of a chain structure is best defined by a unidirectional periodic structure. We assume, as a first approximation, that the structure of the chains are periodic, and perfectly rigid. For polymeric chains in fully ordered crystalline phases, the first part is always fulfilled. The second part can be only approximately true.

SINGLE CHAIN MODELING

In absence of any experimental information, the prediction of stable helical regions which are appropriate for further ordering is not untracktable, eventhough extensive computer time may be required. This issue has been addressed for polysaccharide chains [11], and will not be presented in this work. The problem of finding all the periodic conformations of a polymer chain that are consistent with an experimentally fiber repeat of a given helical symmetry is particularly well understood. Geometric procedures to accomplish this task were used in some of the earliest structural elucidations of stereoregular polymers [12], and have since been actively extended and refined [13,14]. In these calculations, the valence geometry of the repeating unit is fixed, while skeletal torsion angles are allowed to vary over all space. The geometric parameters of the chain (usually the helical parameters) are mapped, and all conformations consistent with the observed fiber repeat are enumerated. Supplemented by estimates of the conformational energy, such calculations have aided in the determination of a very large number of polymer crystal structures.

CHAIN PAIR MODELING

In the following analysis, we assume that the two chains, referred to as chain A and chain B, are regular helices, i.e. that they have screw symmetry, with a repeat distance t_A and t_B, respectively. It is clear that in an ordered periodic arrangement, the periods of the two interacting chains must be commensurate. This requirement of commensurability can be expressed as :

$$p \cdot t_A = s \cdot t_B = t$$

where p and s are small integers.

Chains can be only parallel and antiparallel to one another. A set of 4 interhelical parameters is required, in order to define the geometric orientation of chain A relative to chain B. They are :

μ_A : a rotation of the chain A about its axis, from 0° to 360°.

μ_B : a rotation of the chain B about its axis, from 0° to 360°.

Δ_x and Δ_z are taken as non negative and represent positional shifts normal and parallel to the identity axes respectively. The spatial description of these parameters is shown in Fig. 1.

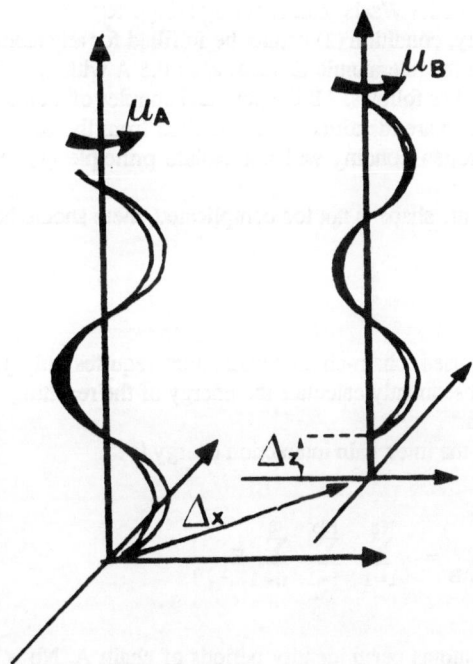

Figure 1. Interhelical parameters required to define the geometric orientation of chain A relative to chain B.

CONTACT PROCEDURE

The minimum energy arrangement of two molecules, with respect to a displacement, will tend to bring the molecules as close as possible without interpenetration. In reality, a small amount of repulsive energy, resulting from interpenetration of some numbers of atom pairs can be compensated by additional attraction from the remaining atom pairs. However, non-bonded "contact" distances deviate by only 10% (0.20 , 0.30 A) in molecular solids.

Since the principle of the contacting procedure has already been described [8], only a brief outline will be given here. The surface of each chain is defined by circumscribing a hard sphere of the appropriate van der Waals radius R around each of the constituent atoms. Then, for a given orientation of the chains (as dictated by rotation angles μ_A, μ_B and increment along the chain axis, Δ_z) one has to find the relative translation which is going to bring the interpenetrating surfaces into a position where they are in contact, with no interpenetration.

In general, the final position of the two initially interpenetrating molecules is characterized by the following conditions :

(1) For at least one atom pair (i,j), the ith atom of chain A is separated from the jth atom of chain B by the sum of the associated van der Waals radii ($R_{ij} = R_i + R_j$). The contact condition, and the atom pair i,j which satisfy this condition is referred to as the *determining contact*.

(2) For all atom pairs in the two separate chains, there is no pair at a distance closer than the appropriate van der Waals radii sum ($d_{ij} > R_i + R_j$).

(3) Obviously, condition (2) cannot be fulfilled for any atom pair (i,j) involved in hydrogen bonding for which the interatomic distance span 0.5 A with no *a priori* optimum value. This limiting case was treated as follows. All the potential couples of atoms eligible to participate in an interchain hydrogen bond are identified, and ommited from the contacting procedure. This implicitely means that hydrogen bonding will not violate principle (1) for the van der Waals bonded atoms.

It is clear that if the molecule shape is not too complicated, there should be only one solution to the interacting problem.

ENERGY CALCULATIONS

If a contacting procedure is used, chain-chain construction requires only geometric information and in principle, one can subsequently calculate the energy of the resulting interactions (E_{AB}) to any degree of approximation.

The formal expression for the interchain interaction energy is :

$$E_{AB} = \sum_{i=1}^{Na} \sum_{j=1}^{Nb} \sum_{n=1}^{\omega} E_{ijn}$$

where Na is the number of atoms per *p* identity periods of chain A, Nb is the number of atoms per *s* identity periods of chain B, and ω is the number of repeating units.

The interaction energy of the two chains is considered to be the sum over all pairwise atom-atom interactions. Such interaction is calculated according to 6-12 potential functions [15]. These terms represent a short-range repulsive interaction and a short-range attractive interactions, respectively. To these, may be added coulombic interactions between the atoms and the on-charges. As for the energy stabilization arising from hydrogen bonding, an extra term has to be included. In the present simulations on polysaccharide chains, none of the hydroxylic hydrogen atoms was considered. Therefore, a simple energy criterium was used, based on the distance between the oxygen and/or the nitrogen atoms which can interact through hydrogen bonding [16].

We limit ourselves to investigating whether the information provided by short-range interactions alone is of utility in identifying structural stable assemblies of polymer chains. In performing the interchain energy calculation, we have used a cut-off distance such that $d_{ij} < 1.5 R_{ij}$, where R_{ij} is the appropriate van der Waals radii sum. The number of interatomic contacts satisfying this particular condition is also computed, and referred to as the number of "close contacts". It represents an indicator of the complementarity of the shapes of the interacting chains, in a particular relative orientation.

PRACTICAL TREATMENT

In practical calculations we have adopted the procedure of mapping Δ_X, E_{AB} and the number of "close contacts" as a function of the structural variables, μ_A, μ_B and Δ_Z. The two orientation angles μ_A and μ_B are bounded between 0 and 360° ; whereas Δ_Z is bounded between 0 and t, which is the fiber repeat. The investigation is performed by rotating μ_A and μ_B over the whole angular range by 10° or 5° increments; the relative translation (Δ_Z) between the two chains is investigated over the length of a whole fiber repeat, typically by 0.5 A increments. For each setting of the chains, as a function of μ_A, μ_B and Δ_Z, the magnitude of the perpendicular off-set (Δ_X) is derived according to the contacting procedure outlined above. Then, the value of the energy E_{AB} and the number of "close contacts", corresponding to each set of chain orientations are computed. Several ways can be envisaged to search for the energy minima. We found that the mapping procedure was quite adequate, particularly in the first steps of the investigation, since it provide a complete overview of the symmetry (or lack of symmetry) of the chain-chain interactions.

EXTRACTING INFORMATION ABOUT THREE-DIMENSIONAL ORGANIZATIONS

The set of interhelical parameters relates directly to the symmetry operations which are found in crystal structures.

* $\mu_A \neq \mu_B$ represents the case where chain A and chain B are not related by any symmetry operation. They would make up the asymmetric unit content, and would be described as being "independent".

* $\mu_A = \mu_B$ represents the situation where chain B is derived from chain A by a pure translational symmetry element.

* $\mu_A = \mu_B + 180°$ and $\Delta_Z = 0$, represents the situation where the two chains are parallel and related by a two-fold operation. For a two-fold screw axis operation, the condition will be $\mu_A = \mu_B + 180°$ and $\Delta_Z = t/2$.

* $\mu_A = -\mu_B + 180°$ represents the situation where the two chains are antiparallel and related by a two-fold or a two-fold screw axis operation.

Similarly, information about the three-dimensional organization of the chains can be obtained, providing that two orientations of chain B with respect to chain A are known i.e.(μ_A, μ_B, Δ_X, Δ_Z)

and $(\mu'_A, \mu'_B, \Delta'_x, \Delta'_z)$. For example, when chain B and chain A are related by a translation, the unit cell parameters $(a, b, c, \alpha, \beta, \gamma)$ can be readily derived in the following way:

$$a\,(A) = \Delta_x{}^2 + \Delta_z{}^2$$
$$b\,(A) = \Delta'_x{}^2 + \Delta'_z{}^2$$
$$c\,(A) = t \text{ (known or assumed)}$$
$$\alpha\,(°) = \text{Arcos}(\Delta'_z / b)$$
$$\beta\,(°) = \text{Arcos}(\Delta_z / a)$$
$$\gamma^*(°) = 180° - (\mu'_A - \mu_A)$$
$$\gamma\,(°) = \text{Arcos}(\cos\alpha.\cos\beta - \sin\alpha.\sin\beta.\cos\gamma^*)$$
$$V\,(A^3) = a.b.c.\sin\alpha.\sin\beta.\sin\gamma^*$$

Results and Discussion

POLY(1,4-TRANS-CYCLOHEXANEDIYL-DIMETHYLENE TEREPHTHALATE)

Poly(1,4-trans-cyclohexanediyl-dimethylene terephthalate) or poly(t-CDT) is a related member of the poly(oligomethylene terephthalates) series of polyesters. A schematic of the chain of poly(t-CDT), along with the labeling of the atoms is shown in Fig. 2.

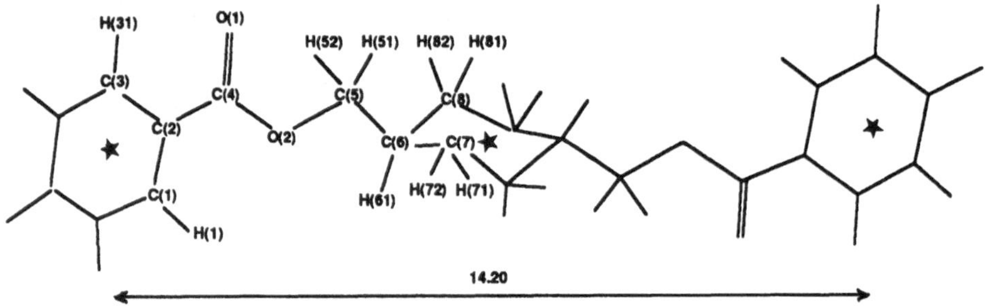

Figure 2. Schematic representation of the repeating unit of poly(t-CDT). (*) indicates the centers of symmetry within the chain.

X-ray fiber diffraction patterns recorded on the poly(t-CDT) indicated that the fiber repeat is 14.2 A. Previous study, involving model compounds has shown that an all *trans* conformation of the repeating unit reproduces the observed fiber repeat of the parent polyester (Fig. 2). From this observation, which was corroborated through conformational analysis a chain was constructed and used as such in the elucidation of the crystal structure of the polymer [17]. The polymer crystallizes in a triclinic unit cell, space group P1 with parameters: $a = 6.46(1)$ A, $b = 6.65(1)$ A, $c = 14.2(2)$ A (fiber axis), $\alpha = 89.4(5)°$, $\beta = 47.0(3)°$ and $\gamma = 114.9(5)°$.

In the present work the same chain model of poly(t-CDT) was used for *a priori* modeling of chain associations without any further experimental data. A list of the atomic coordinates corresponding to two contiguous residues is given in Table 1.

TABLE 1. Cartesian coordinates (A) of the repeating unit used for poly(t-CDT) (oriented along the c axis).

Label	x	y	z	Label	x	y	z
O-1	2.6254	-0.5856	-1.0583	O-1'	-2.6254	0.5856	8.4598
O-2	0.9289	0.3419	0.0568	O-2'	-0.9289	-0.3419	7.3447
C-1	-0.6749	0.2755	-2.1681	C-1'	0.6749	-0.2755	9.5696
C-2	0.6608	0.0073	-2.2518	C-2'	-0.6608	-0.0073	9.6533
C-3	1.2400	-0.2630	-3.4896	C-3'	-1.2400	0.2630	10.8911
C-4	1.5611	-0.0834	-1.0443	C-4'	-1.5611	0.0834	8.4458
C-5	1.6594	0.1424	1.3002	C-5'	-1.6594	-0.1424	6.1013
C-6	0.6679	0.4000	2.4401	C-6'	-0.6679	-0.4000	4.9613
C-7	-0.5109	-0.5403	2.4372	C-7'	0.5109	0.5403	4.9643
C-8	1.4430	0.3051	3.7780	C-8'	-1.4430	-0.3051	3.6235
H-1	-1.1241	0.4850	-1.2096	H-1'	1.1241	-0.4850	8.6110
H-3	2.2971	-0.4753	-3.4667	H-3'	-2.2971	0.4753	10.8682
H-51	2.4218	0.9020	1.3925	H-51'	-2.4218	-0.9020	6.0089
H-52	1.9172	-0.9063	1.2991	H-52'	-1.9172	0.9063	6.1023
H-61	0.2558	1.3984	2.4231	H-61'	-0.2558	-1.3984	4.9784
H-71	-1.0616	-0.4646	1.5113	H-71'	1.0616	0.4646	5.8901
H-72	-0.1767	-1.5673	2.4263	H-72'	0.1767	1.5673	4.9751
H-81	2.2520	1.0202	3.8040	H-81'	-2.2520	-1.0202	3.5974
H-82	1.9011	-0.6701	3.8537	H-82'	-1.9011	0.6701	3.5478

In a first step, all the possible arrangements occurring between two poly(t-CDT) chains were examined. This was performed by rotating μ_A and μ_B over the whole angular range from 0° to 360° by 10° increments. The relative displacement of the two chains was investigated over the whole length of the fiber repeat by 0.5 A increments. For each setting of the chains, as a function of μ_A, μ_B, and Δ_Z the magnitude of the perpendicular off-set, Δ_X, was computed according to the contact procedure outlined above. The value of the energy corresponding to each set of chain orientations, as defined by the set of 4 interhelical parameters, was evaluated. Within the three-dimensional (μ_A, μ_B, Δ_Z) space, the search for the energy minima was performed. The results indicated that the significant energy minima occurred for values of μ_A and μ_B such that $\mu_A = \mu_B$. The coupling of μ_A and μ_B rotation angles suggests that the relative packing of neighboring chains of poly(t-CDT) is best achieved through pure translational symmetry. From now on, the investigation was conducted assuming $\mu_A = \mu_B$. This allows for a straigtforward two-dimensional study. The contour maps calculated as a function of the translation Δ_Z, along the fiber axis and the coupled rotation angles $\mu_A = \mu_B$, are shown on Fig. 3. Fig. 3a is a representation of the variations of the perpendicular off-set (Δ_X). Fig. 3b is a representation of the variations of the interchain energy corresponding to the optimum perpendicular off-set, Δ_X, as a function of translation, Δ_Z, along the chain direction and coupled rotations of μ_A and μ_B. These

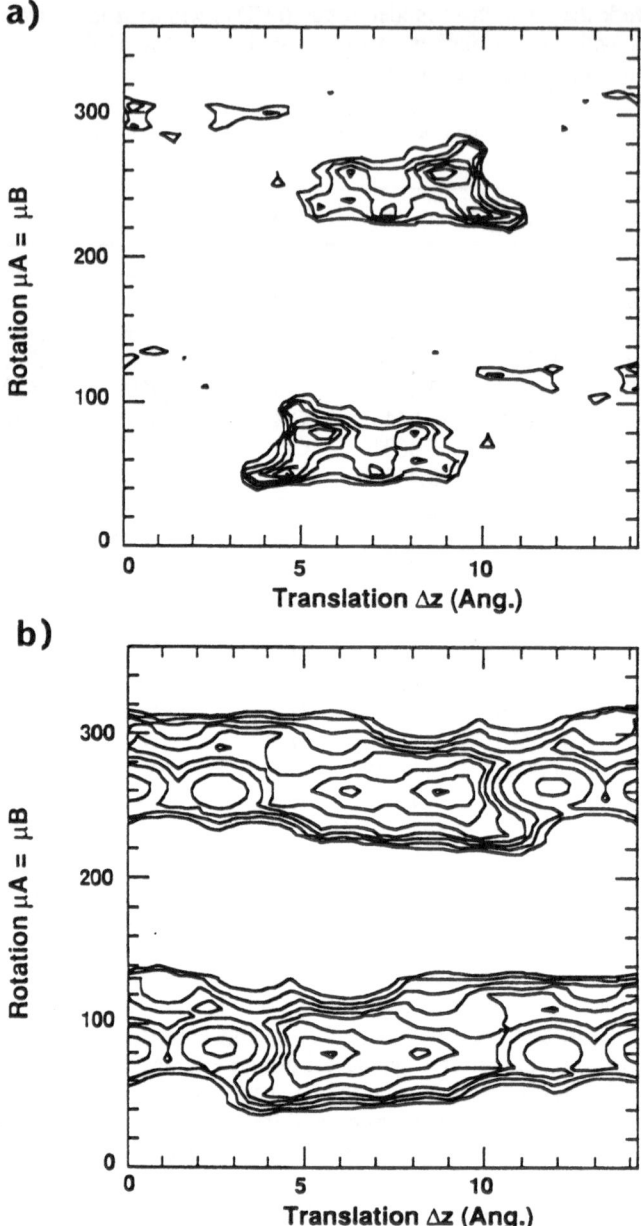

Figure 3. Contour maps as a function of the translation: Δ_Z, along the fiber axis of poly(t-CDT), and the coupled rotation angles $\mu_A = \mu_B$. a) Variations of the perpendicular off-set (Δ_X). Contours correspond to 4.1, 4.3, 4.5, 4.7, 4.9, 5.1, 5.3 A. b) Interchain energies calculated for the corresponding Δ_X. Contours correspond to -11, -10, -9, -8, -7, -6 kcal/mol.

maps exhibit an obvious symmetry which is due to the centers of symmetry within the repeat unit of the polymer chain. Consequently, only one section needs to be described. Domains are found, which correspond to a somewhat restricted range of μ_A values and an extended range of Δ_z values. Within these domains, several energy minima are found, which characteristics are listed in Table 2. In this table, the interchain arrangements observed in the crystal structure [17] are also reported.

TABLE 2. Optimum values for the chain-chain interactions for poly(t-CDT).

		Calculated				Observed [17]		
A chain	B chain	μ (°)	Δz (A)	Δx (A)	Energy Kcal/mol	μ (°)	Δz (A)	Δx (A)
t-CDT	t-CDT	79.0	5.6	3.97	-12.0			
t-CDT	t-CDT	50.0	4.4	4.44	-11.2	51.5	4.4	4.72
t-CDT	t-CDT	50.0	7.0	4.68	-10.1			
t-CDT	t-CDT	78.0	8.0	4.14	-10.2			
t-CDT	t-CDT	111.0	13.9	4.35	-9.5			
t-CDT	t-CDT	110.0	11.6	4.27	-8.1			
t-CDT	t-CDT	111.0	2.2	4.58	-7.4			
t-CDT	t-CDT	134.0	1.3	5.02	-7.2			
t-CDT	t-CDT	18.0	2.9	5.87	-6.5			
t-CDT	t-CDT	180.0	0.2	6.56	-3.7	177.3	0.1	6.64

The occurrence of several energy minima in the vicinity of μ_A = 70 +/- 40°, for values of Δ_z ranging from 4 to 10 A indicates that there are many stable ways to pair parallel chain of poly(t-CDT). Obviously, such situations may occur in the amorphous phase of the polymer. The global energy minimum of -12 kcal/mol occurs at μ_A = 79°, Δ_z = 5.60 A and Δx = 3.92A. It corresponds to an interchain vector of 6.86 A. The second lowest energy minimum occurs at μ_A = 50°, Δ_z = 4.40 A and Δ_x = 4.44 A; it corresponds to an interchain vector of 6.25 A. Therefore, the density is higher than in the previous situation. This minimum which has an energy of only 0.8 kcal/mol higher than the previous one corresponds almost exactly to one of the relative orientations of the chains in the crystal lattice: μ_A = 51.5°, Δ_z = 4.40 A and Δ_x = 4.72 A. A secondary energy minimum of -3.7 kcal/mol is found for μ_A = 180°, Δ_z = 0.2 A and Δ_x = 6.56 A. This calculated chain pairing is close to the other chain orientation found in the crystal (μ_A = 177.3°, Δ_z = 0.07 A and Δ_x = 6.64 A). Combining the relative chain orientations which correspond to these two calculated energy minima yields the following triclinic unit cell: a = 6.25 A, b = 6.56 A, c = 14.2 A (fiber axis), α = 89.4°, β = 47.0° and γ = 117.9°.

POLYETHYLENE TEREPHTHALATE

A schematic of the chain of polyethylene terephthalate (PET) is shown in Fig. 4. Previous studies, involving model compounds of PET have shown that the elucidation of the crystal structure of ethylene glycol dibenzoate [18] provides highly accurate information about the molecular geometry; also, the all *trans* conformation of the methylenic part reproduces the observed fiber repeat of the parent polyester [19]. The structural data derived from crystallographic studies on model compound along with conformational analysis [20] provide the basis for the construction of a chain of PET repeating every 10.70 A. This chain geometry has been used in the present study. A list of the atomic coordinates corresponding to three contiguous motifs is given in Table 3. Results are compared with the three-dimensional crystal structure which is well established [21,22].

TABLE 3. Cartesian coordinates (A) of the repeating unit used for PET (oriented along the *c* axis).

Label	x	y	z	Label	x	y	z
C-11	-0.0002	-0.6988	-7.0467	C-11	-0.0066	-0.4711	3.7456
C-12	0.0174	0.6674	-7.2408	C-12	0.0095	-1.3644	2.6601
C-13	0.0023	1.1470	-8.5150	C-13	0.0234	-0.8983	1.3808
C-14	-0.0496	0.2763	-9.6016	C-14	0.0405	0.4535	1.1704
C-15	-0.0598	-1.0812	-9.4124	C-15	0.0173	1.3425	2.2455
C-16	-0.0376	-1.5744	-8.1392	C-16	-0.0038	0.8871	3.5336
C-17	-0.0032	-1.2466	-5.6892	C-17	-0.0032	-0.9405	5.1338
C-18	0.0104	-0.7653	-3.3691	C-18	-0.0182	-0.3313	7.4269
C-27	0.0434	0.9876	-0.1920	C-27	-0.0548	1.5934	10.5063
C-28	0.0296	0.4820	-2.5027	C-28	-0.0395	0.9606	8.2238
O-11	-0.0039	-2.4256	-5.4532	O-11	0.0275	-2.1020	5.4355
O-21	0.0642	2.1569	-0.4366	O-21	-0.0651	2.7456	10.1965
O-12	0.0124	-0.2970	-4.7190	O-12	-0.0204	0.0603	6.0509
O-22	0.0276	0.0245	-1.1498	O-22	-0.0374	0.5798	9.6022
H-12	0.0632	1.3466	-6.3752	H-12	-0.0091	-2.4308	2.8374
H-13	0.0224	2.1778	-8.6518	H-13	0.0288	-1.5784	0.5692
H-15	-0.0311	-1.7570	-10.2793	H-15	-0.0383	2.4052	2.0641
H-16	-0.0360	-2.6573	-7.9908	H-16	-0.0323	1.5978	4.3892
H-18	-0.7897	-1.3219	-3.1858	H-18	-0.8034	-0.8985	7.6368
H-18'	0.8638	-1.3445	-3.2148	H-18'	0.8501	-0.8753	7.6185
H-28	-0.7934	1.0698	-2.6579	H-28	-0.8777	1.5141	8.0310
H-28'	0.8723	1.0471	-2.6670	H-28'	0.7881	1.5384	8.0328

In a first step, all the possible arrangements occurring between two PET chains were examined, using the procedure described above for poly(*t*-CDT). The results indicated that the significant energy minima occurred for values of μ_A and μ_B such that $\mu_A = \mu_B$. The coupling of μ_A and μ_B rotation angles suggests that the relative packing of neighboring chains of PET is best achieved through pure translational symmetry. From now on, the investigation was conducted assuming $\mu_A = \mu_B$. This allows for a straigtforward two-dimensional study. Fig. 5a is a representation of the variation of the interchain energy corresponding to the optimum perpendicular off-set, Δ_X, as a function of translation, Δ_Z, along the chain direction and coupled rotations of μ_A and μ_B. This map exhibits an obvious symmetry and only one section needs to be described. Essentially only two domains are found, which correspond to a somewhat restricted range of μ_A values and an extended range of Δ_Z values.

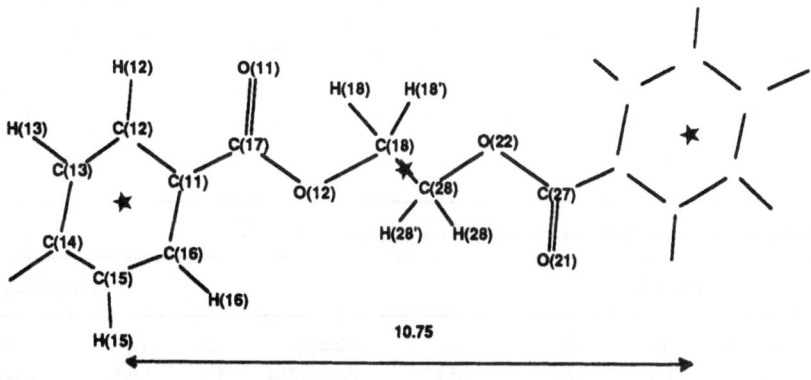

Figure 4. Schematic representation of the repeating unit of PET. (*) indicates the centers of symmetry whithin the chain.

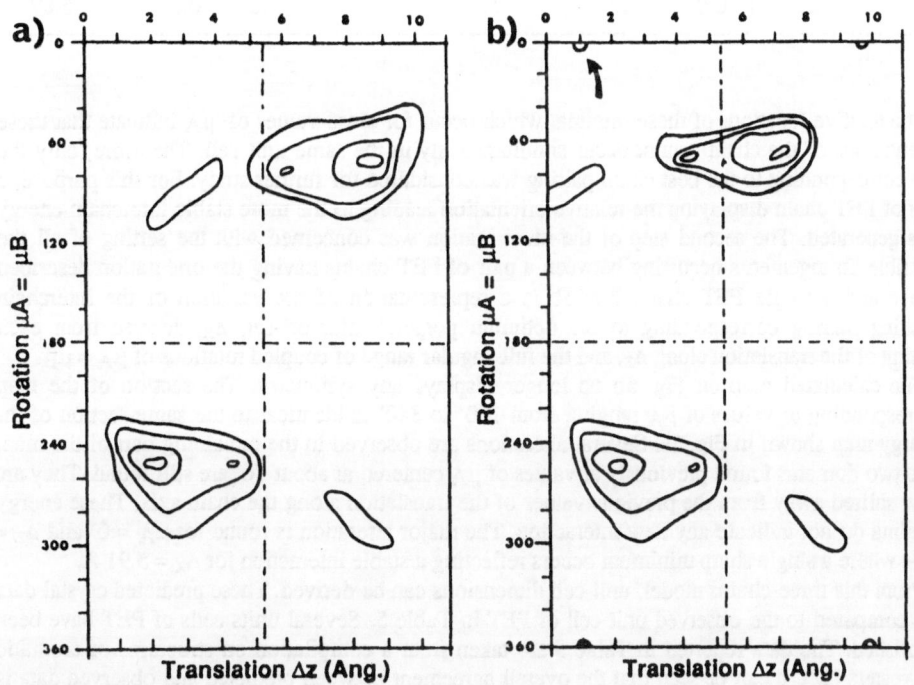

Figure 5. Iso-energy contour maps calculated as a function of the translation Δz, along the fiber axis of PET, and the coupled rotation angles $\mu_A = \mu_B$ for the corresponding Δx. a) between two single chains of PET, b) between one single chain of PET and a pair of PET chains having the orientation $\mu_A = 249°$, $\Delta z = 1.90$ A and $\Delta x = 3.49$ A.

Within these domains, three energy minima are found, which characteristics are listed in Table 4. In this table, the best interchain arrangement observed in the crystal structure [21] is also reported. It is clear that the minimum occurring for : $\mu_A = 249°$, $\Delta_z = 1.90$ A, $\Delta_x = 3.49$ A corresponds to very favorable pairing, not only in terms of energy (-10.7 kcal/mol) but also in terms of density, since the distance between the center of mass of the two chains is 3.97 A.

TABLE 4. Optimum values for the chain-chain interactions for PET.

		Calculated				Observed [21]		
A chain	B chain	μ (°)	Δ_z (A)	Δ_x (A)	Energy Kcal/mol	μ (°)	Δ_z (A)	Δ_x (A)
PET	PET	249.0	1.90	3.49	-10.75	241.6	2.14	4.02
PET	PET	255.0	4.49	3.62	-9.05			
PET	PET	110.0	2.25	3.92	-8.70			
PET	PET-PET	0.0	1.0	5.91		2.2	0.87	5.87

The relative locations of these minima, which occur for close values of μ_A indicate that these orientations of the chain cannot occur simultaneously in the same unit cell. Therefore, only the one corresponding to the best chain pairing was considered for further study. For this purpose, a pair of PET chain displaying the relative orientation leading to the more stable interchain energy was generated. The second step of the investigation was concerned with the setting of all the possible arrangements occurring between a pair of PET chains having the orientation described above and a single PET chain. Fig. 5b is a representation of the variation of the interchain packing pairing corresponding to the optimum perpendicular off-set, Δ_x, derived from each setting of the translation along Δ_z, and the full angular range of coupled rotations of $\mu_A = \mu_B$.

The calculated map on Fig. 5b no longer displays any symmetry. The section of the map corresponding to values of μ_A ranging from 160° to 340° is identical to the same section of the energy map shown in Fig. 5a. Several alterations are observed in the remaining part of the map. The two domains found previously at values of μ_A centered at about 70° are still found. They are now shifted away from the previous values of the translation along the chain axis. These energy minima do not indicate any new interaction. The major alteration is found for $\mu_A = 0°$ and $\Delta_z = 0$ A; where a single sharp minimum occurs reflecting a stable interaction for $\Delta_x = 5.91$ A.

From this three-chains model, unit-cell dimensions can be derived. These predicted crystal data are compared to the observed unit cell of PET in Table 5. Several units cells of PET have been published. The data reported in Table 5 are taken from a compilation on structures of aromatic polyesters [22]. It can be seen that the overall agreement between predicted and observed data is good. The strongest discrepancy occurs for the value of parameter a which is derived from the PET-PET interaction at $\mu_A = 249°$ and $\Delta_z = 1.90$ A. The calculated pairing exhibits a greater lateral packing feature than the one found for the corresponding interactions in the crystal structure.

TABLE 5. Calculated versus Observed Data for PET.

	Calculated	Observed [21,22]				
a (A)	3.97	4.56	4.48	4.52	4.44	4.52
b (A)	5.99	5.94	5.85	5.98	5.91	5.92
c (A)	10.75	10.75	10.75	10.77	10.67	10.70
α (°)	99.6	98.5	99.5	101.0	100.1	99.8
β (°)	119.4	118.0	118.4	118.8	117.0	117.5
γ (°)	111.4	112.0	111.2	111.0	111.8	111.4
V (A)^3	187.6	219.0	210.6	216.0	210.6	215.5

CHITIN

Chitin is the 2-acetamido derivative of cellulose and serves as the fibrous component of skeletal tissues in many lower animals. Two polymorphic forms of chitin have been recognized, of which the α and the β-forms are the best characterized. Both have apparently the same 2_1 helical conformation. Highly crystalline β chitin is obtained from pogonophore tubes and the spines of certain diatoms. The monoclinic unit cell (a = 4.86 A, b = 9.26 A, c = 10.38 A (fiber axis), β = 97.5°, space group P2$_1$) contains a disaccharide unit of a single chain and (Fig. 6), hence the structure is an array of parallel chains [23].

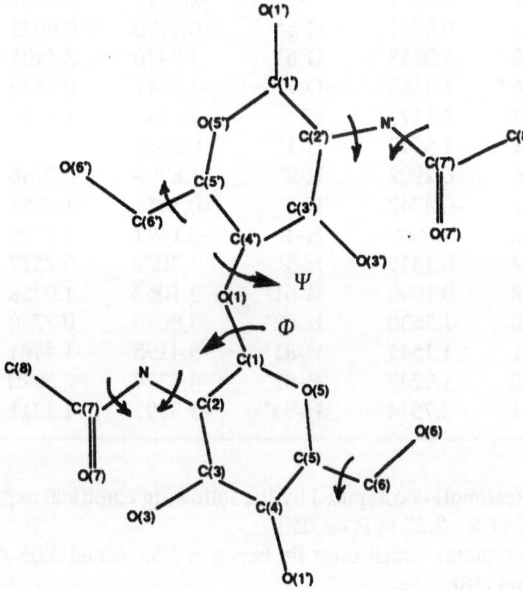

Figure 6. Schematic representation of the disaccharide repeating unit of chitin.

In the present work, the following parameters were taken into account to construct one chitin chain:

- rotation of O(6) about the exo-cyclic C(5)-C(6) bond,
- rotation of the entire amide side chain about the C(2)-N bond,
- rotation of the acetate group about the N-C(7) bond.

The magnitude of these torsion angles was assessed through conformational analysis using the PFOS program [24, 25]. The coordinates of the hydrogen atoms were determined using a C-H bond length of 1.08 A and a bond vector related appropriately to the C-C and C-O bond vectors. Hydroxylic hydrogen atoms were not considered. A list of the atomic coordinates of the disaccharide moiety representative of the chitin chain is given in Table 6.

TABLE 6. Cartesian coordinates (A) of the disaccharide used for chitin (oriented along the c axis).

Label	x	y	z	Label	x	y	z
C-1	0.3055	0.1684	1.3828	C-1'	-0.3055	-0.1684	6.5728
C-2	1.4588	0.2143	0.3967	C-2'	-1.4588	-0.3142	5.5867
C-3	0.9807	0.6913	-0.9735	C-3'	-0.9807	-0.6913	4.2165
C-4	-0.1955	-0.1582	-1.4406	C-4'	0.1955	0.1582	3.7494
C-5	-1.2731	-0.2074	-0.3611	C-5'	1.2731	0.2074	4.8289
C-6	-2.4211	-1.1259	-0.7244	C-6'	2.4211	1.1259	4.4656
C-7	3.6642	0.5423	1.4450	C-7'	-3.6642	-0.5423	6.6350
C-8	4.6699	1.5644	1.9225	C-8'	-4.6699	-1.5644	7.1125
O-4	-0.7602	0.3982	-2.6239	O-4'	0.7602	-0.3982	2.5661
O-3	2.0512	0.6031	-1.9181	O-3'	-2.0512	-0.6031	3.2719
O-5	-0.7170	-0.6922	0.8741	O-5'	0.7170	0.6922	6.0641
O-6	-1.9470	-2.4405	-1.0358	O-6'	1.9470	2.4405	4.1542
O-7	3.8447	-0.6616	1.5385	O-7'	-3.8447	0.6616	6.7285
N	2.5282	1.1039	1.5177	N'	-2.5282	-1.1039	6.7077
H-1	-0.1671	1.1241	1.5549	H-1'	0.1671	-1.1241	6.7449
H-2	1.8799	-0.7746	0.2902	H-2'	-1.8799	0.7746	5.5802
H-3	0.5464	1.6751	-0.8742	H-3'	-0.5464	-1.6751	4.3158
H-4	0.1497	-1.1506	-1.6902	H-4'	-0.1497	1.1506	3.4998
H-5	-1.7002	0.7577	-0.1319	H-5'	1.7002	-0.7577	5.0581
H-61	-3.1062	-1.0758	0.1090	H-61'	3.1062	1.0758	5.2990
H-62	-3.0043	-0.7750	-1.5630	H-62'	3.0043	0.7750	3.6270
H-81	5.4195	1.4461	1.1541	H-81'	-5.4195	-1.4461	6.3441
H-82	4.3374	2.5920	1.9242	H-82'	-4.3374	-2.5920	7.1142
H-83	5.3793	1.4711	2.7314	H-83'	-5.3793	-1.4711	7.9215

The hydrogen bond energy (kcal/mol) is computed by the following empirical expression :

$$V_{HB} = 33.14 \, (R - 2.55) \, (R - 3.05)$$

where R is the distance between atoms which must lie between 2.55 A and 3.05 A in order for this term to be taken into account [16].

a)

b)

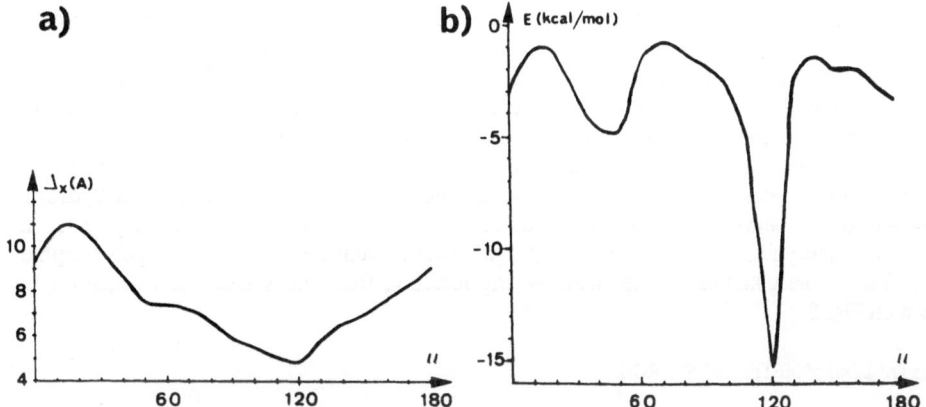

Figure 7. a) Representation of the variation of Δ_X as a function of μ from 0 to 180° and $\Delta_Z = 0$, for chitin chains. b) Representation of the interchain energy as a function of μ from O to 180° and $\Delta_Z = 0$, for chitin chains.

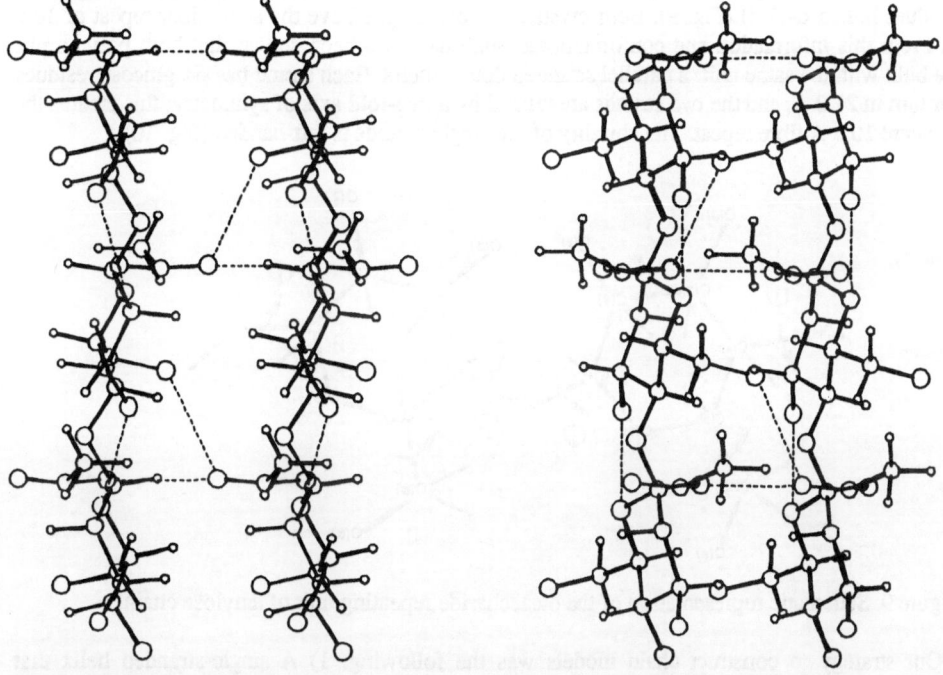

Figure 8. Molecular drawings (PITMOS, [28] of the best parallel pairing of chitin chains.

As in the previous example, all the possible arrangements for two parallel chitin chains were examined. It became clear that the significant energy minima were occurring for μ_A and μ_B and Δ_Z such that $\mu = \mu_A = \mu_B$ and $\Delta_Z = 0$. This was an indication that with their given conformations, packing of the chitin chains is best achieved through translation symmetry operations. This allows for a straightforward one dimensional study. Fig. 7a is a representation of the variation of Δ_X, as derived from the contacting procedure, as a function of rotation of μ from 0 to 180° and Δ_Z = 0. Fig 7b represents the variations of the interchain energy. Essentially only one energy minimum is found at μ = 120°, Δ_X = 4.88 A, and Δ_Z = 0 A. One of the relative orientation of two neighboring chitin chains in the crystal structure of the β-polymorph is characterized by μ = 119.3°, Δ_X = 4.85 A and Δ_Z = 0 A. Therefore, the agreement between observed and calculated interchain orientations is excellent. Close examination also reveals that interchain hydrogen bonds are formed between O-6 and O-7 (2.967 A), and N and O-7 (2.811 A). They reproduce, in a very satisfactory fashion, the observed hydrogen bonding scheme found in the β-polymorph of chitin. Two representations of the chain pairing resulting from our scheme of calculation are shown on Fig. 8.

CRYSTALLINE MOIETIES OF STARCH

In its native forms, starch, an energy reserve for green plants, exhibits two different diffraction patterns depending on the botanical origin: A-type in cereal starches and B-type in tuber starches. In both, diffraction is thought to arise mainly from the short chains that are connected at branch points of the amylopectin component of starch. These short chains have 12 to 20 D-glucose residues linked α-(1-4) (Fig. 9). Both crystalline polymorphs have the same fiber repeat of 10.5 A. From this information and conformational analysis it has been shown that both polymorphs are built with the same unit: a parallel stranded double-helix. Each strand has six glucose residues per turn in 21.0 A, and the two strands are related by a two-fold axis of symmetry; this creates the apparent 10.5 A fibre repeat. The chirality of the single strands is left-handed (Fig. 10).

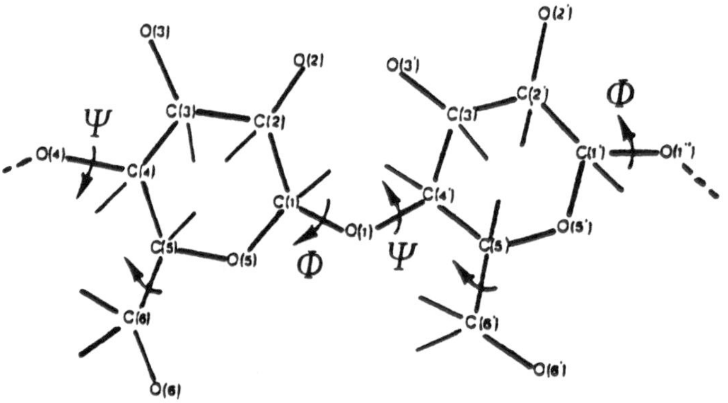

Figure 9. Schematic representation of the disaccharide repeating unit of amylose chain.

Our strategy to construct chain models was the following. 1) A single-stranded helix that repeats in 21.0 A was constructed. 2) The choice for the left-handed chirality was made after a Ramachandran plot of energy vs. glycosidic torsional angles was overlaid with contours of helical parameters. 3) A single strand, having these characteristics, was then used to generate a second strand of a double-helix through two-fold rotation (for every atom with coordinates x, y and z there is a new one at -x, -y and z).

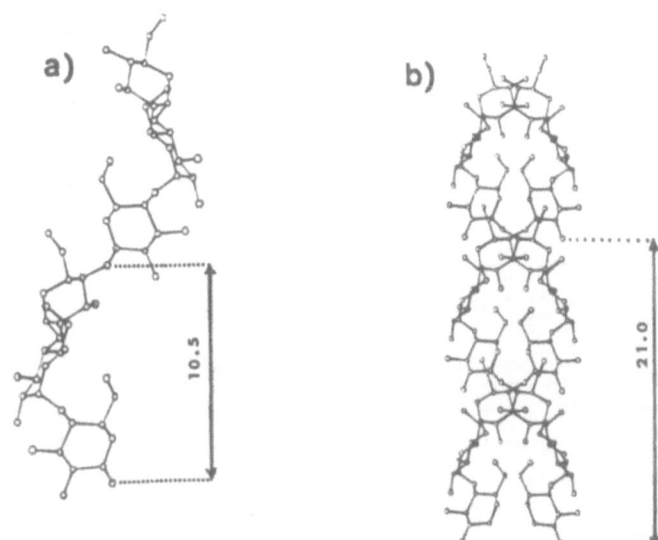

Figure 10. Molecular representation (PITMOS, [28] of (a) one single strand of amylose chain in its left-handed configuration, (b) the assembly of two such strands through two-fold symmetry operation (b).

All the possible arrangements occurring between parallel and antiparallel double-helices were examined. Since the details of these investigations have been reported elsewhere [10], only the essential features will be summarized here. The calculated interchain parameters for the best energy minima for parallel or antiparallel arrangements are collected in Table 7. The corresponding arrangements are shown in Fig. 11. The best pairing in terms of energy is found for a parallel arrangement of the double-helices (PARA 1). Another stable parallel chain-pairing is found which has an interaction energy 6.8 kcal/mol above the previous one (PARA 2). As for the antiparallel case, two stable chain-pairings are also found. The more stable arrangements, for both parallel and antiparallel cases, are characterized by a distance between the center of mass of the two double-helices of 10.77 A.

TABLE 7. Calculated interchain parameters for the best energy minima for parallel and antiparallel of double helices of amylose

	Para 1	Para 2	Anti 1	Anti2
$\mu_A(°)$	11.5	26.0	78.5	47.0
$\mu_B(°)$	11.5	167.5	41.5	13.0
Δ_Z(Ang)	5.25	3.22	7.86	7.39
Δ_X(Ang)	10.77	11.20	10.77	11.16
E(Kcal/mol)	-26.6	-19.8	-23.7	-20.9

50

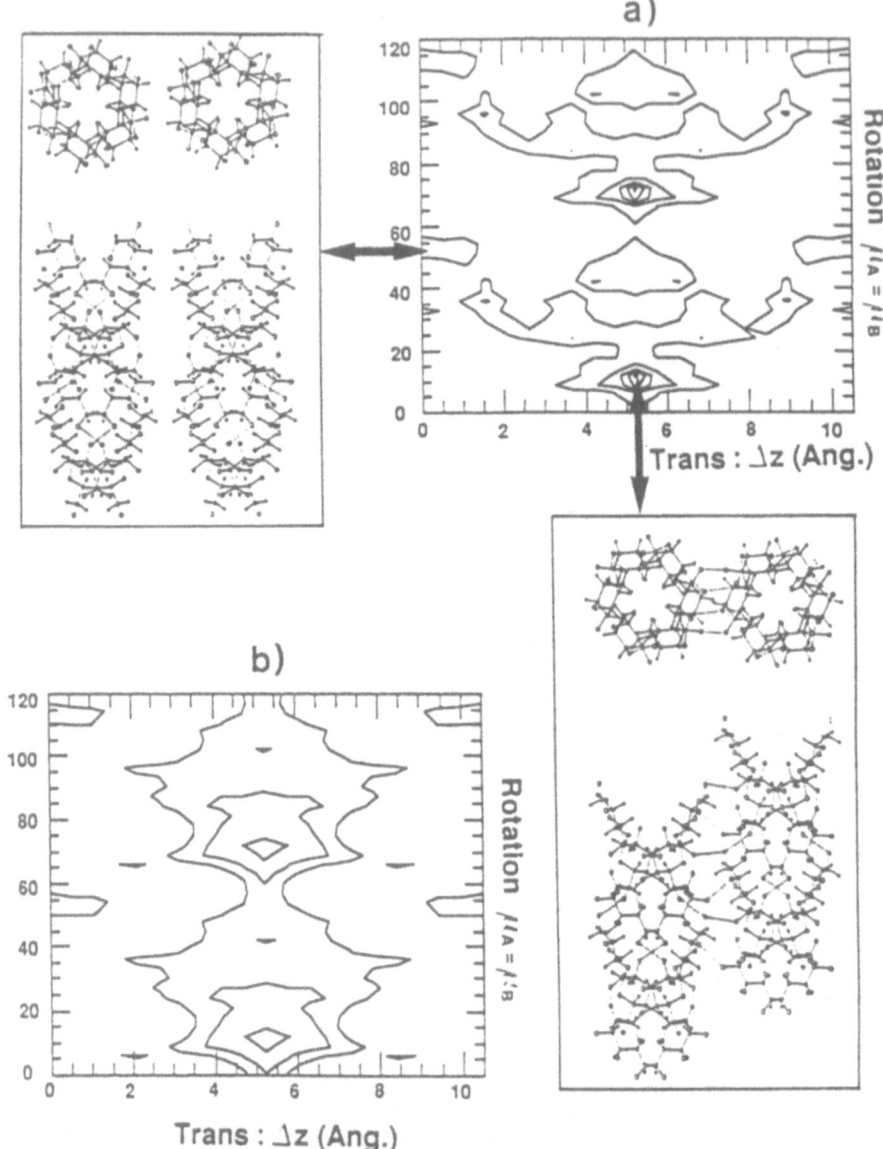

Figure 11. Contour maps calculated for a parallel arrangement of double helices, as a function of the translation Δ_z along the fiber axis and of the coupled rotation angles $\mu_A = \mu_B$. a) Interchain energies calculated for the corresponding Δ_x (see b). Contours correspond to -25, -20, -15, -10 and -5 kcal/mol. b) Variations of the perpendicular off-set (Δ_x). Contours correspond to 11.0, 11.5 and 12.0 Å. The arrangement occurring at $\Delta_z = 5.25$ Å and $\mu_A = 11.5°$ corresponds to the lowest interchain energy and generates the PARA1 model. The one occurring at $\Delta_z = 0.0$ Å and $\mu_A = 55°$ corresponds to a loose interchain interaction as found in the A-type crystal structure.

In the most recent crystallographic studies of the crystalline part of starch [26, 27] the structure of both polymorphs are based on a parallel arrangement of double-helices. In the two observed structures the double-helices are slightly different since small variations away from the perfect six-fold symmetry are found. Nevertheless, they correspond closely to the idealized double-helix studied in this work. The essential result is that in these two observed structures, the closest interactions between two neighboring double-helices correspond closely to the duplex described as PARA 1. In the crystal structures, neighboring double-helices have the same rotational orientation and the same translation of half a fiber repeat as in the PARA 1 model. Only the Δ_x vector is slightly larger in the calculated interactions (10.77 A) than in the observed ones: 10.62 A and 10.68 A in the A type and B type, respectively. This may due to the fact that in the crystal structures, the helices depart slightly from perfect 6-fold symmetry. Also, no interpenetration of the van der Waals surfaces is allowed in the modeling procedure, whereas some of them may occur in the crystal structures. It is interesting to note that the network of inter double-helices hydrogen bonds found in the calculated PARA 1 model reproduces those found in the crystalline structures.

The differences between the two polymorphs occur from other effects. In its crystalline arrangement, the B-type polymorph has an hexagonal symmetry; each double-helix has only three neighbors, corresponding to the interaction described above as PARA 1. The channel created by six double-helices packed in the hexagonal fashion is occupied by a column of water molecules. In the less hydrated A-type structure, each double-helix is surrounded by six neighboring ones. The chain pairing described by the PARA 1 model corresponds to four out of six of these interactions; the two other ones being looser. This type of loose arrangement, which is generated by translational symmetry (Δ_x = 11.72 A, Δ_z = 0 A) is also predicted by the modeling procedure. This arrangement corresponds to a secondary energy minimum, and it is among the low energy chain pairings; it occurs for $\mu_A = \mu_B = 55°$, $\Delta_z = 0$ A, Δ_x = 11.76 A. The agreement between experimental and calculated chain arrangements is found despite the fact that no water molecule was considered in the modeling. This would tend to indicate that these water molecules do not play any driving role on the establishment of the three-dimensional packing of double-helices.

Conclusion

From the knowledge of the stable conformation of a polymer chain it is possible to predict the different ways that this molecule is going to interact with other chain-like molecules. The procedure has been exemplified in the present work, on four different systems of increasing complexity. It has been applied with the same success to other polymeric structures such as Nylon 6,6 or Mannan [29]. The structures derived from this process are good approximations to the experimental ones. Errors in the displacement parameters Δx and Δz, are no more than 0.15 A. Angular parameters do not disagree with the experimental values by more than 7°. In all these cases it is clear that the strongest chain-chain interaction, as present in the crystal lattices, is always predicted. However, the procedure is somehow weak when it comes to find the other chain-pairings which participate in the three-dimensional building of the unit cell. This is partially due to the fact that layer formation undoubtely plays a role in the establishment of lattices. Similarly, space group can be easily predicted only in the case of systems of low symmetry, such as monoclinic or triclinic. Clearly, improvements are needed to automate the procedure.

Despite these deficiencies, useful information can be extracted from the proposed methodology. It offers some insights about how three-dimensional structures of stereoregular polymeric materials are formed. It is quite interesting to note that even in the cases where network of hydrogen bonds may occur between chains, the driving forces remain the van der Waals. Obviously, understanding how the structures are formed provides possible explanations of the polymorphic transitions. Since the present procedure allows for a thorough examination of all possible chain-pairings, many cases are expected to be found. Whereas only a few of these arrangements would correspond to chain pairing capable of generating efficient packing, the other ones may represent situations which are likely to occur in the amorphous state or at the surface of the polymeric materials. Another application extends to the predictions of the strongest chain-chain interactions which may occur in low symmetry systems such as gels. These will be responsible for the formation of the so-called "junction zones". More generally, the proposed methodology seems adequate to investigate the interactions and aggregation phenomena in low-ordered polymeric assemblies.

Acknowledgements

I gratefully acknowledge valuable discussions with Drs. Anne Imberty (INRA-Nantes) and Raymond Scaringe (Eastman-Kodak, Rochester). Appreciation is extended to Dr. Steven Glover (Leeds) for his carefull reading of the manuscript.

References

[1] S. Pérez & H. Chanzy. (1989) 'Electron Crystallography of Linear Polysaccharides'. J. Electr. Microsc. Techn., 11, 280-285.

[2] A.J. Hopfinger, A.G. Walton. (1969) 'Theoretical Calculation of the Conformation and Crystal Structure of Poly-L-hydroxyproline'. J. Macromol. Sci. Phys., B3, 195-208.

[3] A.J. Hopfinger, A.G. Walton. (1970) 'Theoretical Prediction of the Conformations of Poly-(Gly-Pro-Gly)'. J. Macromol. Sci. Phys., B4, 185-199.

[4] A.J. Hopfinger. (1971) ' The Lattice Energetics of Some Polypetide Chains', Biopolymers, 10, 1299-1315.

[5] K. Tai, M. Kobayashi & H. Tadokoro. (1976) ' Conformational and Packing Stability of Crystalline Polymers. VI. Stable Chain Packing of Polyethylene'. J. Polym. Sci., Polym. Phys. Ed., 14, 783-797.

[6] K.C. Chou, G. Nemethy & H.A. Scheraga. (1983) 'Energetic Approach to the Packing of α-Helices. 1. Equivalent Helices'. J. Phys. Chem., 87, 2869-2881.

[7] K.C. Chou, G. Nemethy & H.A. Scheraga. (1984) 'Energetic Approach to the Packing of α-Helices. 2. General Treatment of Nonequivalent and Nonregular Helices'. J. Am. Chem. Soc., 104, 3161-3170.

[8] R.P. Scaringe & S. Pérez. (1987) 'A Novel Method for Calculating the Structure of Small Molecule Chains on Polymeric Templates'. J. Phys. Chem., 91, 2394-2403.

[9] R.P. Scaringe & S. Pérez (unpublished). 'Interchain Interactions in Mixed Small-Molecule/Polymer Systems'.

[10] S. Pérez, A. Imberty & R.P. Scaringe. (1990) 'Modeling of the Interactions of Polysaccharide Chains: Application to the Crystalline Polymorphism of Starch Granules', in "Computer Modeling of Carbohydrate Molecules", A.C.S. Series, A.D. French & J.W. Brady Eds. (in the press)

[11] S. Pérez & C. Vergelati. (1985) 'Unified Representation of Helical Parameters: Application to Polysaccharides'. Biopolymers, 24, 1809-1822.

[12] T. Miyazawa. (1961) 'Molecular Vibrations and Structure of High Polymers. II. Helical Parameters of Infinite Polymer Chains as Functions of Bond Lengths, Bond Angles, and Internal Rotation Angles'. J. Polym. Sci., 55, 213-231.

[13] D. Gagnaire, S. Pérez & V. Tran. (1980) 'Helical Parameters of Infinite Polymers Chains with a Diamond-like Backbone. Application to Homopolysaccharides and Cyclic Oligosaccharides'. Carbohydr. Res., 78, 89-109.

[14] D.Gagnaire, S. Pérez & V. Tran. (1979) 'Polymer Conformation: Analytical Resolution of two Torsion Angles as a Function of Helical Parameters'. Int. J. Biol. Macromolecules, 1, 42-44.

[15] R.A. Scott & H.A. Scheraga. (1966) 'Conformational Analysis of Macromolecules. III. Helical Structures of Polyglycine and Poly-L-Alanine'. J. Chem. Phys., 45, 2091-2101.

[16] S. Pérez & C. Vergelati. (1987) 'Solid State and Solution Features of Amylose and Amylosic Fragments'. Polym. Bulletin. 17, 141-148.

[17] B. Rémillard & F. Brisse. (1982) 'On the Crystal Structure of Poly(1,4-trans-cyclohexanediyl-dimethylene terephthalate)'. Polymer, 23, 1960-1964.

[18] S. Pérez & F. Brisse. (1976) 'The Crystal Structure of Ethylene Glycol Dibenzoate'. Acta Crystallogr., B32, 470-474.

[19] F. Brisse, R.H.Marchessault & S. Pérez. (1979) 'New Insights and Unsolved Problems in the Crystalline Structure of Polyesters' in "Preparation and Properties of Stereoregular Polymers", R.W. Lenz & F. Ciardelli, Eds;, D. Reidel Publishing Company, pp 407-430.

[20] P.R. Sundararajan, P. Labrie & R.H. Marchessault. (1975) 'Conformational Studies on Oligomethylene Glycol Derivatives. 3. Ethylene Glycol Diacetate'. Canad. J. Chem., 53, 3557-3562.

[21]. R. de P. Daubeny, C.W. Bunn & C.J. Brown. (1954) 'The Crystal Structure of Polyethylene Terephthalate'. Proc. Roy. Soc., A226, 531-542.

[22] I.H. Hall. (1984). 'The Determination of the Structures of Aromatic Polyesters from their Wide-Angle X-Ray Diffraction Patterns', in "Structure of Crystalline Polymers", I.H. Hall Eds., Elsevier Appl. Science Pub., pp 39-78.

[23] K. Gardner & J. Blackwell. (1975) 'Refinement of the Structure of β-Chitin'. Biopolymers, 14, 1581-1595.

[24] S. Pérez, (1978), DSc. Thesis, University of Grenoble, France.

[25] I. Tvaroska & S. Pérez. (1986) 'On the Conformational Energy Calculations of Oligosaccharides. A Comparison of Methods and Strategy of Calculations'. Carbohydr. Res., 149, 389-410.

[26] A. Imberty, H. Chanzy, S. Pérez, A. Buléon & V. Tran. (1988) 'The Double Helical Nature of A-Starch'. J. Mol. Biol., 201, 365-378.

[27] A. Imberty & S. Pérez. (1988) 'A Revisit to the Three-Dimensional Structure of B-Amylose', Biopolymers, 27, 1205-1231.

[28] S. Pérez & R.P. Scaringe. (1986) 'PITMOS. A System of Interactive Computer Programs for Visualization of Crystal Packing'. J. Appl. Cryst., 19, 65-66.

[29] S. Pérez. (unpublished results).

SPECIMEN ORIENTATION AND ENVIRONMENT

J. N. Turner, D. P. Barnard, P. McCauley, and W. F. Tivol
Wadsworth Center for Laboratories and Research
New York State Department of Health, and
School of Public Health
State University of New York at Albany
Albany, New York 12201-0509

ABSTRACT. Precise knowledge and control of the specimen's orientation relative to the electron beam is critical for the collection of electron diffraction data. Tilting stages capable of $+/- 70^0$ about a single axis or $+/- 65^0$ with two degrees of freedom (double-tilt or single-tilt and rotation) are used to collect data from most of Fourier space but a portion remains unsampled. For many biological specimens this is a serious limitation due to the lack of symmetry, but for most crystals it may not be serious. However, we are investigating cylindrically symmetric stage designs and specimen mounting methods that allow complete sampling of Fourier space. Conventional stages require the specimen to be vacuum compatible, but in-situ experiments in other environments, as well as temperature control, are increasingly important. Environmental chambers provide a gaseous and/or liquid specimen environment while maintaining normal instrument operation. Temperature-controlled stages can operate near liquid helium temperature or up to several thousand Kelvins.

1. OPERATING CRITERIA

Specimen stages are the most sophisticated mechanical subsystems in the electron microscope (Valdre, (1979), Valdre and Tsuno, (1988), Turner et al. (1988), (1989a & b)). Specimen manipulations include translation in two directions, and two degrees of freedom for orientation relative to the optic axis (double-tilt, or tilt and rotation). All motions must be performed with extreme mechanical precision and stability. It is common practice to use 3 mm specimen grids and to require 0.3 nm or better resolution. Since the image must be stable to at least the limit of resolution, the stage must be stable to one part in 10^7 for periods of ten seconds to a minute or more. By any standard, this is an extreme mechanical requirement. However, even at relatively high magnification the field of view is fairly large allowing the requirement for translational positioning to be an order of magnitude less in most practical applications. Precision of the tilt and rotation angles are typically $+/- 0.1,^0$ but some quantitative studies require higher levels (Turner et al. (1988) Frank et al. (1987)).

Specimen stages must be constructed to fit in a volume of a few cubic centimeters for conventional-voltage instruments and a few tens of cubic centimeters for high-voltage electron microscopes (HVEM). They are located inside a column ranging from 20 cm in diameter and 1.5 m high to 60 cm in diameter and two stories high, and are mechanically coupled to and controlled by devices outside the column. The influence of even minor vibrations can be severe, and the instruments are

J. R. Fryer and D. L. Dorset (eds.), Electron Crystallography of Organic Molecules, 55–62.
© 1990 *Kluwer Academic Publishers.*

difficult to mechanically isolate from their surroundings (Valle et al. (1975), Turner and Ratkowski (1982)). Since optimum electron optics requires immersion lenses, specimen stages must operate in a high strength magnetic field that cannot be perturbed to at least one part in $10.^6$ This restricts the choice of materials to those with no magnetic susceptibility to the same degree. In addition, stop-and-go operation is required with the stage often being in the same position for extended periods of time and then moved with precision. This makes the devices particularly susceptible to stick-slip which is the effect of static friction versus dynamic friction, necessitating more starting force than dynamic force. New high-technology, low-friction, vacuum-compatible coatings may provide a solution to this problem.

The physical state of the specimen can be as important as its orientation. For this reason, temperature-controlled stages and environmental chambers have been developed. Temperature-controlled stages operate both above and below ambient. Electron crystallography of biological materials may benefit from temperatures ranging from a few tens of degrees above ambient to very low temperatures (Hui and Parsons (1974), (1975), Heide (1982), Taylor et al. (1984), Zemlin et al. (1985)). Similar studies involving inorganic materials or organics used as structural components (polymers) may require high temperatures (Valdre (1979), Swann (1979), Komatsu et al. (1977)). In addition, conventional electron microscope designs require that the specimen be vacuum compatible, severely restricting experimental conditions and eliminating any application of dynamic in-situ methods. Specimen stages incorporating environmental chambers are used to maintain the specimen in a gaseous and/or liquid environment while not affecting normal operation of the instrument (Parsons et al. (1974), Turner et al. (1989b)).

2. ORIENTATION STAGES

Orientation stages are defined here as those that change and control the angular position of the specimen with respect to the optic axis of the microscope, which in most cases is also the electron beam axis. These stages are of three basic configurations (single-tilt, double-tilt and tilt-rotation), and of two basic types (top-entry and side-entry) (Valdre (1979), Turner et al. (1989b)). Top-entry stages traditionally have higher mechanical stability, but the incorporation of tilting motions is difficult. Furthermore, incorporation of the tilting mechanisms requires a larger upper pole piece bore which degrades the optical properties of the objective lens. Side-entry stages have direct access to the specimen volume via a rod which can be used as a port for transfer of motion, heat, or gases and liquids. However, the rod serves as an excellent transmitter of vibration, and often requires a wider objective lens gap which degrades the instrument's optical properties.

No electron microscope stage samples all of Fourier space, and the different orientation stages (single-tilt, double-tilt and tilt-rotation) sample different portions each with its own geometry. The regions that are unsampled are shown in Figure 1, and the corresponding sampled portions are listed in Table 1 for a set of practical maximum angles.

For accurate data collection with minimum irradiation, a eucentric goniometer similar in function to those used in x-ray diffraction would be ideal. Goniometers, defined as stages with angular readout and accuracy of as least 0.1^0 (Valdre and Goringe (1971)), are readily available in electron microscopy. However, a eucentric stage has never been implemented. Eucentric means that the specimen orientation has two degrees of freedom (double-tilt, or tilt and rotation) and that the specimen's position does not change as a function of orientation. In other words, there is no change in specimen translation or height (position along the optic axis) as a function of orientation with respect to the beam. A stage is referred to as axis-centered if the specimen position is constant for one orientation motion (single-tilt)(Valdre and Goringe (1971)). The term eucentric is often used when axis-centered is meant. Valdre and Tsuno (1988) have developed a new approach to this perennial problem by designing a stage and objective lens as an integral unit, and by incorporating top-entry and side-entry mechanisms into a single design. This allows optimization of both stage and lens performance.

STAGE TYPE	MAXIMUM TILT ANGLE (Degrees)	PERCENTAGE OF FOURIER SPACE SAMPLED
SINGLE-TILT	60	66.7
	65	72.2
	70	77.8
	75	83.3
DOUBLE-TILT	60	83.9
	65	88.1
TILT-ROTATION	60	86.6
	65	90.6

Table 1 Percentage of Fourier space that can be sampled when using the various specimen orientation stages. The angles used are representative of practical operating designs. The double-tilt figures used the indicated angular values for both orientation motions.

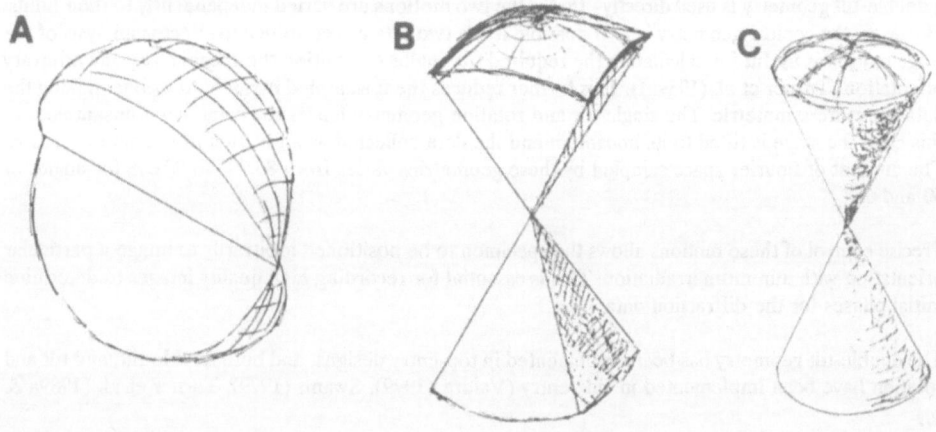

Figure 1 The portions of Fourier space that are unsampled when using electron microscope orientation stages. A) represents the single-tilt geometry, B) the double-tilt case and C) represents tilt-rotation stages.

2.1 SINGLE-TILT

The simplest orientation stage is a single-tilt design. If the specimen holder is made as thin as possible, tilt angles of +/- 70^0 can be achieved (Turner et al. (1986), Frank et al. (1988)), allowing 77.8% of Fourier space to be sampled. If the specimen grid is held in a cantilever fashion with the holder only gripping a portion of the perimeter of the grid, even higher tilt angles are possible (Chalcroft and Davey (1984), Peachey and Heath (1989)). In the latter case, the image and/or diffraction pattern is occluded by the specimen mounting grid, and its geometry determines the maximum tilt angle - typically 75^0 which allows 83.3% of Fourier space to be sampled.

Figure 1A shows the portion of Fourier space that is unsampled by single-tilt stages. The angle of the unsampled wedge decreases as the maximum tilt angle increases, and it is critical to minimize the wedge angle in order to collect the maximum information. The unsampled portion is asymmetric which may be more of a problem than the angular range, as it results in anisotropic resolution, if the diffraction data is used to calculate a structure. However, some stages have unlimited tilt; that is, the specimen can be rotated a full 360^0 thereby accessing all of Fourier space (Chalcroft and Davey (1984), Peachey and Heath (1989)). It is, of course, redundant to tilt more than 180.^{0} The tilt range of these devices is limited by the specimen grid. Thus, we are investigating cylindrically symmetric methods of specimen mounting. These methods have been used to image portions of cells in a patch-clamp pipette. The pipette tip is approximately 1 μm in diameter with 0.25 μm wall thickness and is penetrated by a 1.0 MeV beam (Sachs and Song (1987)). This and other mounting strategies with less amorphous supporting material are being investigated as methods of extending the angular range of tilt.

2.2 DOUBLE-TILT, OR TILT AND ROTATION

Stages with two degrees of freedom can be used to collect a larger and more symmetric distribution of Fourier space. The unsampled wedge of figure 1A is reduced to either the four-sided pyramid or cone shown in figures 2B and C. Double-tilt, and single-tilt with rotation provide equivalent motions given that the mechanical range of particular designs is the same. The four-sided pyramid results when a double-tilt geometry is used directly - that is the two motions are varied independently to their limits. However, the conical geometry is also possible if the two tilts are coordinated. Vector analysis of the stage motion is useful for calculating the required tilt angles to position the specimen at any arbitrary orientation (Turner et al. (1989a)). This further reduces the unsampled pyramid to a cone making the data set more symmetric. The single-tilt and rotation geometry leaves the same cone unsampled. In this case, the stage is tilted to its maximum and the data collected as a function of specimen rotation. The amount of Fourier space sampled by these geometries varies from 83.9% to 90.6% for angles of 60 and 65^0.

Precise control of these motions allows the specimen to be positioned arbitrarily to image a particular orientation with minimum irradiation. This is essential for recording high quality images to determine initial phases for the diffraction data.

The double-tilt geometry has been implemented in top-entry designs, and both double-tilt, and tilt and rotation have been implemented in side-entry (Valdre (1979), Swann (1979), Turner et al. (1989a & b)).

3. HEATING AND COOLING

Control of specimen temperature is important for studying phase transitions in biological and polymer materials, and for minimizing the effects of radiation damage during observation (Zemlin et al. (1985)). This capability further complicates stage design by introducing new materials, large temperature gradients, and more complicated mechanical coupling to exterior mechanisms. If the temperature gradients are large, it is necessary to introduce shielding devices to decrease radiative loses and to act as cryopumps to protect the specimen in the case of a cooled stage. The addition of heat conductive paths and/or electrical components isolated from other stage parts is difficult in the limited space, and results in alteration of the vibration characteristics of the specimen holders. This is especially true for low temperature stages that use an external dewar (Taylor et al. (1984)). If the temperature must be varied both above and below ambient, the design is complicated by requiring both heating and cooling devices. If the operating temperature range is small, a single system such as a Peltier device can be used. However, if a large temperature range is required, it is necessary to incorporate two systems.

Thermally insulating the specimen holder from the other components of the stage or the entire stage from the microscope introduces materials that are electrical insulators. Unless an electrically conductive path with minimum heat conduction is introduced, this may cause charging. The mechanical properties of these insulating materials are very different from those of the metal components of the stage. For example, differences in the coefficients of expansion can cause stage drift as a function of small temperature changes especially if large temperature gradients are used. At extremely low temperatures the properties of some materials are considerably different than at room temperature.

Cooling a stage to extremely low temperatures requires the use of liquid nitrogen or liquid helium. However, direct circulation of liquid through the stage is to be avoided, because as it gains heat while cooling the stage, it boils, introducing vibration (Taylor et al. (1984)). This problem is minimized either by circulating the cold gas, or by using a highly conductive metal braid immersed in the liquid in an external dewar. The braid is connected to an isolated rod of the same material that is in contact with the specimen holder. However, care must be taken not to introduce vibration via the braid and rod due to boiling.

4. ENVIRONMENTAL CHAMBERS

Routine specimen stages operate in the vacuum of the electron microscope column, severely limiting the types of specimens and experimental conditions that can be observed. This is not, of course, the normal environment of most specimens. To perform experiments more closely resembling natural conditions, it is desirable to observe specimens in a gaseous or even liquid environment. Environmental chambers have been incorporated in specimen stages to isolate a small volume surrounding the specimen from the rest of the microscope. This volume is limited either by two continuous thin films (Fukushima et al. (1982), Turner et al. (1989b)), or by a series of small colinear apertures (Parsons et al. (1974), Turner et al. (1981), and (1989b)). The result in both cases is a specimen region controlled from outside the microscope to have any pressure up to atmospheric. Operating pressures elsewhere in the instrument are essentially normal.

Although to date the only liquid used in these chambers has been water, other solvents could be used so long as they do not deteriorate the seals or windows. Many solvents used to prepare polymer

60

specimens as well as mixtures of solvents could be employed. The specimen volume can be purged with gas and/or connected to internal and external reservoirs. It is essential that all parts of the device be at the same temperature to maintain equilibrium vapor pressure. For water, a difference of 1^0C will result in condensation on, or evaporation from the specimen. In spite of this, it is possible to study lipid membranes in-situ and to vary the temperature to document phase transitions (Hui et al. (1974, 1976), Hui and Parsons (1974, 1975)). In addition, high resolution diffraction patterns have been obtained from catalase crystals which disorder irreversibly if exposed to less than 93% relative humidity (Parsons et al. (1974), Matricardi et al. (1972)).

Thin film chambers have been successfully developed and applied by Fukami and co-workers (Fukushima et al. (1982), Turner et al. (1989b)). This device is especially small along the beam direction making it possible to install in the objective gap of a conventional voltage instrument. The windows can routinely withstand atmospheric pressure and are sufficiently large to allow a full field of view. There is little possibility of window rupture or instrument damage. The disadvantages are that the windows may be time consuming to construct, and must be kept clean at all times.

Dynamically pumped chambers have been developed by Parsons and co-workers and have been applied to diffraction studies (for reviews see Parsons et al. (1974) and Turner et al. (1989b)). These chambers can be difficult to align and keep clean, and require extra vacuum equipment. Dynamically pumped chambers are usually too large to install in the objective lens gap - except in high-voltage microscopes. The chamber of Hui et al. (1976) is located above the objective lens of a conventional instrument. Our most recent high-voltage microscope chamber is an improved design with more convenient operation and tilting capability in excess of $+\backslash- 45^0$ (Barnard et al. (1986)).

Specimen damage due to electron irradiation in the presence of water and air appears to be about the same as expected in vacuum at the same temperature (Fukushima et al. (1974)). Thus, the electron beam did not generate free radicals in the gas or liquid that enhanced the effects of the irradiation. A particularly careful comparison was done for polyethylene single crystals (Fukushima et al. (1982)).

5. CONCLUSIONS

Development and application of improved specimen stages should lead to higher quality electron diffraction and correlative image data. Present cold stages make a significant contribution by minimizing the effects of radiation damage (Zemlin et al. (1985)). Eucentric goniometers would result in more precise control of specimen orientation, and in reduced beam exposure while locating and orienting specimens (Valdre and Tsuno (1988)). Environmental chambers should expand the range of specimens examined and provide the opportunity for dynamic in-situ experiments (Turner et al. (1989b)).

6. REFERENCES

Barnard, D.P., Rexford, D., Tivol, W.F. and Turner, J.N. (1986) 'Side-entry differentially pumped environmental chamber for the AEI-EM7 HVEM', in G.W. Bailey (ed.), Proc. 44th EMSA, Albuquerque, San Francisco Press, San Francisco, pp. 888-889.

Chalcroft, J. D. and Davey, C.L. (1984) 'A simply constructed extreme tilt holder for the Philips eucentric goniometer stage', J. Microsc. 134, 41-48.

Frank, J., McEwen, B.F., Radermacher, M., Turner J.N. and Rieder, C.L. (1987) 'Three dimensional tomographic reconstruction in high voltage electron microscopy' J. Electron Microsc. Technique 6 193-205.

Hui, S.W. and Parsons, D.F. (1974) 'Electron diffraction of wet biological membranes' Science 184, 77-78.

Hui, S.W. and Parsons, D.F. (1975) 'Direct observation of domains in wet lipid bilayers' Science 177, 268-270.

Hui, S.W., Parsons, D.F. and Cowden, M. (1974) 'Electron diffraction of wet phospholipid bilayers' Proc. Nat. Acad. Sci. (USA) 71, 5068-5077.

Hui, S.W., Hausner, G.G., and Parsons, D.F. (1976) 'A temperature controlled hydration or environmental stage for the Siemans Elmiskop 1A' J. Phys. E: Sci. Instrum. 9, 69-72.

Heide, H.G. (1982) 'Design and operation of cold stages', Ultramicroscopy 10, 125-154.

Komatsu, M., Sumida, N., Fujita, H. and Hino, H. (1977) 'Very high temperature stage for electron microscope and its application to materials science', in T. Imura and H. Hashimoto (eds.), Proc. 5th Int. Conf. High Voltage Electron Microscopy, Japanese Soc. EM, Tokyo , pp 141-144.

Matricardi, V.R., Moretz, R.C. and Parsons, D.F. (1972) 'Electron diffraction of wet proteins: catalase' Science 177, 268-270.

Parsons, D.F., Matricardi, V.R., Moretz, R.C. and Turner, J. N. (1974) 'Electron microscopy and diffraction of wet unstained and unfixed biological objects', in J.H. Lawrence, J.W. Gofman and T.L. Hayes (eds.), Adv. Biol. and Med. Phys. Vol. 15, Academic Press, New York, pp 162-270.

Peachey, L.D. and Heath, J.P. (1989) 'Reconstruction from stereo and multiple tilt electron microscope images of thick sections of embedded biological specimens using computer   graphics methods', J. Microsc. 153, 193-208.

Sachs, F. and Song, M.J. (1987) 'High-voltage electron microscopy of patch-clamped membranes' in G.W. Bailey (ed.) Proc 45th EMSA San Francisco Press, San Francisco, pp. 582-583.

Swann, P.R. (1979) 'Side-entry specimen stages', Kristall Tech. 14, 1235-1243.

Taylor, K.A., Milligan, R.A., Raeburn, C. Unwin, P.N.T. (1984) 'A cold stage for the Philips EM300 electron microscope', Ultramicroscopy 13, 185-190.

Turner, J.N. and Ratkowski, A.J. (1982) 'An improved double tilt stage for the AEI-EM7 high voltage electron microscope', J.   Microsc. 127, 155-159.

Turner, J.N., Rieder, C.L., Collins, D.N. and Chang, B.B. (1989a) 'Optimum specimen positioning in the electron microscope using a double-tilt stage', J. Electron Microsc. Technique 11, 33-40.

Turner, J.N., Valdre, U. and Fukami, A. (1989b) 'Control of specimen orientation and environment', J. Electron Microsc. Technique 11, 258-271.

Turner, J.N., Barnard, D.P., Matuszek, G. and See, C.W. (1988) 'High-precision tilt stage for the high-voltage electron microscope', Ultramicroscopy 26, 337-344.

Valdre, U. (1979) 'Electron microscope stage design and applications', J. Microsc. 117, 55-75.

Valdre, U. and Tsuno, K. 'A contribution to the unsolved problem of a high-tilt fully eucentric goniometer stage', Acta Crystallogr. A44, 775-780.

Valdre, U. and Goringe, M.S. (1971) 'Special electron microscope specimen stages', in U. Valdre (ed.), Electron Microscopy in Material Science, Academic Press, New York, pp. 208-235.

Valle,R. Gentry, B. and Marraud, A. (1980) 'A new side entry eucentric goniometer stage for HVEM', in P.B. Brederoo and J. van Landuyt (eds.), Electron Microscopy 1980, Vol.4 Proc. 6th Int. Conf. HVEM, Seventh European Congress on EM Found., Leiden, pp 34-37.

Zemlin, F., Reuber, E., Beckmann, B., Zeitler, E. and Dorset, D.L. (1985) 'Molecular resolution electron micrographs of monolamellar paraffin crystals' Science 229, 461-462.

ACKNOWLEDGEMENT

This work was partially supported by United States Public Health Service National Institutes of Health grants RR02984 and RR012219 sponsored by the Center for Research Resources.

ELECTRON DIFFRACTION OF POLYMER SINGLE CRYSTALS

F. BRISSE
Département de Chimie
Université de Montréal
C. P. 6128, Succ. A
Montréal, Québec
H3C 3J7   Canada

ABSTRACT. Single crystals of natural and synthetic polymers may be grown from dilute solutions. These micro single crystals with lateral dimensions of the order of 2 to 10 μm and 50-200 Å thick are suitable for an examination by electron microscopy and transmission electron diffraction.
In order to "solve" the crystal structure of a polymer from electron diffraction intensities one proceeds through a five step process:
1. Syntheses of two to four model compounds of the polymer of interest. The crystal structure of these small molecules is then established by standard X-ray diffraction analyses.
2. Building of potentially acceptable polymer structures using geometrical and conformational information derived from the model compound crystal structures.
3. Determination of the space group and the unit cell dimensions of the polymer.
4. Packing each of the potential structures within the polymer's unit cell, taking the space group symmetry elements into consideration.
5. Selection among the various structures of the one for which there is the best possible agreement between observed and calculated electron diffraction structure amplitudes.
Each step of this process will be examined and discussed with some specific examples.

1. Introduction or why use electron diffraction to study polymers ?

1.1. POLYMERS AND X-RAYS.

The physical state of a polymer is such that when exposed to X-rays, one may only record a so-called fiber pattern or rotation photograph. The diffraction spots are usually observed aligned on parallel lines (cylindical film holder) or placed on hyperbolas (flat film cassette). Depending upon the degree of crystallinity and the degree of ordering of the polymer, the fiber pattern will show spots that are arced and elongated. Often the spots are overlapping, thus the indexing and the attribution of a space group are far from straight forward. Furthermore, the evaluation of the intensities is hampered by the irregular shapes and the overlapping of the diffraction spots. However, at least one parameter of interest, the fiber repeat, i.e., the distance separating one or two chemical repeats, may be obtained from the spacing of the layer lines.

63

J. R. Fryer and D. L. Dorset (eds.), Electron Crystallography of Organic Molecules, 63–75.
© 1990 *Kluwer Academic Publishers.*

64

1.2. POLYMERS AND ELECTRON MICROSCOPY.

For some polymers, single crystals may be available.They are usually grown from dilute solutions. They are very small, too small for X-ray diffraction analysis. Their typical dimensions are, across: 5-20 μm, thickness 50-200 Å or 5 to 20 unit cells. (Keller, 1962; Yamashita, 1965; Geil, 1973; Claffey *et al.*, 1974; Taylor *et al.*, 1975; Chanzy *et al.*, 1979a,b).

The very short wavelength of the electron beam in the electron microscope ($\lambda = 0.037$Å for 100 kV) is well matched to the dimensions of the polymer micro-crystals.

An electron diffractogram usually has more and better defined spots than a fiber pattern.

For a long time only 2D electron diffraction data (hk0 section) were recorded, thus by combining X-ray fiber data (the fiber repeat) and the electron diffraction section one can arrive at the unit cell dimensions.

Systematic absences are easier to recognize on the hk0 section than on the zero layer line of the fiber pattern.

A tilting stage allows for the recording of 3D electron diffraction data, thus one can get the complete unit cell dimensions from polymer single crystals. However, mechanical constraints limit the tilt angle to ±60°.

In any structure factor calculation, the H atoms must be included since their relative contribution is more important in electron diffraction than with X-rays.

Presumably one could now solve a crystal structure from electron diffraction data, if the number of observed reflections were sufficient (Dorset, 1990).

1.3. POLYMERS STUDIED.

We have undertaken to establish the crystal structures of three families of synthetic polymers. These are: the poly(oligomethylene terephthalates) (Poulin-Dandurand *et al.*, 1979; Rémillard and Brisse, 1982); Brisse *et al.*, 1984a,b; Palmer *et al.*, 1984; 1985, the poly(oligomethylene thioterephthalates) (Leblanc and Brisse, 1990) and the poly(oligomethylene terephthalamides) (Brisson and Brisse, 1985, 1989a,b). The discussion reported below uses them as examples and they are shown in figure 1.

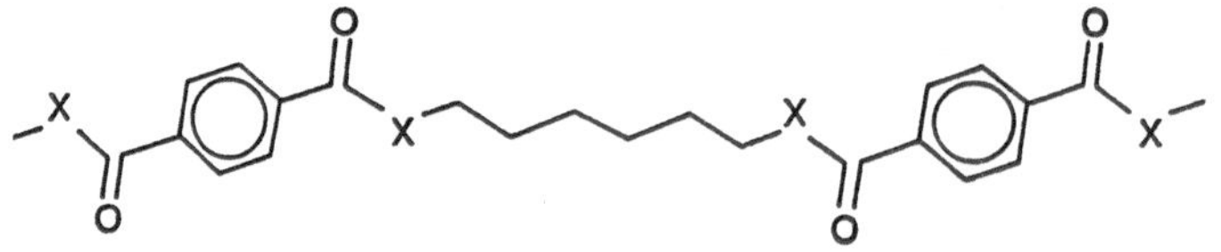

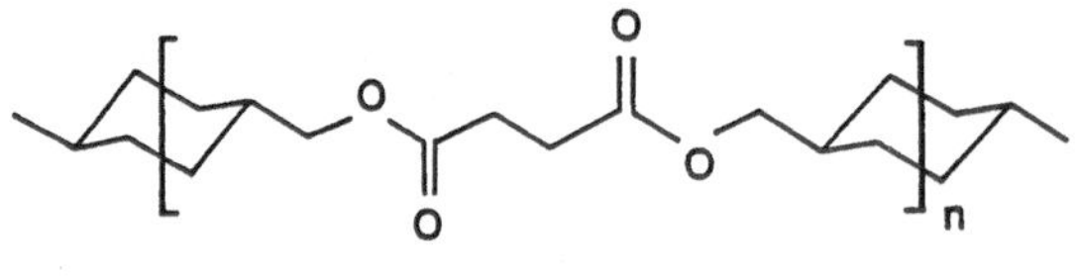

Figure 1. The polymers studied:
A. Poly(oligomethylene terephthalates) (X=O)
   Poly(oligomethylene thioterephthalates) (X=S)
   Poly(oligomethylene terephthalamides) (X=NH) and
B. Poly(1,4-*trans*-cyclohexanediyldimethylene succinate).

2. Why use model compounds ?

Whether one deals with an X-ray fiber pattern or with 3D electron diffraction data, the number of diffraction spots is usually not sufficiently large to allow for a direct-method solution of the crystal structure. Thus one goes around the phasing problem by proposing reasonably acceptable structures and trying to fit one of them to the electron diffracted intensities in the best possible way. The flow chart of figure 2 describes the steps that are followed to this end.

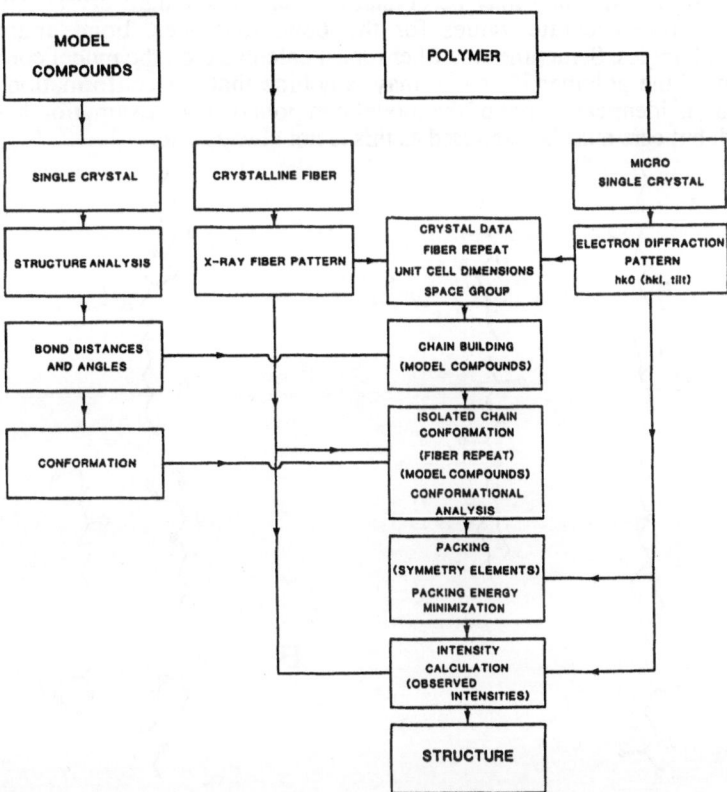

Figure 2. Flow chart showing the various steps required to establish the crystal structure of a polymer.

2.1. CONFORMATIONAL ANALYSIS.

In order to calculate the most likely conformations, one requires the building of a model of the polymer chain for which accurate bond distances and angles are needed. Then one calculates the energy of the chain as its conformational angles are varied. If there are n torsion angles to be varied, one obtains an n-dimension energy map from which a list of energy minima and the corresponding conformations are extracted. When only two or three torsion angles need to be varied, one may have an overall view of the

66

conformational map from which one can pick the conformational angles. In any case, one cannot decide upon inspection which is the actual conformation of the polymer. Each of the most promising conformations must be examined in turn.

Model compounds are, in a way, an alternate approach to obtaining the polymer's conformational angles.

2.2. MODEL COMPOUNDS.

Model compounds are small molecules chemically and hopefully structurally related to the polymer (figure 3). They may be crystallized and a standard crystal structure analysis will yield accurate values for the bond distances, bond angles and conformational angles. Sometimes the fiber repeat calculated on the model compound matches that of the polymer.Thus one may conclude that the conformation of the polymer chain is identical to that of the model compound. This assumption has often been verified, but care must be exercised as this is not always true.

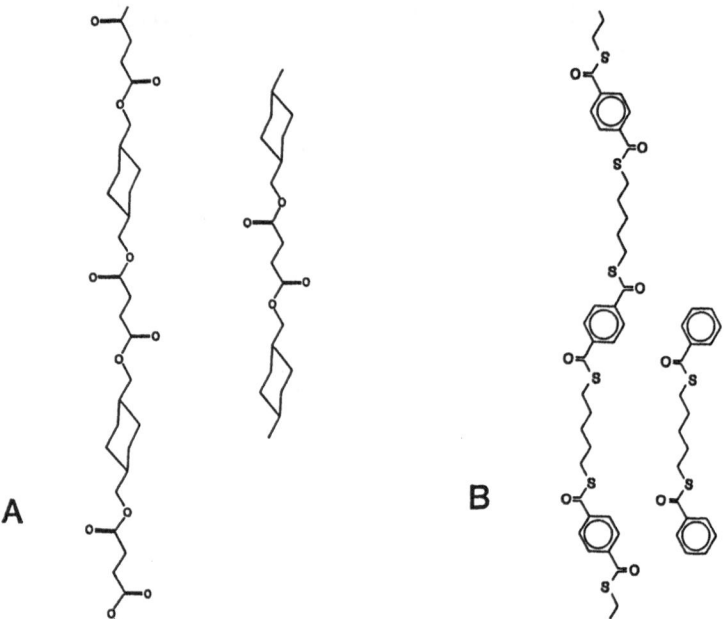

Figure 3. Polymers and related model compounds.
A. Poly(1,4-*trans*-cyclohexanediyldimethylene succinate) and bis-*(trans*-4 methylcyclo-hexylmethylene) succinate.
B. Poly(pentamethylene thioterephthalate) and pentamethylene dithiobenzoate.

2.3. CONFORMATION FROM A SERIES OF MODEL COMPOUNDS.

In a series of chemically homologous model compounds one may recognize specific conformational features (figure 4). It does not follow that all possible or energetically favorable conformations will be observed among the model compounds. It may well be that the polymer adopts a conformation that may not be observed in small molecules. This may occur when there are forces other than van der Waals (H-bonds) or solvent molecules (water channels) which are not present in the model compound or vice-versa. Side chains may also be affecting the polymer differently than its parent model compound.

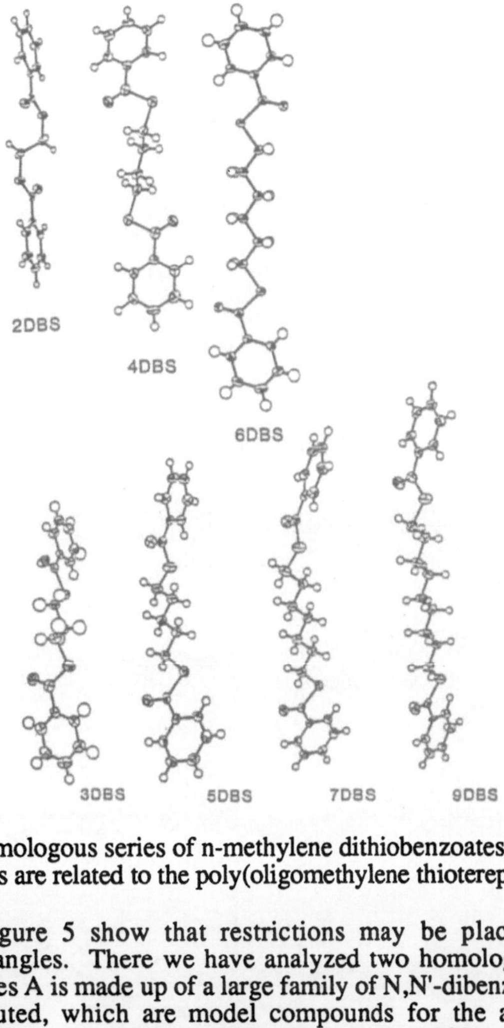

Figure 4. The homologous series of n-methylene dithiobenzoates, for n = 2 to 9. These model compounds are related to the poly(oligomethylene thioterephthalates).

The plots of figure 5 show that restrictions may be placed on some of the conformational angles. There we have analyzed two homologous series of model compounds. Series A is made up of a large family of N,N'-dibenzamido alkanes, some of them substituted, which are model compounds for the aromatic nylons, the poly(oligomethylene terephthalamides) (Forest, 1990). The n-methylene dithiobenzoates in the B series are the model compounds for poly(oligomethylene thioterephthalates) (Leblanc and Brisse, 1990). These polymers are constituted of a semi-rigid group, the terephthalamide or the thioterephthalate group and a flexible methylenic sequence. They may be characterized by only three conformational angles, ε, τ and ϕ. ε is the dihedral angle between the plane of the aromatic ring and that of the amide or the thiocarboxylic group. τ is the first and last torsion angle of the methylenic sequence of atoms, the torsions around the NH-CH$_2$ or S-CH$_2$ bond. The ϕ torsion angles are those around the other C-C bonds of the methylenic sequence.

68

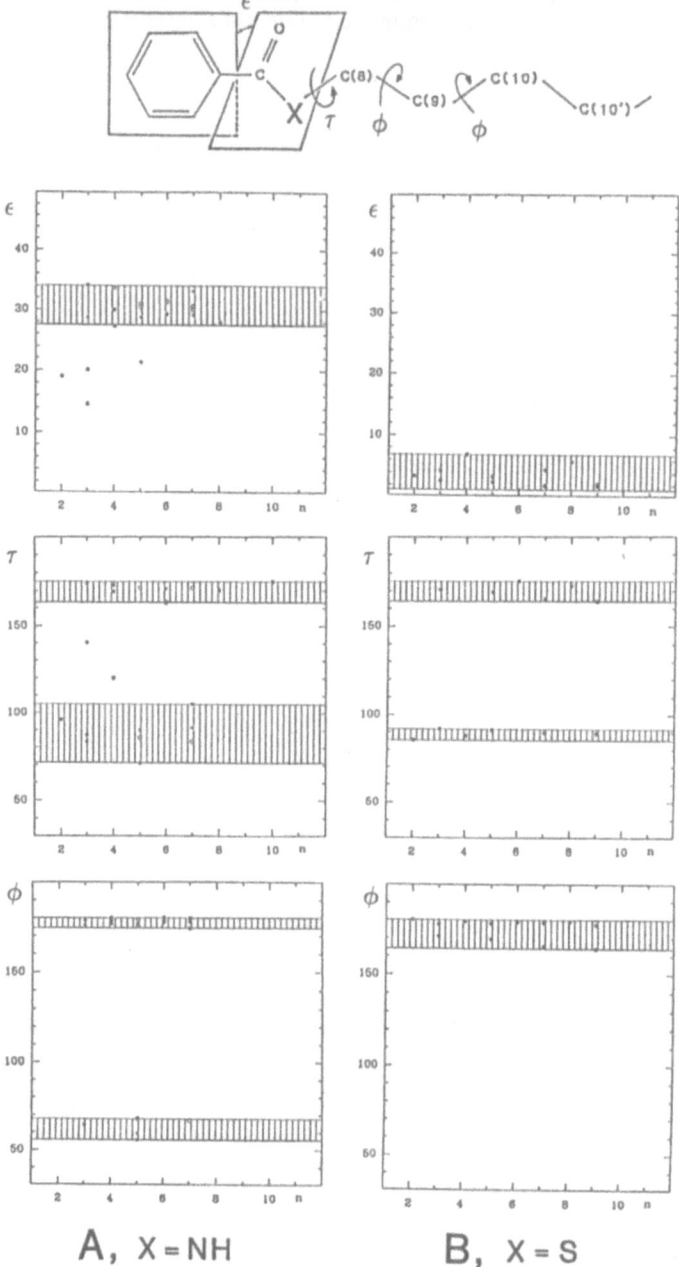

Figure 5. Variation of the dihedral angle ε, the torsion angle τ and the φ torsion angles with the number n of CH_2 groups in the aliphatic segment in A, the N,N'-oligomethylene dibenzamides and B, the n-methylene dithiobenzoates. Note: the influence of the H-bonding, ε is around 30° in A while in B (no H-bonds) ε=0-10°. The φ torsion angles are equally distributed in two ranges 60-70° and 170-180° in A, but only in 170-180° for B.

Taking into account the above constraints one may propose a limited number of acceptable conformations. For example, based on the data presented in figure 5, only three such possibilities should be considered for the structure of poly(oligomethylene thioterephthalates) (figure 6).

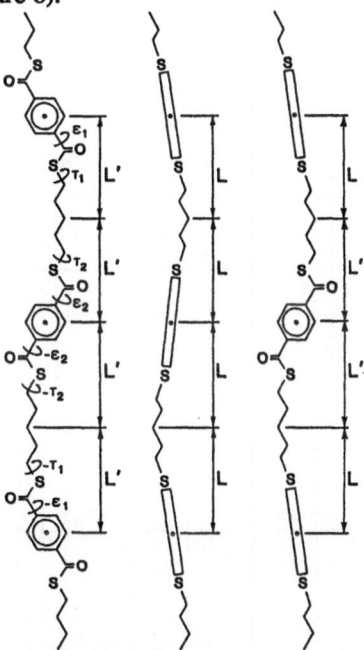

Figure 6. Three potentially acceptable structures for poly(pentamethylene thioterephthalate). These differ only by the value of the τ torsion angle which, as seen in Table 5, may take only the values of 90° or 180°.

3. Interpretation of electron diffractograms

Compared to X-rays, the wavelength associated with the electrons is very short, consequently the radius of the sphere of reflection is very large. The Fourier transform of the extremely thin crystal is a spike which extends significantly above and below the reciprocal lattice plane. Thus the sphere of reflections passes through a large number of lattice planes of the hk0 level and higher levels as well.

3.1. IDENTIFYING THE SECTIONS AS THE CRYSTAL IS TILTED.

When tilting the crystal around one of its crystallographic axes, one should realize that the appearance of a new zone (or section) is to be detected on the outer edge of the diffraction pattern. This is not always clearly recognized since when the tilt angle is small, the pattern will contain prominently-displayed diffraction spots in the hk0 section. This is still present, because of spiking, in tilts of up to 15-20°. Usually only one, or at the most, two rows of diffraction spots are recorded. As the tilt angle increases, the sections are more clearly recognized but are not always free of interference from other ones (figure 7).

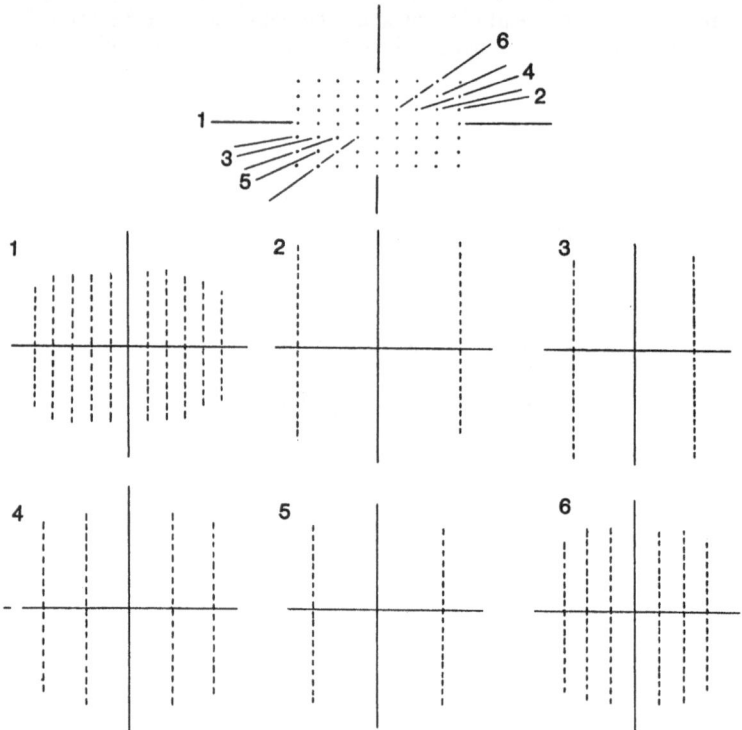

Figure 7. Recognizing new sections of diffraction as the tilt angle increases.

After weeks of examining a given batch of crystals one is able to quickly recognize their features so orienting them to the appropriate tilt axis may be achieved relatively quickly. After having identified the tilt angles in preliminary surveys, one may record a sequence of tilted electron diffractograms from the same micro single crystal within the lifetime of the crystal.

For example, electron diffractograms of the monoclinic form of poly(6-GT) were recorded after tilting of the crystal around the a* axis. In this case all the sections are rectangular. Five sections, from hk0 to 0kl were obtained from the same crystal within less than 3 minutes. This polyester was particularly resistant to the electron beam. At first only the hk0 section and two others which are relatively free from overlap are considered to interpret the diffractograms. The distance between each row of lattice points, parallel to a* (the tilt axis), and the origin is measured on enlargements of each diffractogram. These distances are plotted on a graph, along lines drawn at the proper tilt angle with respect to the hk0. The 0kl section usually clearly emerges from this sketch. One then analyses the other tilted sections in terms of the just recognized 0kl section ensuring that all rows of lattice points form a coherent set (figure 8).

The unit cell dimensions are obtained by a least-squares procedure once the distances between centrosymmetrically related spots are measured and converted to d values, using the Au rings as an internal standard. Au was decorated onto some crystals prior to their examination in the microscope (Brisse and Chanzy, 1990).

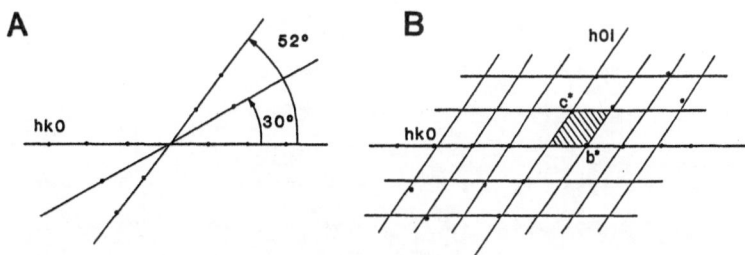

Figure 8. Steps followed to graph the 0kl section.
A. Use of hk0 and two other diffractograms
B. Plot of the 0kl reciprocal lattice

3.2. EVALUATION OF THE DIFFRACTED INTENSITIES. INTER-SECTION SCALING.

Intensities of each diffractogram are either visually estimated using an intensity scale or measured with a film scanner then processed on a computer (see for example: Miller and Brannon, 1980). The intensities of the reflections along the tilt axis, which are common to each section, may be used for inter-section scaling as long as the crystal axis coincides with the tilt axis. The problem that one faces resides in the fact that, in most cases, each section is obtained from a different crystal. Since the shape and the dimensions of the crystals needed are different, the diffraction spots differ from film to film. Furthermore, the shape of the reflection changes from nearly circular to elongated as the tilt angle increases (figure 9). All this causes the inter-section scaling to be less reliable.

Once the intensities are evaluated, the structure factors are obtained by the relation $|F(hkl)|^2 = I(hkl)$ proposed by Dorset (1980). When dealing with polymer single crystals one is fortunate that they are very thin (d < 150 Å) and are constituted of light atoms (C,H,N,O) so that the diffraction data conform to the kinematic approximation (Cowley and Moodie, 1962).

3.3 PACKING OF THE CHAIN

The chain in its chosen conformations must be positioned and oriented in the unit cell. This is done by minimizing the interactions between this chain and all those that surround it. The latter are generated by the symmetry elements of the polymer's space group. Since the chain is aligned with or parallel to the c axis, only four adjustable parameters are required. These are the translations parallel to the a, b and c axes and the rotation of the chain around its axis. This number is often reduced when the chain symmetry and the space group are taken into account.

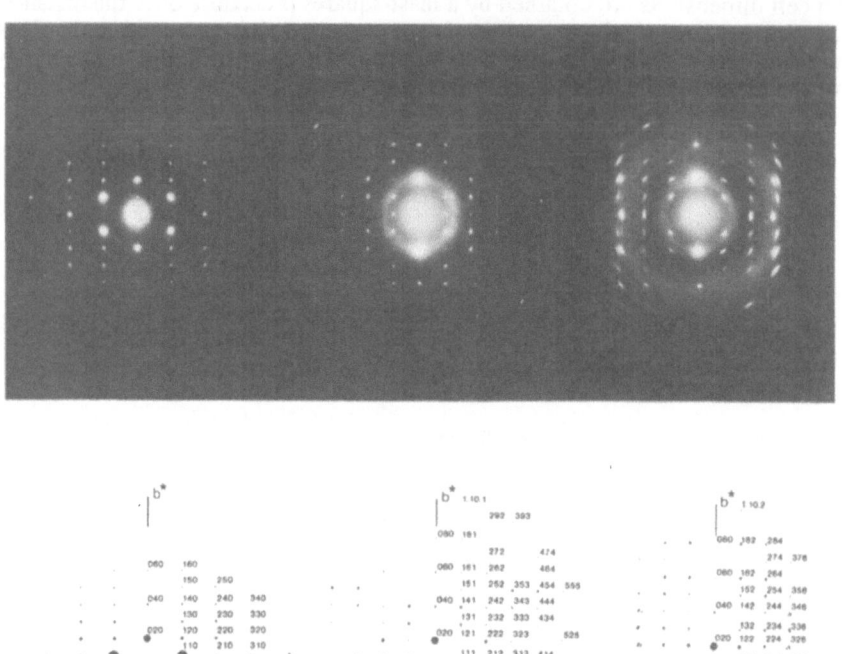

Figure 9. Electron diffractograms obtained from single crystals of poly(1,4-*trans*-cyclohexanediyl dimethylene succinate). hk0 level and two other sections, each issued from a different crystal. The tilt axis is b*. The intensities of the 020, 040, 060 and 080 reflections present in the three diffractograms served to place all the diffraction spots on a common scale.

The orientation of the chain in the unit cell is obtained by minimizing the nonbonded repulsion energy by using the simple function $E = \Sigma w(d_o - d_{ij})^2$ described by Williams (1969) and implemented in his PACK5 program. The function E measures the degree of interaction between atoms i and j in adjacent chains. d_o is a reference distance, d_{ij} is the actual distance between atoms i and j while w is a weighting parameter. The function E is minimized with respect to the positional and orientational parameters of the chain. Once this procedure is carried out for all conformations one may then select which one packs best.

3.4. CHOSING BETWEEN CONFORMATIONALLY FAVORABLE MODELS.

Once each conformation is placed as best as possible within the polymer's unit cell, one calculates and compares the structure amplitude of each reflection to its observed value. The agreement between calculation and observation is measured in the usual manner by the R-factor, $R = \Sigma||F_o|-|F_c||/\Sigma|F_o|$. In most cases, the model which has the lowest packing index, E, also has the lowest R-factor. It is this model which is then taken to represent the structure of the polymer. At this stage the R-factor is of the order of 0.25. It may be lowered by a least-squares refinement procedure. However, care must be taken as the number of observations is low. Only small displacements of the chain, taken as a rigid body, should be applied.

3.5. LIMITATIONS.

3.5.1. *Crystal bending*. Polymer micro single crystals are very thin, so that when deposited on the microscope grid they are not always evenly supported and thus tend to bend somewhat. Due to this distortion, the unit cells are no longer seen parallel to one another but are "separated" as if a wedge had been inserted between them (parallel to the c* or chain axis). Dorset (1979) has analyzed this bending of crystals and noted that it was usually most pronounced in crystals with a long axis parallel to the electron beam.

3.5.2. *Dynamical scattering*. This phenomenon is well-known in electron diffraction as the scattering cross section for electrons is much larger than for X-rays. The amplitudes of the diffracted beams may be computed using the Cowley-Moodie multi slice technique (Cowley, 1976; Goodman and Moodie, 1974).
These two corrections require a structural model, i.e., the structure must be established first as crystal bending and multiple scattering cannot be estimated a priori.

Summary of refinements and corrections for poly(1,4-*trans*-cyclohexane diyldimethylene succinate). In this particular case, only an overall isotropic temperature factor and a scale factor were refined. The polymer chain itself was kept rigid at all times.

Zone	Number of reflections	R	Bending R	Bend angle (deg)	N-beam R	Thickness (Å)
			Type of correction			
h,k,0	20	18.0	15.5	2	13.2	54
h,k,h	38	22.5	20.9	2		
h,k,2h	28	20.2	19.9	3		
h,k,l	86	19.9				

3.5.3. *Inter-section scaling*. Another problem that has been encountered with structure factor and least-squares calculations occurs when intensity data of sections other than the hk0 is included. In other words one often obtains a satisfactory agreement between $|F_o|$'s and $|F_c|$'s when hk0 data alone is used. Attempts to include data from other sections (the tilted ones) usually dramatically worsens the agreement. This has been observed in the case of natural polysaccharides (Chanzy, 1989). There may be a number of reasons for this behavior. The first one could be that the intensities from the hk0 level and the other sections are not properly scaled. This may be the result of the fact that more than one crystal is used, or that the tilt-axis is not perfectly aligned with a

74

reciprocal axis of the crystal or that the shape of the diffraction spot is very different when tilted.
It is also possible that the model is not sufficiently close to the actual structure or that the polymer has side chains whose orientations in space are not well established or again that the polymer itself is too flexible.
In our case, the two synthetic polyesters for which the good $|F_o|$-$|F_c|$ agreement on the hk0 data could be extended to other sections are very rigid. That is, they do not have side chains but most importantly, because of the space group symmetry, the chain is constrained to pass through certain crystallographic centers of symmetry thus reducing the number of degrees of freedom to one, the rotation around the c or chain axis.

ACKNOWLEDGEMENTS.

We wish to thank the National Science and Engineering Research Council of Canada and the Fonds FCAR (Québec) for their financial support to this project. The contribution and hospitality of Dr. H. Chanzy, CERMAV, Grenoble, France, is also gratefully acknowledged.

REFERENCES

Brisse, F., and Chanzy, H. (1990) Electron diffraction analysis of the monoclinic form of poly(hexamethylene terephthalate). *Can. J. Chem.* To be published.
Brisse, F., Palmer, A., Moss, B., Dorset, D., Roughead, W.A., and Miller, D.P. (1984a) Poly(hexamethylene terephthalate)-II. The crystal structure of form I and form II from electron and X-ray diffraction, and packing analyses. *Eur. Polym. J.*, 20:791-979.
Brisse, F., Rémillard, B., and Chanzy, H. (1984b) Poly(1,4-*trans*-cyclo-hexanediyldimethylene succinate): A structural determination using X-ray and electron diffraction. *Macromolecules*, 17:1980-1987.
Brisson, J., and Brisse, F. (1985) Étude comparative dans la série des dibenzamido alcanes, composés modèles des poly(oligométhylènes téréphtalamides). *Can. J. Chem.*, 63:3386-3389.
Brisson, J., Gagné, J., and Brisse, F. (1989a) Model compounds of aromatic Nylons (3): The conformation of 4T Nylon, using X-ray diffraction, solid-state ^{13}C NMR spectroscopy and IR spectroscopy. *Can. J. Chem.* 67: 1293-1297.
Brisson, J., and Brisse, F. (1989b) Model compounds of aromatic Nylons (2): N,N'-trimethylenebis(*p*-methoxybenzamide), N,N'-pentamethylenebis(*p*-methylbenz-amide) and N,N'-heptamethylenebis(*p*-methylbenzamide) by X-ray diffraction, IR spectroscopy and ^{13}C CP/MAS NMR spectroscopy. *Macromolecules*, 22: 1974-1981.
Chanzy, H., Dubé, M., and Marchessault, R.H. (1979a) Structural polymorphism of $(1 \rightarrow 4)\beta$-D-xylan. *Polymer*, 20:1037-1039.
Chanzy, H., Dubé, M., Marchessault, R.H., and Revol, J.-F. (1979b) Single crystal and oriented crystallization of ivory nut mannan. *Biopolymer*, 18:887-898.
Chanzy, H. (1989) Private communication.
Claffey, W., Gardner, K., Blackwell, J., Lando, J., and Geil, P.H. (1974) Structure analysis of polymer crystals by electron diffraction. *Philos. Mag.*, 80:1223-1232.
Cowley, J.M., and Moodie, A.F. (1962) The scattering of electrons by thin crystals. *J. Phys. Soc. Jpn.* 17 (Suppl. BII): 86-91.

Cowley, J.M. (1976) Diffraction Physics, North Holland Publishing Co., Amsterdam, ch. 11.

Dorset, D.L. (1979) Evaluation of the mozaic model for polymer single crystals. *J. Polym. Sci., Polym. Phys. Ed.*, 17:1797-1806.

Dorset, D.L. (1980) Electron diffraction intensities from bent molecular organic crystals. *Acta Crystallogr.*, A36:592-600.

Dorset, D.L. (1990) Prerequisite for successful *ab initio* electron diffraction structure analysis of organic crystals. Communication #1. Electron Crystallography. Nato Advanced Workshop, Erice, Italy.

Forest, H. (1990) Étude de la structure de dibenzamido-alcanes substitués par spectroscopie infrarouge et par résonance magnétique nucléaire. Mémoire de maîtrise. Département de chimie, Université de Montréal.

Geil, P.H. (1963) Polymer Single Crystals. *Interscience*, New York.

Goodman, P., and Moodie, A.F. (1974) Evaluation of n-beam wave functions in electron scatering. *Acta Crystallogr.*, A30:280-290.

Keller, A. (1962) Polymer single crystals. *Polymer*, 3:393-421.

Leblanc, C., and Brisse, F. (1990) Structural studies in the oligomethylene dithiobenzoate series, nDBS, where n, the number of methylene groups is odd. *Can. J. Chem.* In press.

Miller, D.P., and Brannon, R.C. (1980) Accurate fiber X-ray diffraction data from films. Data array calculations. In: Fiber Diffraction Methods. ACS Monograph #141. A.D. French and K.H. Gardner, eds. The American Chemical Society, Washington, pp. 93-112.

Palmer, A., Poulin-Dandurand, S., and Brisse, F. (1985) On the structure of poly(tetramethylene terephthalate). Structural, infrared and raman studies of three tetramethylene glycol dibenzoate derivatives, models for poly(tetramethylene terephthalate). *Can. J. Chem.*, 63:3078-3088.

Palmer, A., Poulin-Dandurand, S., Revol, J.-F., and Brisse, F. (1984) Poly(hexamethylene terephthalate)-I. Interpretation of its diffraction patterns. *Eur. Polym. J.*, 20:783-789.

Poulin-Dandurand, S., Pérez, S., Revol, J.-F., and Brisse, F. (1979) The crystal structure of poly(trimethylene terephthalate) by X-ray and electron diffraction. *Polymer*, 20:419-426.

Rémillard, B., and Brisse, F. (1982) Poly(1,4-*trans*-cyclohexanediyldimethylene adipate). Its polymorphism and the structure of form I, using X-ray diffraction. *Polymer*, 23:1029-1033.

Taylor, K.J., Chanzy, H., and Marchessault, R.H. (1975) Electron diffraction of hydrated crystalline biopolymers. Nigeran;*J. Mol. Biol.*, 92:165-167.

Williams, D.E. (1969) A method of calculating molecular crystal structures. *Acta Crystallogr.*, A25:464-470.

Yamashita, Y. (1965) Single crystals of poly(ethylene terephthalate). *J. Polym. Sci.*, A3:81-92.

THE PREDICTION OF THE CRYSTAL PACKING OF ORGANIC MOLECULES:
GEOMETRY AND INTERMOLECULAR POTENTIALS

A. Gavezzotti
Dipartimento di Chimica Fisica ed Elettrochimica e Centro CNR,
Universita' di Milano, Milano, Italy

Abstract. Methods for the statistical analysis of packing
geometries for organic molecules are described; some rough
predictions on overall packing modes are possible. A procedure for
the systematic analysis of the most likely spatial symmetries in
three dimensional lattices is proposed, producing a number of
possible crystal structures and, in favourable cases, after
optimization of the packing potential energy, a completely ab
initio crystal structure prediction.

1. OUTLINE

In recent times, the growing interest for the solid state
properties of organic compounds, both on the theoretical and the
applications side, has fostered the search for trends in packing
modes, with the hope of finding the rules that govern the
phenomenon of crystallization. In fact, an entirely new branch of
organic solid state chemistry, called crystal engineering, has
emerged (Desiraju, 1989). It signifies the ability of predicting
the crystal structure of an organic compound, given the structure
of the free molecule, in order to be able to engineer a desired
physical property of the solid.

In section 2 of this paper, methods for the statistical
analysis of the existing information on organic crystal structure,
using the Cambridge Structural Database (CSD), will be reviewed.
In Section 3, a more radical approach will be presented, namely,

J. R. Fryer and D. L. Dorset (eds.), Electron Crystallography of Organic Molecules, 77–83.
© 1990 Kluwer Academic Publishers.

the systematic analysis of possible three-dimensional molecular arrangements and the calculation of their packing potential energies. This technique does not rely on new physical models of the solid state; rather, it is an efficient procedure for the screening of favourable packing patterns, based on simple concepts of symmetry, close-packing and empirically modelled intermolecular interactions.

2. GEOMETRY

The CSD contains, in coded, computer-accessible form, a detailed description of the molecular structure, and all the information on intermolecular interaction geometries, for over 70,000 organic compounds. The basic idea of the so-called "geometrical" approach to crystal structure analysis is to define or calculate a number of molecular structure indicators, describing somehow the chemical composition, the size and the shape of the molecule, and to correlate them with crystal structure parameters. Examples of the former are: molecular stoichiometry (the ratio of carbon to hydrogen in hydrocarbons, the ratio of polar to non-polar parts of the molecule, and so on); permanent (or supposedly so) charge separation (size and orientation of molecular dipoles); donor-acceptor capabilities of atoms or fragments; ratios of moments of inertia, roughly describing the mass distribution; molecular van der Waals surface and volume. Examples of crystal descriptors are: the packing coefficients; the distance between centers of mass or the angles between main molecular planes in neighbouring molecules; the mutual orientation of dipoles; distances and angles in hydrogen bonds; the space group and cell parameters (although one should remember that their choice is always arbitrary); and the packing potential energy, which for many crystals can be approximated by lattice summations of atom-atom pairwise potentials of the Lennard-Jones or Buckingham type.

Since space forbids to give here an adequate coverage of the literature, only a few examples will be presented. The reader may find in the corresponding references the necessary perspective and background, as well as cross-references to other relevant work in the area.

a) General methods. A number of statistical techniques to analyze intermolecular interactions have been described (Gavezzotti, 1982, 1989). It is for instance possible to predict the packing energy of any hydrocarbon molecule from its size. Rigid molecules, having one moment of inertia much smaller than the others, are defined as cylindrical; for such molecules, nearest-neighbours in crystal are always parallel.

b) Planar aromatic hydrocarbons. The molecular surface is divided into core and rim surface; the former promotes stacking, while the latter favours a T-shaped (herringbone) arrangement. It has been shown (Desiraju and Gavezzotti, 1989) that for a rim/core surface ratio greater than 1.8, the molecules always choose a herringbone arrangement.

c) Populations. For organic compounds, six space groups account for 78.9% of the observed crystal structures; for hydrocarbons, the same space groups account for 84.3 of the observed structures, but centrosymmetric space groups are more populated. only 19% of the reported hydrocarbon crystal structures are for molecules with an odd number of carbon atoms. (Gavezzotti, 1990a). For C=O or C≡N containing molecules, there is a significant increase in the population of space group $P2_12_12_1$.

d) Dipoles. For crystals of molecules containing C=O or C≡N groups, the bond dipoles will never stack at short distances in a parallel fashion, nor point to each other with the electronegative atom first. The preferred arrangement at short dipole-dipole distances is an antiparallel stacking.

e) Other forces. Hydrogen bond patterns have been analyzed in leading papers by Leiserowitz and Hagler (1983) and Taylor and Kennard (1982). The contact radii of several atoms in crystals have been studied by Nyburg et al. (see for instance Nyburg and Faerman, 1985); interactions involving chlorine or sulphur atoms have been reviewed by Sarma and Desiraju (1986), and those involving other halogens by Ramasubbu et al. (1986).

3. FORCES

The cohesive forces acting between molecules in organic crystals are essentially of the dispersion (i.e. between induced dipoles) or electrostatic (i.e between permanent charges) type. The distinction between the two is not straightforward. Many empirical schemes for the calculation of the associated potentials have been proposed over the last decades (see Pertsin and Kitaigorodski, 1987, for an exhaustive discussion).

The most complete approach to the prediction of the crystal structure for an organic compound of known molecular structure is then the calculation of the packing potential energy (PPE) for possible structures, and its optimization to reach the most stable one. In spite of the outstanding progress achieved by computing facilities, the task of generating possible structures in all space groups is still far beyond the practical stage. From the analysis of the most likely spatial symmetries, sketched in the foregoing section, it emerges that the essential symmetry operators are, in descending order of importance, the screw operator (S), the inversion center (I), and the glide operator (G). It is therefore possible to build up couples of molecules related by S, I or G (nuclei), and to judge their stability by computing the interaction potential energy (IPE). For the most promising nuclei, one can calculate the full 3-dimensional PPE after adding pure translation (T), according to the following scheme:

Pure T	P1 space group
S+T	$P2_1$ space group
G+T	Pc space group
I+T	$P\bar{1}$ space group.

A promising nucleus may then be subjected to a second operator, thus obtaining a supernucleus (in which also product symmetries appear); adding again pure translation, one gets:

(S+I)+T or (G+I)+T $P2_1/c$ space group

(S+S)+T $P2_12_12_1$ space group

(G+G)+T or (S+G)+T $Pca2_1$ or $Pna2_1$ space groups.

Any supernucleus composed of S and G operators, when operated upon by I, produces the Pbca space group. Care must of course be taken to preserve the appropriate reciprocal orientation of these symmetry operators for each space group. In many cases, these relationships produce directly some or all of the cell translation vectors.

The space groups thus generated account for nearly 80% of the observed organic crystal structures. A computer program for the automatic scanning of all these possibilites is being developed. It would be a rather easy task to add other operators (mirror, twofold axis, centering) thus giving access to many other, although less populated, space groups.

The IPE of the nuclei is calculated using the same potential energy functions used in the calculation of PPE; since these have been calibrated on crystal structures, they are not fit to the prediction of the most stable dimer structures. Conversely, the crystal need not contain the most stable molecular dimer structure. Therefore, the IPE's are just convenient indices of possible dimer structures, but success in crystal structure prediction can only be judged after a full 3-dimensional PPE calculation. This is an inherent shortcoming of the method; in general, more than one promising nucleus can and must be exploited in the crystal structure search. When multiple operators are used, the number of trial structures is the product of the number of promising nuclei or supernuclei at each stage. But this is still much smaller than the number of possible structures in a blind search over all molecular orientations and space groups. In some favourable cases, however, the most stable nucleus and the main couple found in the crystal coincide: this happens frequently for the I operator. In such cases the complete crystal structure prediction follows immediately in very short computing times (even a few minutes).

After a plausible trial structure has been obtained, a final computing stage is entered, in which a fine optimization of cell parameters and molecular orientation parameters is performed. This is accomplished by using program PCK83 (Williams), which embodies the best potential energy parameters and a very efficient optimization algorithm.

In about 50% of the trials so far conducted, using known hydrocarbon crystal structures as a test, the true structure was succesfully reached after a small number of attempts. Besides, quite a number of other structures for the same compounds were obtained, with energies differing from that of the true structure by 0.5-2.0 Kcal/mole. Therefore, this method affords a straightforward route to crystal polymorphism, being able to predict the most stable crystal structure for a given compound in any preselected space group (with the proviso that at present the interactions between the molecular and crystal fields are ignored).

As mentioned before, the statistics on existing compounds allows the prediction of packing energies and of many other packing parameters for each class of compounds; thus, one knows what to expect for a given molecule, and is able to judge whether a postulated crystal structure conforms to the standards, and is therefore acceptable.

Preliminary accounts of this method have been presented at the Mendel D. Cohen Memorial Symposium (The Weizmann Institute of Science, Rehovot, february 1990) and at the Symposium on Chemistry and Structure (Schweizerische Gesellschaft fur Kristallographie, Zurich, march 1990). A detailed description of the relevant procedures, with a critical assessment of the results, is being prepared for publication (April 1990).

REFERENCES

Desiraju,G.R. (1989). Crystal Engineering, The Design of Organic Solids, Elsevier, Amsterdam.

Desiraju,G.R. and Gavezzotti,A. (1989). Acta Crystallogr. Sect.B,45,473.

Gavezzotti,A. (1982) Nouv.J.Chim.6, 443.

Gavezzotti,A. (1989) J.Amer.Chem.Soc.111, 1835.

Gavezzotti,A. (1990a). Acta Crystallogr., Sect.B, in the press.

Gavezzotti,A. (1990b). J.Phys.Chem., in the press.

Leiserowitz,L. and Hagler,A.T. (1983). Proc.Roy.Soc. A388,133.

Nyburg,S.C. and Faerman,C.H. (1985). Acta Crystallogr. Sect.B, 41,274.

Pertsin,A.J. and Kitaigorodski,A.I. (1987). The Atom-Atom Potential Method, Springer-Verlag, Berlin.

Ramasubbu,N.,Parthasarathy,R., and Murray-Rust,P. (1986). J.Amer.Chem.Soc. 108,4308.

Sarma,J.A.R.P. and Desiraju,G.R. (1986). Acc.Chem.Res. 19,222.

Taylor,R. and Kennard,O. (1982). J.Amer.Chem.Soc. 104,5063.

Williams,D.E. PCK83: A Crystal Molecular Packing Analysis Program, Quantum Chemistry Program Exchange no. 548, Indiana University, Bloomington.

A THEORETICAL TECHNIQUE FOR LAYER STRUCTURE PREDICTION

RAYMOND P. SCARINGE
Life Sciences Research Laboratories
Eastman Kodak Company
Rochester, NY 14650
USA

ABSTRACT. A new technique for the prediction of molecular layer structures is described. The essential information required is the structure of the molecule. The method takes advantage of the fact that molecular assemblies can be described as close-packed. In effect, this allows an analysis of the 9-dimensional potential hypersurface to be performed by searching a 3-dimensional, close-packed subspace. Use is made of geometric concepts, construction techniques, potential energy functions, and data base analysis to arrive at a quantitative method. Predictions are in the form of an all-atom, atomic-scale description of the structure. The methodology and underlying theory are discussed in detail. Predictions for 12 compounds of known structure are presented and compared to the experimental results. Molecular displacements can usually be predicted to within a few tenths of an Angstrom and the molecular orientation to within about ten degrees. The results indicate that the method is capable of accurate predictions even for systems with relatively strong interlayer interactions.

1. Introduction

Molecular layers are the building blocks of more complex condensed phases. Interest in these structures spans the diverse fields of Biology, Chemistry, and Materials Science. Membranes, Langmuir-Blodgett (LB) films, epitaxial films, the smectic phases of liquid crystals, and molecular crystals, all provide examples of layered structures. Typically, such materials are highly organized and often display diffraction maxima characteristic of a periodic arrangement. However, with the exception of molecular crystals, experimental difficulties in preparing single domain samples, beam sensitivity, low intensities, and relatively few observable intensities also characterize these materials. Hence, a completely experimental determination of the structure is rarely possible. There are many intriguing diffraction patterns awaiting an atomic-level interpretation (for some examples of LB films, see [1,2], for model membranes [3,4], for liquid crystals [5-7], and for epitaxial films [8]).

The theoretical prediction of the structure of molecular assemblies is confronted with many difficulties related to the complexity of the potential hypersurface and the sheer size of the problem. Even the *a priori* prediction of a molecular crystal structure, frequently a routine proposition for experimental diffraction techniques, is beyond the scope of current theoretical methods and computational resources.

In this work we report progress made in our laboratory toward a practical theoretical method of predicting the intermolecular organization of layers. The goal is to develop a model that is as general as possible, but still computationally feasible to investigate. The

J. R. Fryer and D. L. Dorset (eds.), Electron Crystallography of Organic Molecules, 85–113.

method makes use of close-packing principles, construction techniques, potential energy functions, and data base analysis. An exhaustive account of the layer constructions and numerical techniques is not possible here. However, a fairly thorough discussion of symmetry principles and a detailed account of some constructions will be presented. Results take the form of a short list of minimal potential structures among which the correct structure should occur with a high probability. Calculations for 12 compounds with known structures will be presented and compared to the experimental structures. For some molecules, there may be one minimum that is much deeper than all of the others, but this cannot normally be expected. It should be possible to utilize statistical mechanical calculations to estimate which structure from the list exhibits the lowest free energy. However, such calculations are difficult, and in any event, will usually require consideration of experimental conditions as well as the three-dimensional phase in which the layer resides. At present, the best method of selecting the correct structure from the list is to have the positions of a few reciprocal lattice points. Fortunately, this is just the information that diffraction experiments usually provide.

2. Preliminaries

2.1 BACKGROUND

For the purposes of this report, the term layer should be taken to mean an assembly of molecules with all mass centers lying roughly in a plane. In particular, we do not restrict ourselves to flat molecules lying in a plane. The term periodic layer should be taken to mean a 2-periodic assembly of 3-dimensional molecules in the same way that a crystal is a 3-periodic assembly of 3-dimensional molecules.

As is well known, the equilibrium structure of a molecular assembly represents a minimum free energy arrangement. However, the calculation of the free energy for a molecular assembly is still a matter of some difficulty [9,10]. A very readable discussion of the origin of this difficulty has been given by Owicki and Scheraga [11]. Current statistical mechanical techniques rely on the use of reasonably accurate starting structures from which to initiate a simulation. Put another way, the results of practical simulations are strongly biased by the starting structure; the number of barriers that can be surmounted and hence the number of potential wells that can be sampled is limited by computer resources. In the case of molecular solids, it is well known [12] that any observable structure must also represent one of the lowest potential energy configurations. However, the thermal contribution to the free energy can be large enough that the structure corresponding to the absolute, or global minimum of the potential surface, may not correspond to the observed structure at finite temperatures. Hence, molecular solids are frequently polymorphic and one must examine all structures of sufficiently low potential energy if the goal is to find an approximation to the observed structure for some set of experimental conditions.

The basis for such a method can be understood in terms of the following expression for the lattice potential of a formally infinite layer composed of identical molecules.

$$\Phi^S = f(a,b,\gamma,x,y,z,\phi_1,\phi_2,\phi_3); \ G^S \tag{1}$$

The potential energy of the layer is a function of at most nine structural parameters. There are six translational degrees of freedom, where a, b, and γ are the lattice constants, and, x, y, and z are the mass center coordinates of the molecule. The angles, ϕ_1, ϕ_2, and ϕ_3 fix the orientation of the molecules with respect to the cell axes. The symbol, G^S, specifies any one of the 80 crystallographic layer groups [13,14]. The function, f, that relates the structural parameters to the lattice potential depends on the method of calculating the intermolecular potential and the level of rigor employed for evaluating lattice sums. Since the method to be described makes no explicit assumptions concerning these points, numerical techniques will be discussed later when quantitative results are presented. Although (1) is quite general, there are restrictions inherent in its use, and discussion of these is deferred to the next section.

Given the above, the problem of finding all observable layer structures is reduced to the technical problem of constructing a 9-dimensional potential surface so that all deep depressions can be located. For each molecule of interest, it would be necessary to repeat the process 80 times, once for each layer group. Also, some of the structural parameters may be fixed, depending on the layer group symmetry. As pointed out previously [15], such a calculation cannot be considered practical. For the case of crystals, a direct assault on the 3-periodic version of (1) has only been attempted in a few cases [16-18]. Practical use of (1) requires a method of reducing the number of free structural variables (the parameter problem) and reducing the number of layer groups (the symmetry problem). In the remainder of this report, we describe an approach to these problems and present some quantitative results to provide an assessment of its efficacy.

2.2 SCOPE AND LIMITATIONS

Here, we discuss the main limitations inherent in the use of the lattice potential as parameterized in (1). The first is that the layer can be described as a 2-periodic array of 3-dimensional molecules. For substances that display several diffraction maxima such a description seems appropriate, at least as a first approximation. Recent experimental results [19] clearly indicate that even the "liquid analogous" phases of LB films can display order over the range of 100 intermolecular separations. Simulations on atomic systems have even prompted discussions of the "inherent structure" of liquids [20]. Hence, even in the absence of true long-range (i.e., semi-infinite) order, a lattice description of the intermolecular organization of layers seems an appropriate starting point for experimental comparisons, and also for more detailed statistical mechanical investigations.

A second restriction is that there should be only one symmetry independent molecule per layer cell (hereafter referred to as minimal Z_S layers). Experimental data available from single crystal structure determinations has allowed us to assess the validity of this assumption. From a survey of about 5000 structures [21] it has been reported that over 92% of all molecular structures crystallize with only one independent molecule per cell (hereafter, minimal Z_X). From this it seems likely that 92% of all layer structures should display minimal Z_S layer cells. If one considers structures such as those for p-hydroquinone [22] and tolane [23] that contain two symmetry independent molecules, but are still composed of minimal Z_S layers (in these structures the two layers are not related by a symmetry element) it would seem that the above statistic represents a lower limit. If one further considers the tetracene structure [24] which closely approximates that of naphthalene, but nonetheless is triclinic with two symmetry independent molecules per layer, it is clear that some low symmetry structures can closely approximate minimal Z_S packing in a higher symmetry layer group. In crystals, such arrangements are said to exhibit super-symmetry. In a survey of compounds with non-minimal Z_X packings [25], it has been reported that 80% of these exhibit super-symmetry between the crystallographically "independent" molecules. Hence, one would estimate that perhaps 98% of all compounds exhibit at least approximate minimal Z_S packing. It will also become apparent that the method described herein can be extended to include genuinely non-minimal Z_S layers, but this will be discussed in Section 3.2.

The remaining limitations concern the parameterization of the molecules in the layer. In particular, (1) can only be used if the molecular structure is known beforehand. Fortunately, it is often possible to determine the structure of the molecule in the crystalline state before the study of experimentally less convenient layered phases commences. Also, given the rapid advances in molecular mechanical techniques [26], one can expect reasonably accurate theoretical estimates of the gas-phase molecular structure of most modest sized organic molecules. As is well known [12], with the exception of rotations around single bonds, the field produced by the surrounding molecules in the condensed phases produces only minor deformations of the gas-phase molecular structure. It is also tacitly assumed that the molecule has only one stable conformation, but this is only a formal restriction. If the molecule has several stable conformations then each one can be treated as a separate molecule as far as the use of (1) is concerned. Molecules with broad conformational minima (e.g., biphenyl) present a somewhat more difficult problem, but it should be possible to use several representative conformations and explore the packing of each one independently.

A final point concerns the fact that layers serve only as the building blocks of three-dimensional phases; they are not isolated in real systems. Is it in fact reasonable to seek a potential minimum for just part of a system? In general, the answer to this question must certainly be negative, but in the present case there are further considerations. First, it is not unusual that the intralayer interactions are somewhat stronger than the interlayer interactions. Obviously, there must be a point at which interlayer interactions are no longer strong enough to materially disturb the free-space minimum potential structure of the layer. Such 3-dimensional structures can be described as stratified. It would seem that many LB films, smectic liquid crystals, and epitaxial layers fall into this category. However, as discussed below, it is even common in 3-dimensional crystal phases. For the purposes of this discussion, let us approximate the lattice potential by considering only nearest-neighbor interactions (this approximation is described in Section 3.3). In its monoclinic form [27], we find that layer potential of the crystalline paraffin, n-$C_{36}H_{74}$, is about -57 kcal/mol, while the crystal potential is approximately -59 kcal/mol. In other words, the interlayer interactions contribute only 3% to the crystal potential. This is an extreme case made possible by the long chain aspect of the paraffin molecule. In crystalline naphthalene the intralayer and interlayer contributions are approximately 0.76 and 0.24, respectively. Here, there is less stratification, but as we will show in Section 4.1, the in-crystal layer structure is very close to that predicted for the free layer.

In the following we will always assume that the full three-dimensional system to which the layer belongs is sufficiently stratified that searching for the potential minima of (1) is logical in the hierarchical sense. However, it is not clear that such a restriction is necessary. Using the method described below we have successfully predicted the structure of layers [28] in crystals that cannot be considered stratified (i.e., the intra- and interlayer interactions are identical). It should also be pointed out, that we have had some previous success in partitioning packings that are chain-like in character [29].

2.3 THE GEOMETRIC MODEL AND CONSTRUCTION TECHNIQUES

In order to discuss the symmetry and parameter problems in detail, it is necessary to introduce the geometric model of a molecule and some of the ideas of close packing. The geometric model is inspired by the fact that many experimental observations can be understood if molecules are assumed to have shape and therefore occupy volume. Many criteria have been suggested for calculating these presumed molecular properties [30-32], but the one used here is based on the familiar idea of the van der Waals radius, or intermolecular radius of an atom. The shape of a molecule is then described by representing each atom in the molecule by a sphere of the appropriate radius. The volume is calculated by summing the constituent sphere volumes and applying a correction for overlaps. This volume and its associated surface can be correlated to a variety of observed and calculated condensed phase properties [30,33-36]. For our purposes, the advantage of this description is that it is possible to develop a set of intermolecular radii [37,38] such that the structures of molecular solids can be considered packings of the molecular shape. That is to say, the spatial arrangement of molecules in solids is such that the volumes of the molecules do not overlap. This in turn leads us to the idea of close-packed arrangements, or packings that cannot be made more dense for a given molecular shape.

The use of sphere packings in the description of crystal structures was the subject of many early investigations. Patterson and Kasper [39] have collected many of the results of these studies. For molecules, a quantitative measure of the packing density is given by the packing coefficient, k, defined as the ratio NV_m/V_c, where N is the number of molecules in the unit cell; V_m and V_c are the molecular and unit cell volumes, respectively. As first pointed out by Kitaigorodsky [40], k is usually between 0.65 and 0.77 for molecular solids, a range quite comparable to the value of 0.74 for close-packed spheres.

A second important aspect of a packing is the coordination number, defined as the number of molecules touching an arbitrarily chosen reference molecule in the structure. For spheres in a plane-lattice the maximum number is 6, in a space-lattice the maximum number is 12. For molecules in crystals, the number is generally between 10 and 14, with 12 being the most common. However, the coordination number can vary depending on the geometric model used [31,32]. Also, as the molecular shape becomes more complicated, very high coordination numbers can occur. Apparently, the observed coordination number in a crystal not only depends on the packing pattern (i.e., disposition of mass centers), but also on the details of the molecular shape. This is also true of layer packings of 3-dimensional molecules where coordination numbers as high as 10 have been observed [41].

For our purposes, it would be convenient to have a number that is characteristic of a packing, but does not depend on the molecular shape. Generally speaking, this should be the coordination number of an ellipsoid array with the same packing pattern as the one under consideration. In the following, we will refer to this number as the convex coordination number, but the concept also holds for plane packings of flat objects of a more general shape. For flat molecules lying in a plane, Zorkii and Porai-Koshits used the method of trajectories, and introduced the idea of a shape with no sharp corners [42,43]. Hence, a square in a square lattice would have, in our terminology, a convex coordination number of four instead of eight; one arrives at the former value by rounding the corners of the square. It was reported [43] that any flat molecule whose shape is determined by intermolecular radii has the desirable characteristic of displaying only a small number of possible coordination numbers.

The reason this is not true for layer packings of 3-dimensional molecules is illustrated in Figure 1. In Figure 1, we see a packing that we interpret as an array of molecular cross-sections. From this point of view, the molecule is 4-coordinate and this is another possible way of defining the convex coordination number for a layer packing. The lines attached to the molecular cross-sections represent intermolecular contacts above or below the layer plane. These are made possible by the 3-dimensional shape of the molecule. In this example, the actual coordination number is 8, but for arbitrarily complex molecules the coordination number can be indefinitely large. In that the convex coordination number avoids such complications, we will make extensive use of it in analyzing and systematizing layer packings. It should be clearly understood that this will in no way limit the actual coordination number of predicted layer packings. Since the actual molecular shape will be used in all numerical calculations, the actual coordination number will be unrestricted. For example, using the techniques to be described in Sections 3.1 - 3.3, we have recently constructed a hypothetical molecular layer for which the convex coordination number is 4 but the actual coordination number is 10 [28].

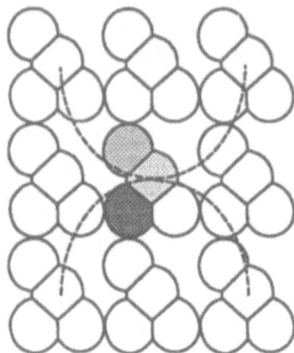

Figure 1. Layer packing with molecular cross-sections displaying 4-coordination. Intermolecular contacts above the layer plane shown as curved lines.

In the following we will need to construct hypothetical packings of molecules in a general way. Here we describe two useful techniques that can be combined to produce more complex arrangements.

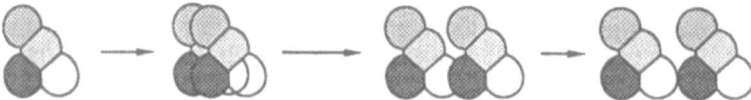

Figure 2. A displacement search is shown. Two molecules are initially superimposed. The moving molecule is displaced in a predetermined direction until an optimum separation is reached.

In Figure 2 we depict a displacement search. Initially, we start with a pair of identical molecules with their mass centers in coincidence. We then move one of them along some predetermined direction (usually along a cartesian axis) until their mutual positioning is in some sense optimal. For the geometric model, we would interpret optimal as just touching, but other criteria are possible as well. Note that if the two molecules in question are identical and in the same orientation, that a displacement search determines the magnitude of a lattice vector in the displacement direction. Therefore, the displacement search can determine the positions of an entire chain of molecules.

A line search is depicted in Figure 3. Here we start with a pair of molecules that define a line segment, such as the pair that resulted from the displacement search of Figure 2. A third molecule is initially placed with its mass center coincident with one of the other two molecules. This third molecule is not necessarily in the same orientation as the other two. A displacement search is then performed along a direction normal to the line segment

91

defined by the first two. This determines one possible position for the third molecule. The third molecule is then returned to the starting position and moved along the line segment by some predetermined increment (Δx_1 in Figure 3). From this position another displacement is performed thus determining another possible position for the third molecule. The process is repeated until the next increment would take the third molecule beyond the end of the line segment. Having determined some number of possible positions for the third molecule, we need a resolution criterion, that is, a way to decide which position is optimal. Such criteria will be discussed in Section 3.3. Note that resolution of the line search not only places the third molecule but also determines positions of an entire chain of molecules parallel to the line segment direction.

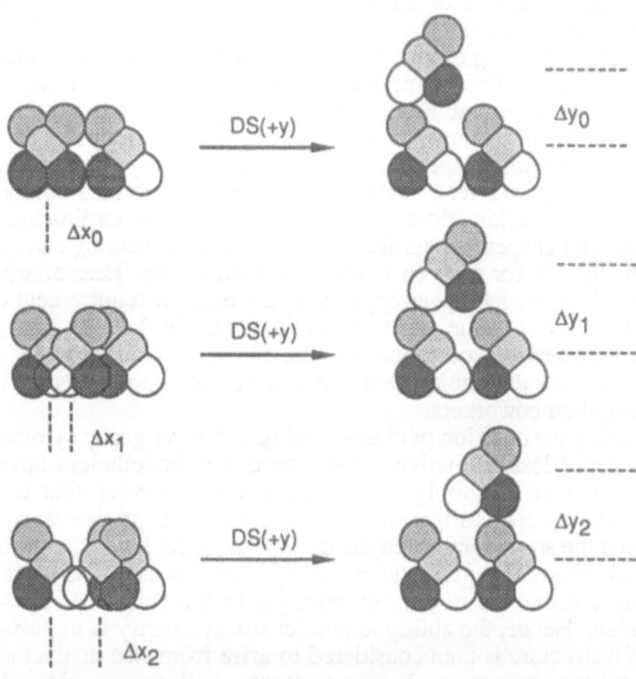

Figure 3. A line search is shown. (top) The moving molecule is initially superimposed with the reference molecule. A displacement search along the +y direction determines the value of Δy_0. (middle) The moving molecule is initially displaced from the reference molecule by the increment Δx_1. A displacement search along +y determines the value of Δy_1. (bottom) Same as above but starting with initial displacement Δx_2. The process determines a set of trial coordinates, $\{x_n, y_n\}$, for the moving molecule. The magnitude of the increment, Δx_n, is bounded by the positions of the two stationary molecules (i.e., by a lattice vector).

3. Description of the Method

3.1 THE SYMMETRY PROBLEM

Each of the 80 crystallographic layer groups occur in one or more of the 230 crystallographic space groups. Hence, if one examines space-group frequencies [44] in molecular crystals it is possible to estimate the relative importance of particular layer symmetries for generating the building blocks of three-dimensional structures. The main conclusion from such a survey is that layer symmetries higher than orthorhombic are very rare. This follows from the fact that fewer than 1% of all organic structures occur in the higher symmetry space groups. Neglecting the tetragonal, trigonal, and hexagonal layer groups still leaves 48 of the original 80 layer groups.

A more detailed analysis of these can be made in terms of close-packing theory. Kitaigorodsky [40] first demonstrated the remarkable fact that not all of the 48 remaining groups allow close-packing of an arbitrarily shaped molecule. In particular, he was able to show that the presence of mirror planes normal to the layer, or diads lying in the layer precluded the attainment of 6-coordination. By imposing a number of additional requirements, he found that only three of the original 80 groups were suitable for close-packing. The main result of the analysis is a prediction of preferred space groups, and in this respect the theory was very successful. However, as pointed out by Zorkii and Porai-Koshits [43], closest-packing does not necessarily imply 6-coordination when considering complex molecular shapes. In particular, a 4-coordinate packing can be more dense than any 6-coordinate one for a given shape. This raises the issue of whether the general requirements of close-packing, as opposed to the explicit requirement of 6-coordination, impose any restrictions on the symmetry of the layer. In the following, we will point out some differences between our approach and those of Kitaigorodsky and Zorkii, but it should be made clear that the method was inspired by the pioneering work of these two researchers and their coworkers.

In re-examining the question of close-packing and layer group symmetry, we have tried to develop a set of less restrictive requirements that nonetheless have some predictive capability. The idea is simply to formulate a set of rules that are consistent with observation and then to investigate their consequences. In particular, we have adopted the philosophy that the site symmetry of the molecules in the layer should be predicted rather than assumed. From the practical point of view, Belsky [45] has shown that site symmetry preferences are not strong enough, in the statistical sense, to be of much predictive value. Hence, the ability to predict site symmetry is important. The observed symmetry of a structure is then considered to arise from two distinct sources, molecular symmetry and lattice symmetry. Lattice symmetry is always considered to relate different molecules in the layer and all molecules occupy formally general positions. In addition, the molecule may impose part or all of its symmetry on the lattice rather than the more common view that the lattice can impose symmetry on the molecule. There are some subtleties arising from this viewpoint that will not be discussed here. A similar comment on the origin of symmetry elements was made by Mackay [46].

As an immediate consequence of this view, all layer groups with mirror planes that coincide with the layer plane are considered unsuitable; the molecule would be obliged to occupy them. Also, since we are interested in layer structures in anisotropic environments, polar layer symmetries should be considered as viable options, contrary to work of Kitaigorodsky [40] for layers in crystals. Hence, a less restrictive set of close-packing rules that still seem consistent with experimental observation can be stated as follows:

(1) For a given molecular orientation, the layer should be as dense as adjustment of the lattice constants and molecular displacements allow.

(2) The symmetry elements of the layer should not constrain the orientation of the molecules nor impose unfavorable intermolecular contact points.

(3) The symmetry elements of the layer should not constrain (either implicitly or explicitly) the convex coordination number to be less than the maximum (in layers, 6-coordination).

The first rule is identical to that used by Kitaigorodsky [40] and refers to the fact that molecules in solids are not "suspended in air" or interpenetrating. This rule is most useful in the context of layer constructions (Section 3.2).

Aside from the obvious question of site symmetry discussed above, the second rule also relates to the observation that molecules pack bump-to-hollow implying that particular types of intermolecular contact points are important elements of close-packing. For this to be possible, the orientational degrees of freedom cannot be constrained by the lattice symmetry. An energetic interpretation of this is also possible. Anisotropic forces (e.g., dipole-dipole interactions) can only be optimized if molecular orientation is unconstrained. Moreover, other forces (e.g., hydrogen bonds) require a particular complementarity in the intermolecular contacts. Also, intermolecular dispersion forces are highly anisotropic even if it is assumed that the interatomic dispersion forces are isotropic and additive.

It turns out that glide planes and screw axes always satisfy the second rule, but some point symmetry elements will violate it to a greater or lesser extent. If two molecules are brought together across an inversion center, they must pack bump-to-bump (or hollow-to-hollow) at a single point in space. But this imposes no constraint on the packing density since any molecule that is packed bump-to-hollow can exhibit bump-to-bump packing across a point. On the other hand, a two-fold axis imposes bump-to-bump packing along a line parallel to it, and a mirror plane imposes bump-to-bump packing across an entire plane; generally either situation would result in a decrease in packing density. Hence, this rule gives us the qualitative idea that inversion centers present no conflict with close-packing, diads cause some conflict, and mirror planes even more. This does not, however, tell us how much generality might be lost by ignoring the layer groups with the offending point symmetry elements. For this information we must resort to a statistical analysis of observed structures.

In Table 1, we report our preliminary results for a survey of the Cambridge Crystallographic Database. In this survey we only use space group data for organic compounds with coordinates in the data file and no reported disorder. Also, we have surveyed only the 20 most frequently occurring space groups; these represent about 95% of reported structures. Of the six column headings in Table 1, only the last two are not self-explanatory. The column labeled N(t) is the total number of occurrences for a particular space group and the space group ranking is based on this number. The column N(g) is the number of occurrences subject to the additional restriction that all molecules occupy general positions in the cell.

The following conclusions can be drawn from the survey. Structures with inversion centers comprise over 50% of the entries, those with two-folds less than 10%, while only 2% are found with mirror planes. This is in agreement with the qualitative ideas presented above, and also similar to the results of Mighell and coworkers [44], but an even more direct analysis is possible. In our scheme, there are two kinds of point symmetry elements, those occupied by molecules and those that effect packing between molecules. From the standpoint of finding the layer groups most suitable for close packing we are only interested in the latter kind, since the former are considered to arise as a consequence of molecular symmetry.

If we examine the entries under N(g) for the first three space groups containing inversion centers, we find that these are still 85-90% of their values under N(t). These groups do not owe their popularity to structures in which the molecules occupy inversion centers. At the other extreme, if one analyses the two space-groups containing mirror planes, reductions of at least 69% occur. The result for Pnma is particularly striking; only 10% of the structures for this space group display packings with no molecules occupying mirror planes. This clearly indicates that mirror planes are not an effective means of bringing about commonly observed packings. The situation with diads is between these two extremes, N(g) is about 60% of N(t) on average.

TABLE 1. Observed frequencies for organic compounds in the 20 most common space groups

Rank	No.	Symbol	Pt. Sym.[a]	N(t)[b]	N(g)[c]
1	14	$P2_1/c$	-1	9056	8032
2	19	$P2_12_12_1$	none	4415	4415
3	2	P-1	-1	3285	2779
4	4	$P2_1$	none	2477	2477
5	15	C2/c	2,-1	1371	802
6	61	Pbca	-1	1180	1064
7	33	$Pna2_1$	none	445	445
8	1	P1	none	370	370
9	5	C2	2	275	225
10	62	Pnma	-1,m	266	33
11	29	$Pca2_1$	none	220	220
12	60	Pbcn	-1,2	205	94
13	9	Cc	none	196	196
14	11	$P2_1/m$	-1,m	127	40
15	18	$P2_12_12$	2	123	89
16	92	$P4_12_12$	2	92	46
17	43	Fdd2	2	88	51
18	7	Pc	none	82	82
19	56	Pccn	-1,2	82	55
20	148	R-3	3,-3,-1	79	40

[a] Point symmetry of special positions
[b] Total number of occurrences
[c] Number with molecules in general positions only

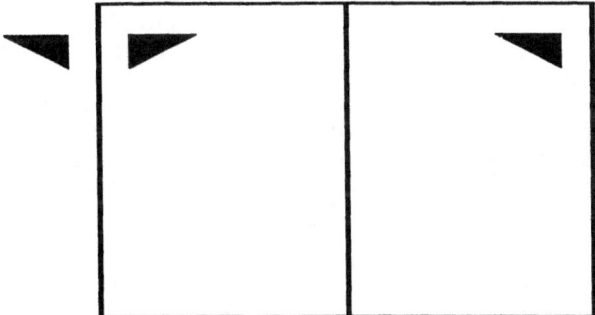

Figure 4. Symmetry elements and object transformations in layer group P1m1.

Regarding the third rule, a thorough analysis has already been given by Kitaigorodsky [40]. In the following discussions we will assign symbols to some of the layer groups. They are generally similar to those used in space group notation except that the direction of the layer normal is underscored. Consider the layer group P1m1 (Figure 4) that contains mirror planes normal to the b axis, and c is taken as the layer normal. Starting with the molecules in any arbitrary orientation (Figure 5), we perform a displacement search along the a axis resulting in a chain of molecules. This line is then replicated by the action of the mirror plane and brought into contact with the initial chain, in accordance with rule 1. It is clear that independent of the assumed orientation, the coordination number will always be four and hence violates rule 3. This is what we mean by an explicit symmetry-imposed constraint on the coordination number. Had we considered P121 that contains diads parallel to the b axis in the layer plane, the result would have been the same.

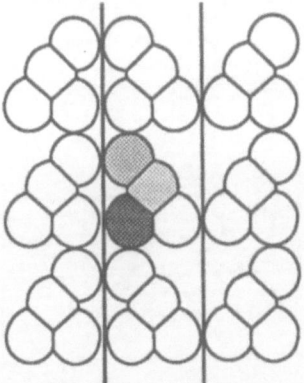

Figure 5. Illustration of 4-coordination in layer group P1m1.

A different situation applies to P112 (Figure 6), which contains diads normal to the layer. If we again start with a line of molecules parallel to the a axis (Figure 7) and replicate any one of them with the diad, we can then move it freely in the plane in accordance with rule 1. In this case, the resulting convex coordination number can be either four, five, or six, in accordance with rule 3, but molecules that are brought in contact by the diads will exhibit bump-to-bump contacts along a line normal to the layer. If the coordination number is six, there will be four such lines; for 5-coordination there will be three such lines and for 4-coordination there will be two such lines. Hence, the attainment of a higher coordination number can be achieved only at the expense of additional breaches of rule 2. This is what we mean by an implicit restriction on the coordination number. Finally, if we consider group P11-1, as an example of packing with the aid of an inversion center, it is analogous to P112 in that convex coordination numbers of four, five, and six are possible but this does not result in bump-to-bump packing.

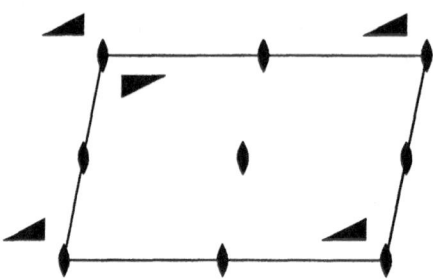

Figure 6. Symmetry elements and object transformations in layer group P112.

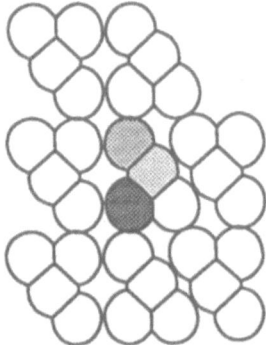

Figure 7. Illustration of 5-coordination in layer group P112. Packing with 4-coordination and 6-coordination are also possible in this layer group.

As a first approximation, it seems that all layer groups with diads and mirror planes can be discarded. This leaves only 7 of the 48 low symmetry layer groups and these are shown in Figure 8.

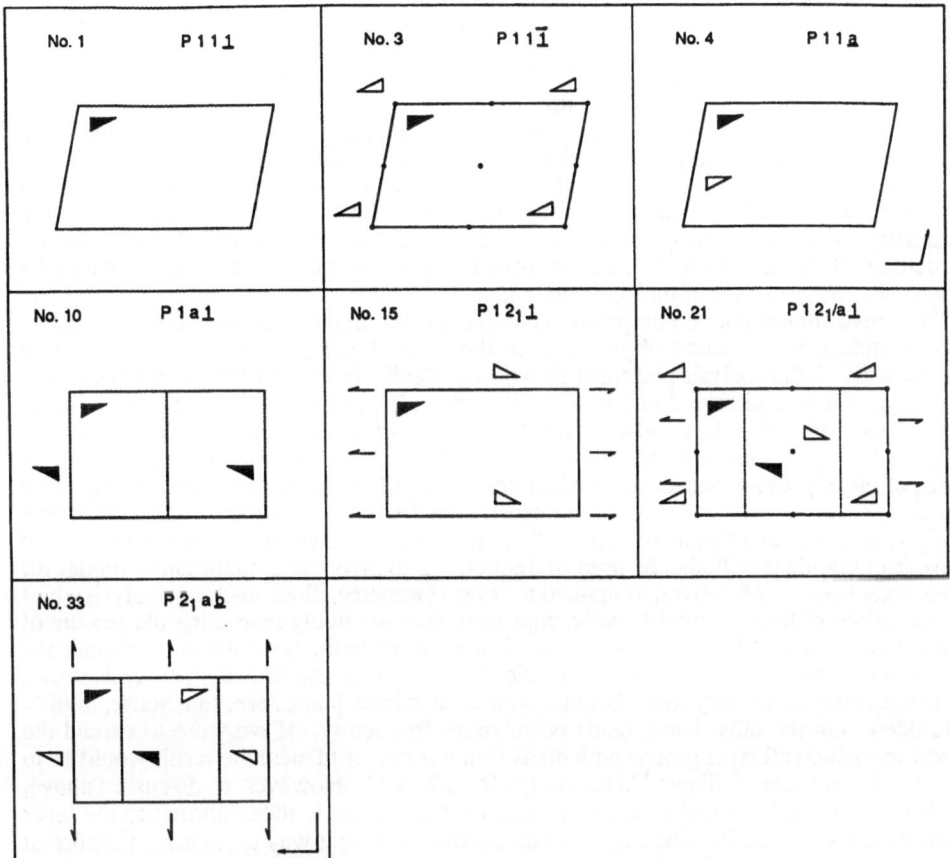

Figure 8. Symmetry elements and object transformations in 7 close-packed layer groups.

The numbering system is that of Kitaigorodsky [40], and the style of showing object transformations with the aid of polar scalene triangles was used by Weber [13]. The above analysis is not sufficient to show that these 7 groups allow structures consistent with rules (1) - (3), but rather, that the other 39 are, to a greater or lesser extent, in conflict with these rules. Here we simply state that these 7, which will be referred to as close-packed layer groups, do in fact satisfy all three of the close-packing requirements. Complete support for this statement for each of the seven will be presented elsewhere, but the method of proof will be illustrated in the next section for the particular case of layer group $P12_11$ (No. 15).

The list in Figure 8 is somewhat longer than that of Kitaigorodsky [40] and we would like to comment on the reasons for these differences. In his model, only groups 3, 4, and 15 were allowed and these can also be found in Figure 8. Layer groups 1 and 10 were rejected by Kitaigorodsky on the basis of being polar, but for our purposes polar layers are clearly of interest. Finally, groups 21 and 33 were rejected on the basis of not allowing 6-coordination without restrictions, but as stated above, this is not in accord with our results. Also, our interpretation of the layer groups is different. In our model, these are appropriate groups for packing molecules of any symmetry, while molecules with symmetry were treated separately by Kitaigorodsky [40]. Hence, in our model, the symmetry of the layer group turns out to be the minimal symmetry of the structure. Structures of higher symmetry, and therefore, higher symmetry layer groups can re-emerge as a consequence of molecular symmetry.

The above should not be interpreted as an indication that all observed layer structures can be understood in terms of packings in the seven layer groups of Figure 8. On a statistical basis, the analysis presented above would indicate that perhaps 94% of observed structures can be understood in this way. However, in practical terms, we have reduced the number of viable layer groups from 80 to 7, which translates into an order of magnitude in computational effort. At this point, one may ask whether the analysis in terms of close packing was even worthwhile. After all, the same conclusions regarding the allowed layer groups could have been reached solely on the basis of a statistical analysis of observed crystal structures. First we point out that the geometric model and close-packing ideas will also be used in Section 3.2 to arrive at a quantitative model for layer organization. Moreover, as applied to layer symmetry, close-packing analysis gives us indications of how the model can be improved without unduly increasing the amount of computational work. For example, mirror planes normal to the layer not only produce the worst violation of rule 2 but also violate rule 3; accordingly, the statistical analysis shows their frequency to be very low. Groups with such mirror planes are, in a sense, doubly forbidden. On the other hand, diads occur more frequently. If we were to extend the model to include all layer groups with diads (but not mirror planes) the result would be to increase the number of allowed layer groups from 7 to 21. However, as discussed above, diads lying in the layer plane violate both rules 2 and 3 while those normal to the layer plane violate only rule 2. This suggests that the most efficient way to improve the current model would be to add only those groups displaying diads normal to the layer; there are only 4 such groups. Finally, there are the 32 high symmetry (tetragonal, trigonal, and hexagonal) layer groups that we have dismissed solely on statistical grounds. An analysis of these in terms of close packing might indicate that only a few of these could possibly improve the model.

3.2 THE PARAMETER PROBLEM

The main purpose of this section is to demonstrate that the qualitative ideas presented in the previous sections can be used to obtain a quantitative model for layer organization. In the process, we will also illustrate the method for ensuring that the seven layer groups of Figure 8 adhere to the close-packing rules of Section 3.1. In reference to the expression for the lattice potential (1), it should be clear that an arbitrary choice of values for the 9 structural parameters will usually give rise to a structure that is inconsistent with what is generally known from experiment. In particular, most arbitrary choices for the lattice parameters will result in interpenetrating molecules or molecules suspended in air, contrary to the rules of close packing. This implies that if close-packing rules are imposed, not all of the parameters of (1) are independent and therefore a systematic search of some close-packed subspace may be computationally feasible. We will use the layer group $P12_11$ to illustrate the method.

We begin by noting that there are three important characteristics of a packing: the convex coordination number, the symmetry composition of the coordination shell, and the orientation of the coordination shell. For close-packed layers the only convex coordination numbers possible are 4, 5, and 6. Strictly speaking, 3-coordination is also possible but it is unlikely that its neglect will lead to conflict with experiment. As pointed out by Zorkii and Porai-Koshits [43], a packing with a coordination number of 2, or even 0, can be more dense than a packing with a coordination number of 6. However, such packings can always be made more dense by suitable adjustment of the translational degrees of freedom, at which point they will exhibit one of the coordination numbers mentioned above.

For the particular case of $P12_11$, only coordination numbers of 4 and 6 are possible (for the layer groups of Figure 8, 5-coordination occurs only in groups with an inversion center). Each member of the coordination shell must either be in the same orientation as the reference molecule, denoted (1), or related by a twofold screw symmetry, denoted (2_1). Thus, the symmetry composition of a shell with two neighbors related by a translation and two neighbors related by a screw would be $(1)_2(2_1)_2$. Also note that in this layer group all entries must occur in pairs. We can now go on to enumerate, in the combinatorial sense, all of the possible symmetry compositions for each possible coordination number; the results are collected in Table 2. It can be shown that not all of the shells listed in Table 2 are physically possible. We will not prove this statement here, but note that this conclusion can be reached by considering the results reported by Zorkii and Porai-Koshits [43]. On this basis one finds only three feasible coordination shells in Table 2, namely, the two 4-coordinate shells, $(1)_2(2_1)_2$, and $(2_1)_4$, and the 6-coordinate shell $(1)_2(2_1)_4$.

TABLE 2. Coordination shell compositions for $P12_11$

K	(1)	(2_1)	CP
4	4	0	no
4	2	2	yes
4	0	4	yes
6	6	0	no
6	4	2	no
6	2	4	yes
6	0	6	no

We now wish to develop a definite algorithm for constructing layer models that exhibit the coordination shells mentioned above. To simplify the discussion, we will assume that the molecular mass centers lie in the layer plane. Allowing small deviations from the plane requires a bit more computation but does not add anything new to the discussion. We choose the y axis as the screw direction, and place an arbitrarily oriented reference molecule in the xy plane (Figure 9) with cartesian mass center coordinates (x_0, y_0). We generate the first molecule in the coordination shell by performing a displacement search in the +x direction. This results in the final mass center coordinates (x_1, y_1), and determines the magnitude of the lattice constant, a. The second molecule in the shell is initially generated with mass center coordinate $x_2 = x_0$ and $y_2 = y_0$, but is orientationally related to the reference molecule by the action of a two-fold axis parallel to the y axis. In order to determine the magnitude of the screw displacement, a line search bounded by the interval, $x_0 \leq x_2 < x_1$, is performed. Upon resolution, we have the mass center coordinates x_2, y_2, and therefore $b = 2(y_2-y_0)$. The placement of these two molecules completely determines the structure. The lattice constants are given by $a = x_1-x_0$, and $b = 2(y_2-y_0)$. Finally, if we take the cell origin to lie on the screw we can arbitrarily set the mass center coordinate, y, to be zero. However, the x coordinate is determined by the perpendicular distance from the screw, hence, $x = -(x_0 + x_2)/(2a)$, where division by the lattice constant was necessary to obtain the customary fractional coordinate. Note that the coordination number is determined by resolution of the line search. If molecule 2 comes to rest in contact with both molecules 0 and 1, the structure is 6-coordinate. If only one of the two molecules is contacted, 4-coordination is the result. Hence, a single construction can give rise to either 4- or 6-coordinate layers of the desired shell compositions. Note also that all of the structural parameters of (1) have been determined except the three orientation angles. Since we have taken the orientation angles of the reference molecule to be arbitrary, they must be determined in another way. This is discussed in Section 3.3.

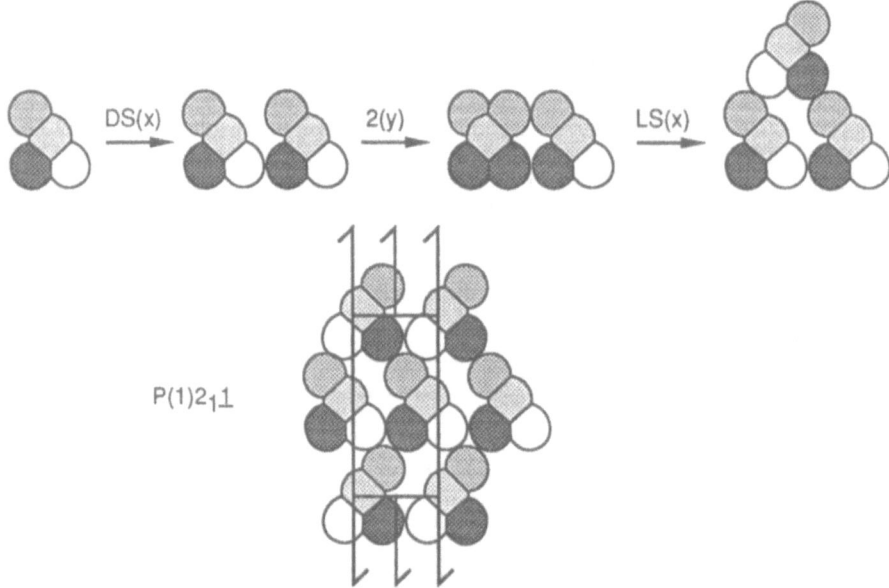

Figure 9. A layer construction in group $P12_11$ is illustrated. (top) A displacement search along +x determines an intermolecular separation and the lattice constant, a. A third molecule is superimposed on the first by the action of a twofold axis parallel to the y direction. A line search bounded by the first two molecules fixes the position of the third. (bottom) The positions of all other molecules in the lattice are determined from the above three.

Returning to the construction, note that we arbitrarily began with a displacement search along x. Given the composition of the coordination shell, a displacement search is an obvious start, but the y direction could have been chosen instead. To investigate this possibility (Figure 10), we start with a line search along the +y direction, using a molecule in the same orientation as the reference molecule, and thus determine the coordinate pair (x_1, y_1). Immediately, we have $b = (y_1 - y_0)$ as a lattice constant. As with the previous construction, we next generate the second molecule with $x_2 = x_0$ and $y_2 = y_0$, but the orientations are related by a twofold axis parallel to y. Since b has already been determined, we have $y_2 = y_1 + b/2$, from the translational component of the screw. To determine x_2, we now perform a displacement search along the +x direction. In this case, having two of the molecules of the coordination shell is not sufficient to determine the lattice since the value of the lattice constant, a, is still unknown. A third molecule, orientationally identical to the second and with $y_3 = y_2 = y_1 + b/2$ is generated. A displacement search along the -x direction determines x_3. The lattice constant, a, is now given by $a = x_2 - x_3$. Again taking the origin to lie on the screw, the fractional coordinates of the reference molecule are given by, $x = -(x_0 + x_2)/(2a)$, and $y = 0$. Again, all the structural constants of (1) can be determined except the orientation angles of the reference molecule. For this construction, the layer is always 6-coordinate. This follows from the fact that as molecule 2 is brought in contact with molecule 0 it must also make a physically identical contact with molecule 1, by the definition of 2_1 screw symmetry. By translational symmetry this implies that molecule 2', related to molecule 2 by a lattice translation -b, also makes an identical contact with molecule 0, bringing the coordination number to 4. Similar considerations hold for molecule 3, bringing the coordination number to 6. It is perhaps worth mentioning that none of this implies that the contacts that molecule 3 makes with molecules 0 and 1 are similar, in any way, to those made by molecule 2 with molecules 0 and 1 (see Figure 10).

In comparing Figures 9 and 10 we note that the orientation of the screw with respect to the cartesian axes is the same in both cases as are the arbitrary orientations of the reference molecules; also, the symmetry composition of the coordination sphere is identical for the two. The structures are nonetheless different. The difference lies in the fact that in Figure 9, the line of molecules related by translations runs normal to the screw while in Figure 10, the translation chain runs parallel to the screw. This is what is meant by the orientation of the coordination shell, and it clearly has nothing to do with the orientations of the molecules themselves. It is convenient to use a packing symbol that somehow includes this information. Accordingly, we designate the construction in Figure 9 as $P(1)2_1\underline{1}$, and that in Figure 10 as $P1(2_1)\underline{1}$, where the parentheses indicate the direction of the translation chain. In the work of Kitaigorodsky [40], a construction similar to that described above for $P(1)2_1\underline{1}$ was used to show the feasibility of 6-coordination, but the possibility of $P1(2_1)\underline{1}$ was not mentioned. Finally we draw attention to the fact that neither construction will ever result in symmetry composition $(2_1)_4$. Using methods similar to those introduced above, such a construction is easily devised. However, for a given molecular orientation, it seems clear that such a construction can never lead to a lattice constant, a, as small as that for construction $P(1)2_1\underline{1}$, or a lattice constant, b, as small as that for $P1(2_1)\underline{1}$. In some sense, these are the limiting cases. Moreover, we have not yet found an experimental structure composed of only such layers (similar compositions occur in $P1a\underline{1}$ and $P12_1/a\underline{1}$ and $P2_1a\underline{b}$). Hence, as a first approximation towards a quantitative model, we have not included such constructions. Should this lead to conflict with experiment, the necessary adjustments are obvious.

In a completely analogous manner, combination tables similar to that just given for $P12_1\underline{1}$ can be generated for the other layer groups of Figure 8, and some number of constructions devised that result in the feasible coordination shells. The resulting construction symbols are presented in Table 3. We also point out that a lattice symbol set off by braces rather than parentheses implies that the orientation of the coordination shell may be determined by a chain of higher symmetry than a translation chain. The general meaning of these symbols should be clear. Each symbol represents a construction that adheres to all three of the close packing rules of Section 3.1 and also determines all of the parameters of (1) except the orientation angles. In particular, no definite coordination number is imposed for any layer group. Shells exhibiting 6-coordination as well as 4-coordination are possible for each layer group (also 5-coordination is possible for those with inversion centers). Taken together, these symbols should represent, with a high probability, most of the close packed layer structures a particular molecule is capable of exhibiting.

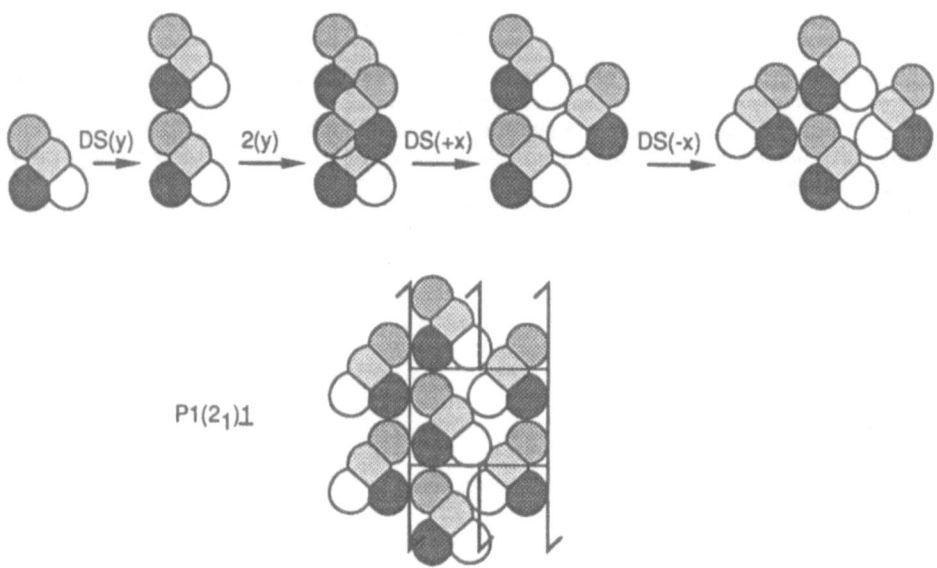

Figure 10. A layer construction in $P12_11$ is shown. (top) A displacement search along +y determines the lattice constant, b. A third molecule, related to the first by a twofold axis parallel to y is placed midway between the first two molecules. A displacement search along the +x direction determines the final position of the third molecule. A fourth molecule is initially located as was the third. A displacement search in the -y direction determines the final position of the fourth molecule. (bottom) The positions of all other molecules in the lattice are determined from the above four.

TABLE 3. Constructions and properties for some close-packed layer groups.

No.	Group	Construction(s)		Properties	
1	P11$\underline{1}$	P(1)1$\underline{1}$		Polar	Chiral
3	P11-$\underline{1}$	P(1)1$\overline{1}$		Non-Polar	Achiral
4	P11$\underline{a}$	P(1)1$\underline{a}$		Non-Polar	Achiral
10	P1a$\underline{1}$	P(1)a$\underline{1}$	P1(a)$\underline{1}$	Polar	Achiral
15	P12$_1\underline{1}$	P(1)2$_1\underline{1}$	P1(2$_1$)$\underline{1}$	Non-Polar	Chiral
21	P12$_1$/a$\underline{1}$	P{1}2$_1$/a1	P1{2$_1$/a}1	Non-Polar	Achiral
33	P2$_1$a$\underline{b}$	P(2$_1$)a$\underline{b}$	P2$_1${a}$\underline{b}$	Non-Polar	Achiral

We would now like to comment on the extension to packings with non-minimal Z_S. The main difference in such a packing is that one needs to consider the molecular composition as well as the symmetry composition of the coordination shell. Even for the case of $Z_S = 2$, this involves considerably more computation, but the principles introduced above still apply. Finally, we would like to mention that the idea of enumerating different types of coordination shells is not unique to the present work. If one examines the ideas used by Zorkii and Porai-Koshits in their investigations of plane packings of flat molecules [42,43], their vector angle plays a role similar to the symmetry specifiers used here.

3.3 NUMERICAL METHODS

In the construction procedures discussed above the molecular orientation was taken arbitrarily and therefore the three orientation angles of (1) are left as undetermined parameters of the theory. Formally, we have transformed (1) into an expression of the form,

$$\Phi^S = f(\phi_1, \phi_2, \phi_3) \tag{2}$$

where the translational degrees of freedom of (1) are now considered implicit "functions" of the molecular orientation. The constructions listed in Table 3 can be considered the functions that determine the translational degrees of freedom from the orientational ones. A straightforward way of determining the most favorable molecular orientations is to construct the 3-dimensional potential surface implied by (2) and to search this surface for all deep minima. The computational effort required for such a calculation depends upon the method used for calculating the intermolecular potential and lattice sums.

In this work we have used the atom-atom potential method [12] where it is assumed that the intermolecular potential can be approximated by a sum over individual atom-atom interactions. There are a great number of parameter sets and functional forms available from the literature [47]. The set used in this study was developed by Williams and Starr [48] and has the following form for the interatomic potential:

$$u_{ij} = A_{ij}\exp(-B_{ij}r) - C_{ij}/r^6 + q_iq_j/r \tag{3}$$

where r is the distance between atoms i and j, and all of the subscripted quantities, on the right hand side, are empirically determined constants. The expression for the intermolecular potential is

$$U_{nm} = \Sigma_{ij}u_{ij} \tag{4}$$

where the atom i is on the nth molecule and the atom j is on the mth. As mentioned earlier, the intermolecular potential can be highly anisotropic even though the interatomic potential, (3), is isotropic. This follows immediately from (3) and (4), and the fact that molecules do not have spherical symmetry. The formal expression for the layer lattice potential per molecule becomes

$$\Phi^S = (1/2)\Sigma_m U_{0m} \tag{5}$$

where the summation runs over all molecules in the lattice and it is therefore assumed that all of the structural parameters of (1) are known. The factor of 1/2 in the expression on the right hand side comes about to avoid counting each intermolecular interaction twice [49]. However, the potential energy of a molecule in the lattice is given by

$$U_S = \Sigma_m U_{0m} \tag{6}$$

which is twice the lattice potential. Either quantity can be used for ranking trial structures. If we truncated the sum in (6) to allow only the nearest neighbors, we can define the coordination potential,

$$U_S = \Sigma_k U_{0k} \tag{7}$$

where the sum is over the K members of the coordination shell. This should be comparable to a truncation radius of about 7 - 10 Å for most compounds considered here. It is well known [38] that such a truncation underestimates the lattice potential by about 10 - 15%, but has little effect on the structural constants in a lattice minimization. Hence, we have investigated the use of the coordination potential as a measure of the lattice potential. However, it should be made clear, that the layer constructions represented in Table 3 yield the lattice constants and the full summation in (5) or (6) could be evaluated if necessary.

It is now possible to define a complete algorithm for the determination of all likely layer structures given only a molecular structure. One first chooses one of the layer constructions from Table 3. Next, a set of trial orientation angles are selected, and the translational parameters of the lattice are determined from the construction procedure. This completely determines the structure for one set of orientation angles and the lattice potential can be computed to any desired level of approximation using (5) - (7). This determines one point on a 3-dimensional potential energy surface. The process is then repeated systematically for other values of the orientation angles until an entire 3-dimensional map of the energy surface is constructed. The unique ranges of the orientation angles are not necessarily the same for all of the constructions of Table 3, but are determined by both the molecular symmetry and the lattice symmetry. The symmetry relationships between otherwise distinct orientations is deduced from the Cheshire symmetry or Euclidian normalizer of the layer group. In the case of crystals, the subject has been investigated extensively [50-52].

Having constructed the entire potential energy surface, it can then be searched for minima and the minima recorded. The process is then repeated for all constructions of interest. The net result is a list of close-packed structures that are ranked by potential energy and which may occur in any of the 7 layer groups of Table 3. There is no reason to believe that the minimum of lowest potential energy will correspond to the observed structure. However, if the experimental conditions are such that the interlayer interactions are weak, we expect that the observed structure will correspond to one of the minima of low potential energy. Support of this statement will be presented in Section 4.2. As a practical matter, this means that one must have some way of selecting the correct structure from the list. Even meager diffraction data would be sufficient for this purpose. For a completely theoretical prediction, free energies would have to be calculated using each of the lowest potential minima as a starting point. It is not clear that current statistical mechanical techniques coupled with current computational resources are adequate for the task, but this would be an interesting line of research.

It should be noted that not all of these layer groups are of interest under all experimental conditions. For example, only groups 1 and 10 are polar and only these should be necessary for the study of amphiphiles in the context of LB films. Similarly, most of these layer groups are achiral. If a resolved chiral compound is being studied, only layer groups 1 and 15 are possible.

For a complete description of the technique, a further discussion of the line and displacement searches (Section 2.3) is necessary. The first point is concerned with the idea that these searches can be resolved (i.e., result in a unique answer). We have already discussed displacement searches in a different context [29] and only note here that although it is possible for a non-unique optimum to exist, the molecular shape must be quite complex for this to become a problem. For line searches the situation is different because the topography of the closed interval over which the search takes place can display considerable complexity owing to the presence of two or more boundary molecules. One could therefore claim with some justification that the variable that steps over the closed interval is an independent parameter of the model, and all of the constructions involving line searches would then leave four parameters undetermined. For our initial studies, the choice is largely arbitrary. Since each increment must be sampled, we could construct the corresponding lattice and store the potentials as a function of four parameters instead of three. The disadvantages to this approach are that it requires the calculation of additional lattice sums, there is more bookkeeping to perform, and a 4-dimensional map must be searched. A possible advantage is that the correct structure might otherwise be missed due to a poor choice of resolution criterion. So far, we have not encountered such a case using even the most simple resolution criteria.

The obvious criteria for resolution of the displacement and line searches involve either energy or geometry. In the geometric resolution of a displacement search, the optimum is defined as the point where the two molecules are in contact but not interpenetrating. For displacement searches, the most obvious idea is to choose the displacement pair (Δx, Δy of Figure 3) that leads to the smallest cell area. In other words, we would choose the most closely packed layer. For the constructions in Table 3, this pair can be chosen even when a single displacement search does not suffice to determine the second cell length (e.g., $P1(2_1)\underline{1}$). Since the intermolecular potential can be calculated with the aid of (3) and (4), resolution of the displacement search could also be defined as the point along the displacement vector where the potential energy is a minimum. For line searches, several such criteria are possible. For example, one could minimize the interaction energy of the moving molecule with those defining the closed interval, or with the entire chain implied by these. In our initial studies, we have confined ourselves to using the geometric resolution criteria defined above. The intermolecular radii necessary to define the molecular shape were taken from Bondi [37].

4. Results

In order to assess the validity of the model presented above, it is desirable to make comparisons with experimentally determined structures. In that the bulk of what is known about the organization of condensed phases is the result of single crystal studies, it would be convenient if this information could be used. However, the layers in crystals are subject to relatively strong interlayer interactions and, as discussed in Section 2.2, there is no guarantee that all layers in crystals represent a minimum of the layer lattice potential. On the other hand, layers of practical interest are always subject to some interlayer interactions, so layers in crystals should provide a very stringent test of the model. In the next two sections we report the results for a number of such tests. The results for naphthalene are presented in enough detail to give the reader an appreciation for the practical aspects of the method.

106

4.1 NAPHTHALENE

Naphthalene crystallizes in space group $P2_1/a$ with two molecules per unit cell [53]. In the coordination approximation, the potential energy of a molecule in the crystal, U_x, is about -33 kcal/mol. The crystal displays a 6-coordinate layer of symmetry $P12_1/a\underline{1}$, $Z_s=$ 2, and all molecules occupy inversion centers. The coordination potential of the molecule in an isolated layer, U_s, would be about -25 kcal/mol. Taking the molecule to be 12-coordinate in the crystal and 6-coordinate in the layer, we would expect the ratio U_s/U_x to have a value of 0.50 for a packing in which each intermolecular potential contribution is equal. Such a packing could be termed isotropic. In the case of naphthalene, this ratio is 0.76, indicating that the intralayer contributions to the crystal coordination potential substantially exceed the interlayer contributions. Following the discussion in Section 2.2, we would characterize the naphthalene structure as being stratified.

TABLE 4. Observed and calculated layer structures for naphthalene in $P1(2_1)\underline{1}$

Rank	ϕ_1	ϕ_2	ϕ_3	a	b	x	$U_s^{(a)}$
obs	55	57	37	8.27	5.97	-0.25	-25.1
1	90	60	90	8.38	5.52	-0.25	-25.0
2	50	60	30	7.98	6.03	-0.25	-24.7
3	90	60	10	9.52	6.52	-0.25	-18.1
4	60	60	150	9.08	7.02	-0.25	-18.1

[a] Coordination potential (kcal/mol)

Using the ideas of Section 3.1, we ignore the fact that the molecule has inversion symmetry, and therefore this packing can arise from either layer group $P1a\underline{1}$ or $P12_1\underline{1}$. In Table 4 we summarize the results obtained for the construction $P1(2_1)\underline{1}$. The top line in Table 4 represents the observed structural parameters for the in-crystal layer. The other entries are the lowest points sampled for each depression located in the free-layer, 3-dimensional, potential surface implied by (2). The surface was constructed using the numerical techniques outlined in Section 3.3. In particular, the orientation angles were sampled at regular 10 degree intervals and the lattice potential was approximated by the coordination potential. Note that there are only four minima on the entire surface. This result is peculiar to naphthalene, most molecules display more. However, note that there are only two minima within a few kcal/mol of the global minimum. We have found this result to be quite typical.

For naphthalene, the global minimum does not correspond to the observed structure whereas the second ranked minimum corresponds quite closely. A visual comparison of the second ranked minimum with the observed structure is given in Figure 11. The structure of the first ranked minimum is shown in Figure 12 for comparison. In that the sampling interval is coarse, and geometric criteria are used in the layer constructions, the small difference in coordination potential between the two lowest calculated structures may not be significant. This point could be checked by subjecting the two structures to a direct lattice optimization but this has not been attempted as yet.

For each minimum in Table 4, the optimum value of the fractional coordinate, x, is predicted to be -0.25, which indicates that the molecular inversion center is located 1/4 of the unit cell from the screw axis. Taking both the molecular inversion center and the

packing symmetry into account, the full symmetry of the layer is predicted to be P12$_1$/a1,
as observed.

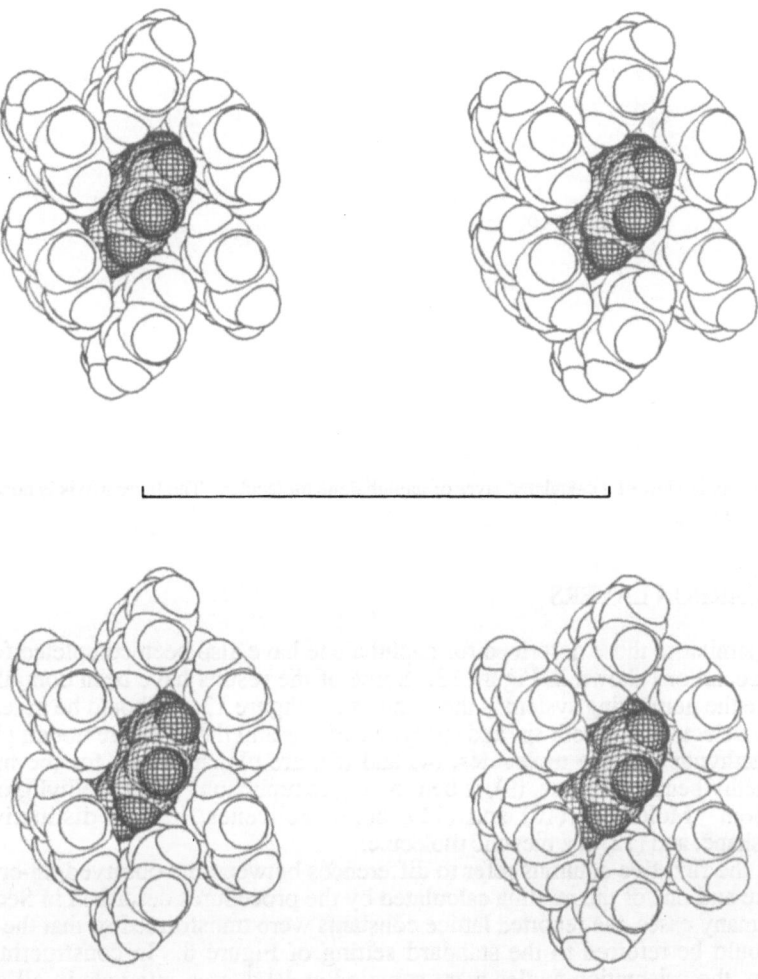

Figure 11. Stereoscopic view of observed (top) and calculated (bottom) structures for naphthalene layers.
The layer a axis is horizontal and the b axis is vertical.

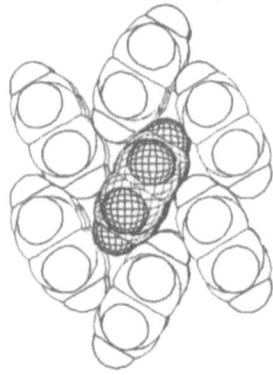 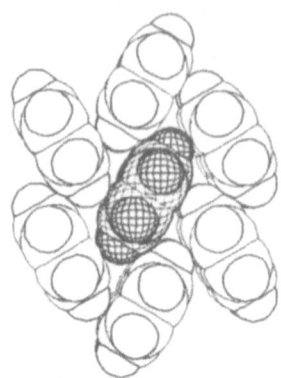

Figure 12. Steroscopic view of a calculated layer of naphthalene molecules. The layer a axis is horizontal and the b axis is vertical.

4.2 HYDROCARBON LAYERS

Calculations similar to those described for naphthalene have also been completed for the 11 other hydrocarbons shown in Figure 13. Some of the results have been compiled in Table 5 where the numbering system is the same as for Figure 13. It should be noted that the molecules selected have both sp^2 and sp^3 type carbons and therefore are not all planar. The two dimethynaphthalene molecules, (4) and (5), are planar except for the methyl groups. Dimethylbenzanthracene, (9), is bent owing to repulsions of the methyl group in the bay region. Molecules (10) and (11), being helicenes, have a distinctive 3-dimensional shape, and (12) is a bicyclic molecule.

In Table 5, the first five columns refer to differences between the observed, in-crystal, layer structure and one of the minima calculated by the procedures described in Sections 3.1 - 3.3. In many cases, the reported lattice constants were transformed so that the layer of interest could be referred to the standard setting of Figure 8. In constructing the potential map, the orientation angles were sampled at 10 degree intervals in all cases. Also, the z coordinate of the molecules was fixed at zero even though, for many of the observed structures, there is no such constraint in the crystal. The intent here was to investigate whether the line search described in Section 2.3 might be sufficient, or if its analog, which includes a bounded interval along z, is necessary. One complete calculation required 2 to 12 hours on a Micro VaxII, depending on the size and symmetry of the molecule and the layer construction under consideration.

TABLE 5. Comparison of observed and calculated layer structures for 12 hydrocarbons[a]

No.	Δa	Δb	Δx	$\Delta\phi_1$	$\Delta\phi_2$	$\Delta\phi_3$	$U_s(o)$	S_x	$U_s(c)$	ΔU
1	0.3	0.1	0.00	5	3	7	-25.1	0.76	-24.7	0.3
2	0.3	0.1	0.00	4	5	8	-38.7	0.80	-38.4	2.0
3	0.1	0.1	0.00	7	2	7	-50.4	0.88	-51.2	0.0
4	0.2	0.4	0.00	6	3	2	-32.8	0.85	-33.4	0.0
5	0.0	0.1	0.00	11	6	20	-27.1	0.73	-28.7	0.0
6	0.9	0.8	0.02	1	12	3	-57.7	0.95	-58.9	0.0
7	0.1	0.0	0.00	10	1	12	-36.0	0.83	-35.7	0.0
8	0.5	0.4	0.00	4	5	5	-49.6	0.87	-49.0	0.0
9	0.3	0.0	0.01	5	9	3	-43.8	0.76	-41.6	0.6
10	1.7	0.2	0.10	10	3	10	-38.3	0.68	-32.6	0.0
11	0.3	2.5	0.01	5	4	10	-42.7	0.76	-39.1	0.0
12	0.2	0.1	0.00	0	0	0	-39.8	0.72	-39.0	0.0

[a]Compounds numbered as in Figure 13, with cell length difference in Angstroms, angle differences in degrees, energies in kcal/mol. See Section 4.1 for definition of stratification index, S_x, and Section 4.2 for discussion.

The calculated minimum selected for comparison in Table 5 was in each case the one whose orientation angles differed least from the observed structure. Most maps displayed a larger number of minima than that of naphthalene. In that we do not present the complete peak lists, the column headed ΔU is of interest. Here, we give the energy difference between the selected minimum and the global (1st ranked) minimum. For example, as discussed earlier in the case of naphthalene, it is the 2nd ranked minimum whose orientation angles are closest to the observed and it lies 0.3 kcal/mol higher than the 1st ranked minimum. Hence, one finds the value of 0.3 kcal/mol under ΔU for compound (1), naphthalene. Here we draw attention to the result that most of the entries under ΔU are zero. This means that it is the global minimum that corresponds to observed structure in most cases. This result is very encouraging, but quite unexpected. The results for compounds (2) and (9) are analogous to those of naphthalene in that it is also the 2nd ranked minimum that corresponds to the observed structure for these molecules. It should be pointed out that calculations were performed only in the observed layer group. The current results indicate that among the calculated structures we always find one that is very similar to the observed. However, the energy difference between this structure and the global minimum may change if calculations (not yet performed) in other layer groups give rise to structures of even lower potential energy.

As can be seen from Table 5, lattice translations are normally predicted to better than 0.3 Å and the fractional coordinates are reproduced even better. The orientation angles are normally predicted to better than 10 degrees. Considering that a sampling grid of 10 degrees was used, the agreement can be considered perfect in many cases. The close approximation of the calculated structures to the observed ones is also reflected in the coordination potentials. The layer coordination potential for the calculated structure, $U_s(c)$, is often within 1 kcal/mol of that for the observed structure, $U_s(o)$. The worst results are clearly for compounds (10) and (11). In both cases, the molecular mass center was displaced from the layer plane by 0.5 Å or more in the experimental structure. For compound (10), dimethylbenzphenanthrene, we have verified that allowing for non-zero z results in a sharp drop of about 6 kcal/mol in the coordination potential and brings the calculated and observed lattice constants into close agreement. It is interesting that even such "ruffled" structures are predicted reasonably well without including the z parameter in the search.

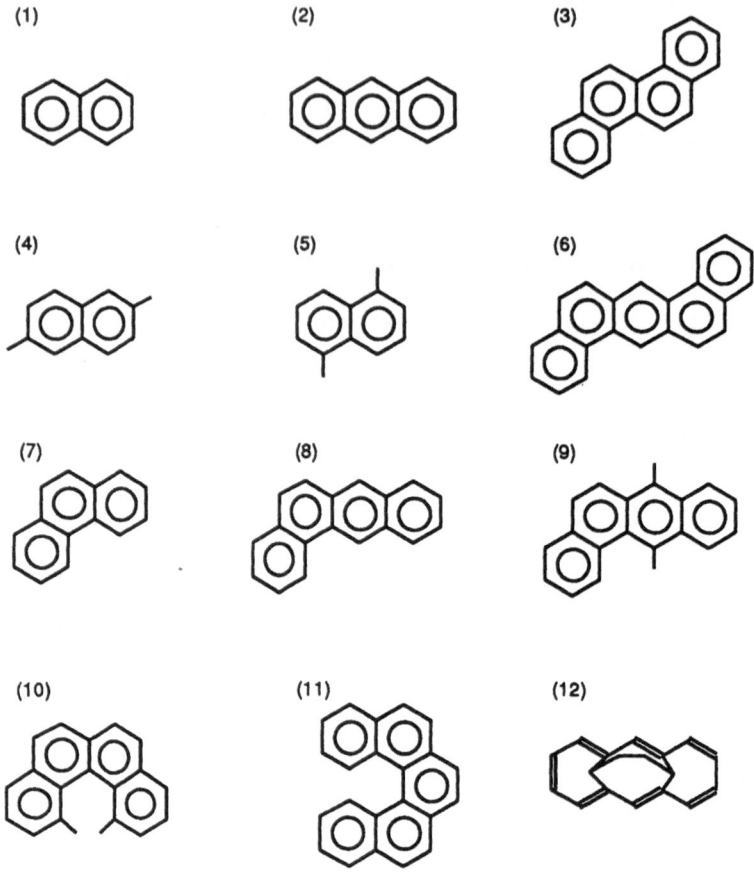

Figure 13. Molecular structures of 12 hydrocarbons considered in this work.

Also included in the table is a measure of the stratification of the crystals from which the experimental layers were taken. The stratification index, S_x, is the ratio of the in-crystal layer coordination potential to the crystal coordination potential ($S_x = U_S(o)/U_x(o)$). It varies from 0.68 to 0.95 for the structures considered here. These are all well in excess of isotropic value of 50. There is no apparent correlation between the stratification index and the precision with which the calculated structure approximates the observed.

It is also possible to extend the method to include molecules capable of hydrogen bonding. A geometric characteristic of hydrogen bonds is that the hydrogen atom usually approaches the receptor atom more closely than would be expected on the basis of their intermolecular radii [54, 55]. This necessitates a slight modification of the geometric model of a molecule. Taking this into account, we have performed calculations on several layers containing hydrogen bonds; the results were comparable to those presented above for hydrocarbon layers.

5. Discussion

In this work we have described a new technique for the prediction of molecular layer structures. The technique seeks to locate all of the important (i.e., observable) minima of the layer lattice potential function while neither constructing the entire 9-dimensional surface nor sampling all 80 of the crystallographic layer groups. The basis of the method is the observation that molecular solids are normally close-packed. This tendency toward maximal density is exploited both to minimize the number of layer groups and to reduce the dimensionality of the potential hypersurface. However, close-packing ideas are not sufficient to put the method on a quantitative basis. For the symmetry problem, it is necessary to make a statistical analysis of experimentally observed structures in order to estimate the probability of an unknown structure occurring in one of the close-packed layer groups of Figure 8. For the parameter problem, a method of calculating the intermolecular potential is required.

The method is quite flexible and numerical results can be obtained in a variety of ways. Some of the options left open to the investigator include: (a) to require resolution of line searches or to include extra parameters in the model, (b) the criteria used in resolving displacement and line searches, (c) the method used in estimating the intermolecular potential, (d) the method used in estimating lattice sums, and (e) the number of layer groups deemed necessary to achieve an acceptably high probability of success. In making such choices one is not using a different model, but rather a different implementation of the same model. In the preliminary work presented above, we have used the least time consuming, and presumably crudest method for each option. In particular, the coordination potential which can be viewed as a very crude approximation to the lattice sum (6), has sufficed for all systems studied so far.

The essential elements of the model are: (a) each layer group of interest is assumed to display a finite number of close-packings and that each can be characterized by its convex coordination number, symmetry composition, and orientation of its coordination shell, (b) that for each such packing, a construction can be devised for a given molecule in an arbitrary orientation, (c) that a suitable model for calculating the intermolecular potential is available, and (d) that the minima of (1) will provide an adequate representation of the layers under the experimental conditions of interest.

Intuitively, one might expect that (d) would impose the greatest limitations on the model. However, the results presented in Section 4.2 clearly indicate that the method is reliable even when interlayer interactions are quite strong. For many of the systems of Table 5, the stratification ratio is only about 0.75. Apparently, interlayer interactions of 1/3 to 1/2 the strength of the intralayer interactions are still insufficient to markedly perturb the ideal structure of the free layer. This is the most important numerical result of the present investigation. It does not seem reasonable that this could be true of any arbitrary potential minimum. However, these structures are not arbitrary, they are all close-packed. It may be that close-packing holds the key to a more fundamental law of molecular organization.

112

6. Acknowledgments

The many discussions with my colleagues, Drs. S. Perez, D. L. Dorset, J. R. Fryer, R. H. Young, J. H. Perlstein, J. P. Spoonhower, P. M. Henrichs, A. Ulman, T. L. Penner, B. H. Rubin, A. F. Corin, R. L. Parton, H. J. Price, and A. P. Marchetti, are gratefully acknowledged. It is also a pleasure to thank Drs. R. S. Eachus, R. C. Baetzold, D. M. Sturmer, C. T. Goodhue, P. J. Grisdale, J. C. Chang, and R. F. Large for their support of this work.

7. References

(1) Fryer, J. R., Hann, R.A., and Eyres, B. L. (1985) Nature, 313, 382-384.
(2) Wolf, S. H., Deutsch, M., Landau, E. M., Lahav, M., Leiserowitz, L., Kjaer, K., and Als-Nielsen, J. (1988) Science, 242, 1286-1290.
(3) Fischer, A. and Sackmann, E. (1984) J. Phys. (Paris) 45, 517-527.
(4) Kjaer, K., Als-Nielsen, J. Helm, C. A., Laxhuber, L. A., and Mohwald, H. (1987) Phys. Rev. Lett., 58, 2224-2227.
(5) Doucet J. and Levelut, A. M. (1977) J. Phys. (Paris) 38, 1163-1169.
(6) Levelut, A. M., Doucet, J., and Lambert, M. (1974) J. Phys. (Paris) 35, 773-779.
(7) Leadbetter, A. J., Mazid, M. A., and Richardson, R. M. (1980) Liquid Crystals, Chandrasekhar, S. (ed.), Heyden, London, pp 65-79.
(8) Strong, L. and Whitesides, G. M. (1988) Langmuir 4, 546-558.
(9) Pertsin, A. J. (1985) J. Phys. Chem. Solids 46, 1019-1024.
(10) Gibbons, T. G. and Klein, M. L. (1974) J. Chem. Phys. 84, 1803-1814.
(11) Owicki, J. C. and Scheraga, H. A. (1977) J. Am. Chem. Soc. 99, 7403-7412.
(12) Pertsin, A. J. and Kitaigorodsky, A. I. (1987) The Atom-Atom Potential Method, Springer-Verlag, Berlin.
(13) Weber, L. (1929) Z. Krist. 70, 309-327.
(14) Alexander, E. and Hermann, K. (1929) Z. Krist. 70, 328-352.
(15) Scaringe, R. P. (1987) Proc. 45th Ann. Mtg. El. Micros. Soc. Am., Bailey, G. W. (ed.), San Francisco Press, San Francisco, pp 472-475.
(16) Kitaigorodsky, K. I., Pertsin, A. J. and Kozlova, I.E. (1976) Sov. Phys. Cryst. 20, 633-634.
(17) Dzyabchenko, A. V. (1984) Russ. J. Struc. Chem. 25, 416-420.
(18) Dzyabchenko, A. V. (1984) Russ. J. Struc. Chem. 25, 559-563.
(19) Dutta, P. (1989) Abstract, 47th Pittsburgh Diffraction Conference, Pittsburgh Diffraction Society, Pittsburgh.
(20) Stillinger F. H. and La Violette, R. A. (1985) J. Chem. Phys. 83, 6413-6419.
(21) Belsky, V. K. and Zorkii, P. M. (1977) Acta Cryst. A33, 1004-1006.
(22) Maartman-Moe, K. (1966) Acta Cryst. 21, 979.
(23) Espiritu, A. and White, J. G. (1978) Z. Krist. 147, 177.
(24) Campbell, R. B., Robertson, J. B. and Trotter, J. (1962) Acta Cryst. 15, 289.
(25) Zorkii, P. M. and Razumaeva, A .E. (1979) Russ. J. Struc. Chem. 20, 390-393.
(26) Allinger, N. L. (1977) J. Am. Chem. Soc., 99, 8127-8134.
(27) Smith A. E., (1953) J. Chem. Phys. 21, 2229-2231.
(28) Scaringe, R. P. (1989) unpublished work.
(29) Scaringe, R. P. and Perez S. (1987) J. Phys. Chem. 91, 2394-2403.
(30) Meyer, A. Y. (1986) Chem. Soc. Rev. 15, 449-474.
(31) Fischer, W. and Koch E. (1979) Z. Krist. 150, 245-260.
(32) Fischer, W. and Koch E. (1980) Z. Krist. 153, 255-263.

(33) Immirzi, A. and Perini, B. (1977) Acta Cryst. A33, 216-218.
(34) Gavezzotti, A. (1982) Nouv. J. Chim. 6, 443-450.
(35) Gavezzotti, A. (1983) J. Am. Chem. Soc. 105, 5220-5225.
(36) Gavezzotti, A. (1985) J. Am. Chem. Soc. 107, 962-967.
(37) Bondi, A. (1964) J. Phys. Chem. 68, 441 - 451.
(38) Kitaigorodsky, A. I. (1973) Molecular Crystals and Molecules, Academic Press, New York.
(39) Patterson, A. L. and Kasper J. S. (1972) International Tables for X-ray Crystallography Vol. II, Kynoch Press, Birmingham, pp 342-354.
(40) Kitaigorodsky, A. I. (1961) Organic Chemical Crystallography, Consultants Bureau, New York.
(41) Fryer, J. R. (1988) private communication.
(42) Zorkii, P. M. and Porai-Koshits (1962) Sov. Phys. Cryst. 6, 529-533.
(43) Zorkii, P. M. and Porai-Koshits (1968) Sov. Phys. Cryst. 12, 863-866.
(44) Mighell, A. D., Himes, V. L. and Rodgers, J. R. (1983) Acta Cryst. A39, 737-740.
(45) Belsky, V. K. (1974) Russ. J. Struc. Chem. 15, 631-634.
(46) Mackay, A. L. (1972) J. Microsc. 95, 217-227.
(47) Timofeeva, T. V., Chernikova, N. Yu. and Zorkii, P. M. (1980) Russ. Chem. Rev. 49, 509-525.
(48) Williams, D. E. and Starr, T. L. (1977) Comput. Chem. 1, 173-177.
(49) Busing, W. R. (1978) J. Phys. Chem. Solids 39, 691.
(50) Hirshfeld, F. L. (1968) Acta Cryst. A24, 301-311.
(51) Dzyabchenko, A. V. (1983) Acta Cryst. A39, 941-946.
(52) Fischer N. and Koch, E. (1983) Acta Cryst. A39, 907-915.
(53) Brock, C. P. and Dunitz, J. D. (1982) Acta Cryst. B38, 2218-2228.
(54) Hamilton, W. C. and Ibers J. A. (1968) Hydrogen Bonding in Solids, Benjamin, New York. pg 16.
(55) Kuleshova, L. N. and Zorkii, P. M. (1981) Acta Cryst. B37, 1361 - 1366.

ELECTRON MICROSCOPY AND MOLECULAR MODELLING OF SOME LANGMUIR–BLODGETT
FILMS

J.R.FRYER,and   C.H.McCONNELL.Electron Microscopy Centre,and G.H.GRANT,
Chemistry   Department,   University of   Glasgow,G12 8QQ.Scotland.U.K.
R.A.HANN,S.K.GUPTA and B.L.EYRES,
Imperial   Chemical   Industries   plc,Imagedata,Brantham,Manningtree,
Essex CO11 1NL.England.U.K.

ABSTRACT.Langmuir–Blodgett films of the lead and copper derivatives of
tetra-t-butyl phthalocyanine,and 9-butyl-anthryl propionic acid have
been examined by electron diffraction,high resolution imaging and
simulated by molecular modelling.The monolayers are not well ordered
crystals,so that detailed crystallographic analysis was not possible,
but a combination of these techniques-and,in the case of the anthracene
derivative-molecular modelling,has   revealed   the   structure of the
monolayers of these materials.

INTRODUCTION.

The principle of casting extensive thin films on a   liquid   surface is
attractive for the large scale fabrication of oriented material.In
particular this is true when the compound involved is an organic
molecule, which cannot easily be crystallised to produce   large
crystals.The current interest in organic[1] crystals,as the basis of
electronic and opto-electronic devices[1],places a   technological
motivatation on the study of such films which was not present when
Langmuir and Blodgett did their initial investigations[2-4]. The
concept of a molecule having a polar group,which is directed   to
a water surface,leads to the conclusion that,as all the molecules have
one axis in common,then all other axes are aligned creating a single
crystal film.This   may be correct in some cases,but significant
deviations have been found,and it is the purpose of this work to use
high   resolution electron microscopy,electron   diffraction   and
molecular modelling   to simulate   structures,and   endeavour   to
elucidate the structure of the partially crystalline monolayers.
 Macrotechniques have tended to confirm the single crystal concept for
multilayer films.A summary of the application of X-ray and neutron
scattering has been published by Pomerantz[5], and spectroscopic methods
have also been used to identify the orientations of particular
functional groups[6]. However,the microstructural observations have
been mainly confined to electron optical methods.

J. R. Fryer and D. L. Dorset (eds.), Electron Crystallography of Organic Molecules, 115–127.
© 1990 *Kluwer Academic Publishers.*

The major difficulty with these methods is that the high energy electrons degrade the specimen. For high resolution information at least 100 electrons/nm^2 are required. The lifetime of an aliphatic hydrocarbon under this irradiation per second is less than 5secs. whilst an aromatic hydrocarbon is only a factor of 10-100 times more stable[7]. In contrast,normal electron microscopy of inorganic crystals and metals employs dose rates of 10^3 to 10^5electrons/nm^2/sec. Electron diffraction reduces this effect because the structural information is averaged over an area from 0.1-1um^2. This technique has been used extensively by Peterson[8-10] to interpret the structure of multilayer L-B films of w-tricosenoic acid and an anthracene derivative. Dorset has shown that the information derivable by electron diffraction is limited by bending of the specimen,and dynamical effects in the intensities of the scattered beams[11,12]. However, for structure and orientation recognition it is a successful method for the areas involved. Direct imaging by both bright and dark field methods has been used to identify the microstructure of films at the molecular level.

Baumeister and Hahn[13] reported images of a thorium hexafluoroacetyl acetone complex L-B film.They did some spatial filtering and the filtered image showed lattice structure consistent with ordered islands of organic complex of 2-5nm in extent.The islands were randomly oriented.Another relatively stable molecule is copper phthalocyanine, for which we obtained lattice images in the monolayer film[14]. The random island structure was similar to that observed by Baumeister. Uyeda et al[15] used dark field electron microscopy to examine monolayers of stearic acid, and also found a heterogeneous structure.

In the present work three materials were used.They were tetra-t-butyl copper phthalocyanine(tb-CuPc), tetra-t-butyl lead phthalocyanine(tb-PbPc) and 9-butyl-anthryl propionic acid (C$_4$ anthracene).Their molecular structures are shown in Fig.1.The first two are molecules which rely on their shape to form a monolayer film .The metal free tetra-t-butyl phthalocyanine had been shown to form films using the LB method[16] as had the copper derivative[14] and tetra-cumylphenoxy lead phthalocyanine[17]. There has been some discussion about the structure of such films. Kovacs et al[18] has suggested that the adjacent phthalocyanine molecules are staggered with a vertical sheer,so that every alternate molecule is aligned totally above its neighbours and has no interaction with the water surface. However,Barger et al[19] proposed that each molecule was flat on the water surface and formed the template for a tilted stack growing from the aqueous substrate.Both of these hypotheses stemmed from interpretation of film pressure data.

The other compound forms a more classical LB film with the film structured by the hydrophilic carboxyl groups oriented to the water surface.The molecular structure of the anthracene derivative was that previously investigated by Peterson et al[10], and its thin film properties had also been examined[20]. The major difference expected between these molecules and the linear chain acid molecules is that the Van der Waal's interaction of the aromatic moieties should affect the symmetry of the structure in the plane of the film.The ability of such molecules to form LB films had been explored by several workers with particular emphasis on anthracene derivatives[21,22,23].

Extensive studies have been made of both their structure and properties[10,20-24].From surface pressure measurements on a more heavily substituted anthracene Stewart[22] had suggested that the hydrocarbon chain was inclined at 55° to the water surface,and that the major axis of the three aromatic rings was at 90°to this tilt angle. Vincett and Barlow[24]assumed the same geometry and, by shallow angle X-ray methods applied to C_4 anthracene multilayers, found a value of 55-65° to the surface normal for the tilt angle of the hydrocarbon chain axis. Peterson et al[10] concluded that the angle was 60° for 11 layers of the same material.The latter author also found a structural solution at 13°,but rejected it on the grounds that for this model,it would have required some molecules not to be in contact with the surface. The work of Steven et al[23] had shown that aromatic hydrocarbons with hydrophilic substituents will form stable layers,provided there is a balance between the hydrophobic and hydrophilic ends of the molecule.Thus the butyl-anthracene is balanced by the propionic acid. Another factor for consideration is that the interactions between the planar aromatic nuclei may be a significant factor in the structure of the film.

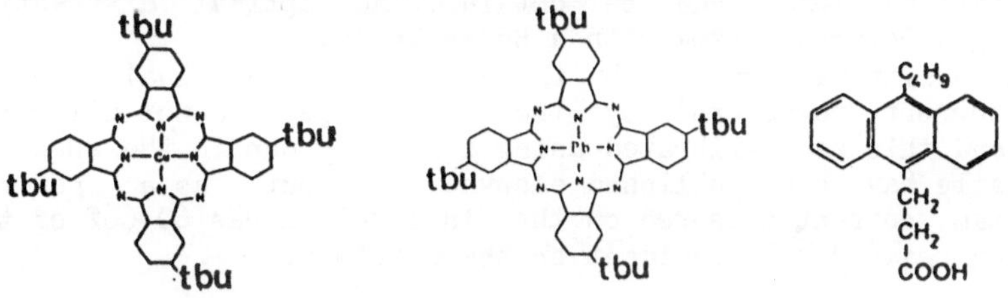

Figure 1.a.Tetra-tert-butyl copper phthalocyanine.
         b.Tetra-tert-butyl lead phthalocyanine
         c.9,butyl-anthryl propionic acid.

EXPERIMENTAL

The compounds were dispersed in xylene(for phthalocyanines)or chloroform and the solutions(1mg/ml) were spread on to the surface of an ICI Langmuir trough[17,24] filled with water deionised to 18M resistivity and Milli-Q filtered.The pH was not adjusted.The barriers were compressed to obtain a surface pressure of 20mNm^{-1}. The monolayer was then deposited on a carbon covered microscope grid by lifting the grid horizontally through the film. The films were examined in a JEOL 1200EX microscope at 120keV. Because of the extreme radiation sensitivity of these materials a rapid exposure technique was devised, so that the specimen exposure time was 20-30ms[25]. Single sided X-ray film (CeaverkenAB,Reflex15), was used for recording the images.
Electron diffraction was done by the selected area technique so that

diffraction patterns were obtained from circular areas 0.2um in diameter.The resulting diffraction pattern represented the periodic structure within the circumscribed area,in terms of the periodicities normal to the incident beam-i.e.only lattice planes lying parallel to the beam direction showed reflections in the diffraction pattern. Therefore,the diffraction pattern showed the orientation of the crystal.In most cases the patterns were arced,rather than the sharp spots expected from a single crystal, indicating a polycrystalline array with some preferred orientation.The electron diffraction technique was used to compare orientations in different areas of the LB film,as well as to provide crystallographic information.

One additional advantage of electron diffraction,is it requires a much lower dose to record information than does the corresponding image. Hence,for radiation sensitive specimens, high resolution information(i.e.reflection intensities), averaged over the 0.2um area,can be obtained without specimen degradation.

Smaller areas than 0.2um have diffraction patterns of proportionally lower intensity,and identification of smaller areas is more difficult. Therefore, it is more convenient to obtain an image of the specimen and carry out optical diffraction of the required areas,provided the specimen is sufficiently stable for images to be obtained. The imaging has all the difficulties described previously,but periodicities from areas as small as 10nm^2 can be obtained.The optical diffraction was done using a Polaron system with a He/Ne laser.

Radiation intensity was measured by the current on the final viewing screen.The current registered by the microscope did not allow for backscatter,and this was calibrated using a Faraday cup in the specimen position.Details have been published elsewhere[25], but as a general rule the beam current measured on the final screen was 60-66% of the total electron beam(120keV) incident on the specimen.

Molecular modelling was done using the QUANTA package run on a Silicon Graphics IRIS 3130 workstation.It did not prove possible to model the phthalocyanines to allow for tetragonal distortion,so the C_4 anthracene was the only one studied in this way.

The initial molecular structure was obtained from the two dimensional construction facility within QUANTA and then refined by three dimensional minimisation of the energy of the isolated molecule.This involved an initial run of 200 steps of steepest descent,followed by 20 steps using the Adopted Basis Newton Raphson method to obtain proper convergence of the energy.

The minimised geometry obtained by the above procedure,was used as a starting point to construct an array of 25 similar molecules in a two dimensional quasi-crystalline array.For molecules with hydrophilic groups,their orientation was conferred by fixing the hydrophilic groups on a plane to represent the water surface.The spacings and orientations of the molecules were set up from the approximate values obtained from the electron micrographs.Each molecule was positioned to reduce non-bonded contact.Each array was allowed to relax without any external constraints,and was subjected to 200 iterations using the steepest

descent method to provide a total interaction energy for the array.
The energies were calculated using the CHARM molecular mechanics
programme [26] within the QUANTA package and considered both inter- and
intra-molecular interactions.For intermolecular interactions a cut off
of 1.2nm was used for the Van der Waals and electrostatic interactions.
This means that only nearest neighbour interactions would be
significant,as has been found from calculations involving proteins[27].
The minimisation method of steepest descent was chosen,because of the
size of the array under consideration.In this method the direction of
geometric change is calculated by determining the vectors of the
force(F),which is acting on each atom due to all potential energy
interactions(V).This force is found from:

$$F = -\nabla V$$

where $\qquad \nabla V_g(x) = (\partial V/\partial x_i) \qquad i=1,3,\ldots\ldots$

This represents the gradient of the energy with respect to
displacement(x) of the atom along any component of its Cartesian
coordinates.These derivatives may be calculated analytically for a
given potential energy function $V_g(x)$.

As with all gradient minimisations,the final structure is the optimum
structure within the potential well defined by the initial structure.
For example,if the initial structure has the molecules tilted,then the
final structure is unlikely to have changed the tilt angle,but has
optimised the molecular stacking for that parameter.Hence,several
configurations were chosen for each array, and the final energy minima
tabulated.The final structure is an idealised situation,and represents
an energy minimum or plateau.The real structure will either be very
close to the computed one,or totally different.For C_4 anthracene,the
hydrophilic interaction with the water surface limits the range of
structures possible.

RESULTS

Copper tetra-tertiary butyl phthalocyanine.
Monolayer films of this compound showed either ring diffraction
patterns or arced diffraction patterns.The 1.9nm reflections arise from
intermolecular separation across the width of the molecules,and the
majority of the 1.9nm lattices,in this selected area,will lie at 90°
to the line which has been drawn intersecting the arcs.Areas up to
100um^2 were mapped for the various lattice directions,using 0.2um
selected area diffraction patterns,with 0.5um between area centres.
Typical results are shown diagrammatically in Fig.2. A single line is
the intersection of a pair of arcs ,and multiple lines indicate that
several pairs of arcs were present.An absence indicates that the
particular area was not analysed.
Electron diffraction of a monolayer only showed the two reflections,
1.9 and 0.35nm,but multi-layers had three dimensional crystallinity
with extensive spot patterns indicating well ordered crystal formation.
Direct imaging of a monolayer also showed differences in orientation,
between periodic areas in the film.Fig.3 shows such an area of 1.9nm

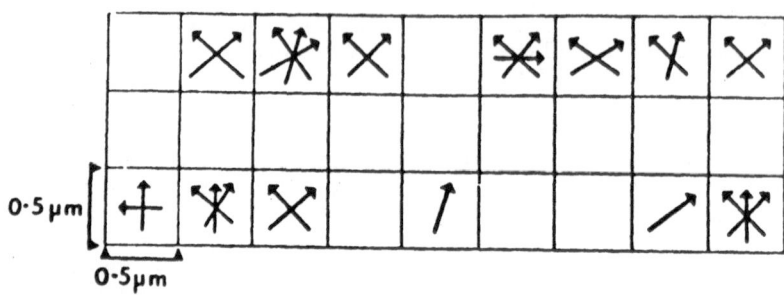

Fig.2.Orientation of domains in tb-CuPc monolayer.

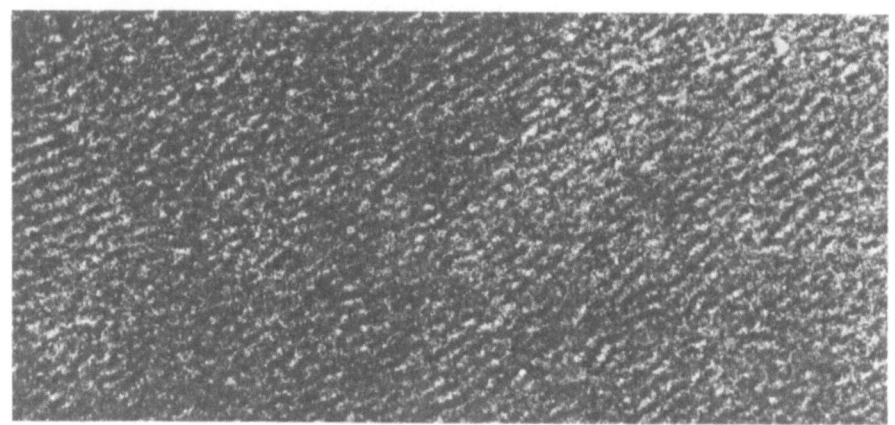

Fig.3.1.9nm lattice fringes in a tb-CuPc phthalocyanine monolayer.

lattice fringes,and again there is no common orientation of the islands of lattice fringes in the monolayer.
In the image it is possible that radiation damage and lack of contrast can create the appearance of an island structure,which was not originally present.However,it has not been known to cause orientation changes.To confirm this the same monolayer specimens were examined at liquid helium temperature,where molecular mobility could not be present.The images were indistinguishable from those taken at room temperature.In particular the varied orientations and domain sizes of areas of continuous lattice were similar.
Multilayer films of tetra-t-butyl copper phthalocyanine exhibit diffraction patterns resembling the crystalline state.Occasionally single crystal spot patterns were observed in a small area of a monolayer film.The intensities of the reflections,and low resolution imaging,indicated that these patterns originated from local three dimensional nucleation and crystallisation on the film.

Lead tetra-t-butyl phthalocyanine.
The diffraction patterns produced by monolayer and multilayer films of tetra-t-butyl lead phthalocyanine were extremely variable.Normally only two sets of reflections were obtained,corresponding to 0.33-0.37nm and 0.41-0.415nm.Often these were arcs and the 0.34nm set were used to map local orientations as was done for the copper compound.A similar degree of disorder was shown.
The two sets of reflections occasionally appeared as spot patterns in a near hexagonal arrangement as shown in Fig.4. This was produced by a monolayer,but some areas in both monolayers and multilayers(up to six layers) produced polycrystalline rings . A perfect hexagonal array occasionally occurred with a spacing of 0.417nm.No other reflections from a monolayer were observed. The large intermolecular spacing-which in the copper analogue is 1.9nm-was not observed.

Fig.4.   Orthorhombic diffraction pattern from a monolayer of lead tetra-tert-butyl phthalocyanine.

9-butyl-anthryl propionic acid(C_4anthracene).
This material was much more radiation sensitive than the phthalocyanine compounds($4e/A^2$ for extinction of first order reflections from a monolayer at room temperature compared to $148e/A^2$ for copper tetra t-butyl phthalocyanine).
A low magnification micrograph is shown in Fig.5.It is evident that the film is not homogeneous,but small crystals are dispersed randomly over its surface.Diffraction patterns taken between the crystals were extremely weak or absent,whilst the crystals gave sharp spot patterns. The spacings agreed(within the accuracy of electron diffraction) with those of bulk C_4 anthracene.
Observations of the monolayer film at low temperature failed to show lattice structure, but using the fast exposure technique some structure was resolved-Fig.6. The poor contrast and irregularity mean that lattice distances are only approximate,but the lattice distances in the background film are between 0.9and 1.1nm. In Fig.7. a crystalline island is shown with strong lattice periodicities of 0.56nm. The greater contrast is probably a result of the increased radiation resistance of the thicker crystal compared to the monolayer.
Modelling the C_4anthracene commenced with an energy minimisation of the molecular shape in three dimensions,and then this shape was used in

five rows,each having five molecules.This is shown in projection in Fig.8a. The major axis of the anthracene nucleus is at 10° to the substrate.The first array configuration was straight rows of molecules as is shown in Fig.8b. The minimised energy of this structure -521.8kcals. is alongside the molecular diagram(NB.this energy only refers to the 25 molecule array).The rows were then staggered to minimise inter-row distances -Fig.8c- but the energy was similar.A big improvement was made by tilting the aromatic plane at an angle of 65° to the water surface,so that the aliphatic chain hydrogens did not interfere -Fig.8d. Finally a herringbone structure was adopted-Figs.8e and f.

The inter-row distance in this final structure is 1.08nm with an energy of 416.5 kcals.

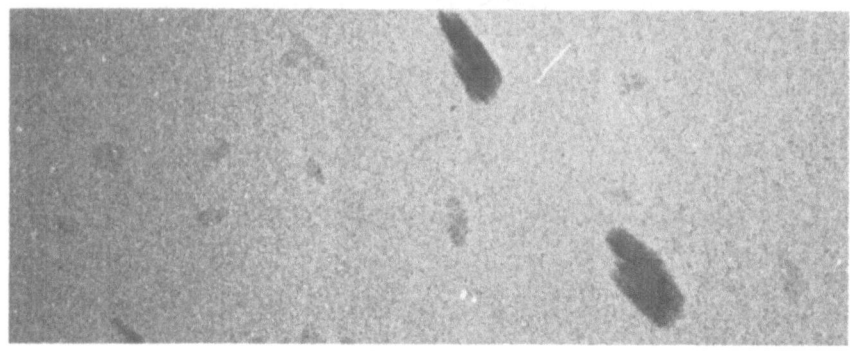

Fig.5. Low magnification micrograph of C_4 anthracene

Fig 6. Monolayer film structure of C_4 anthracene.

Fig.7.Crystalline island on a film of C$_4$anthracene

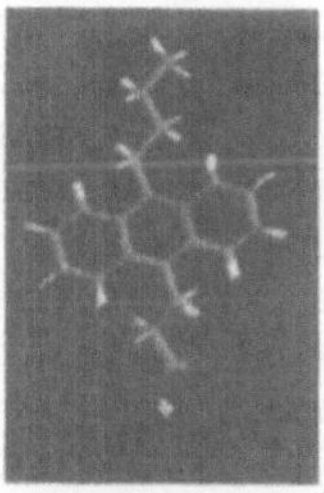

a.Projection of
a single row
of molecules.

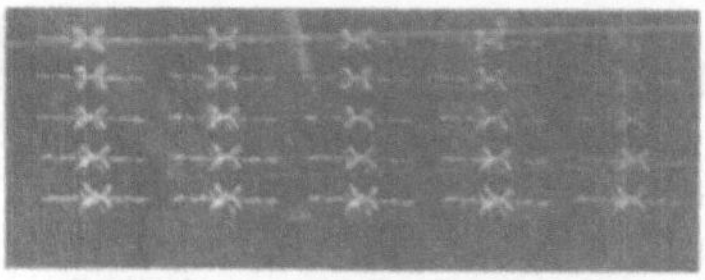

b.521kcals

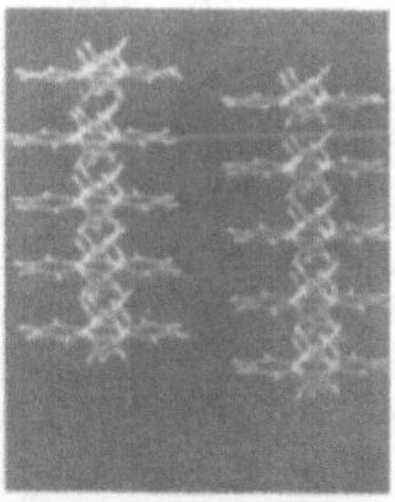

c.517kcals(two rows)

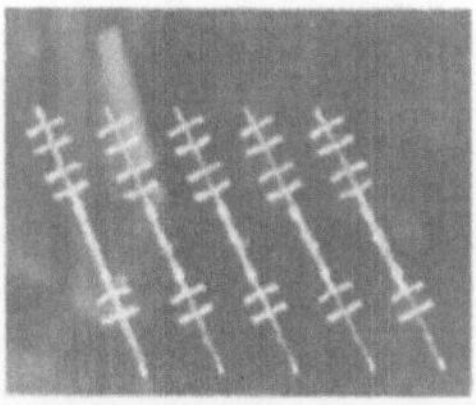

d.430kcals.

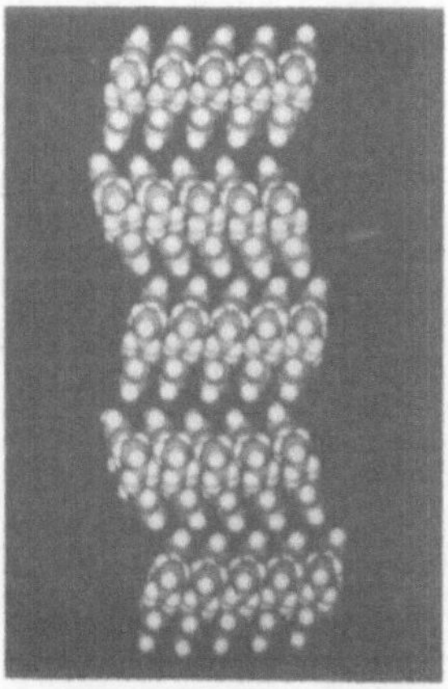

e.Herringbone structure
(untilted)

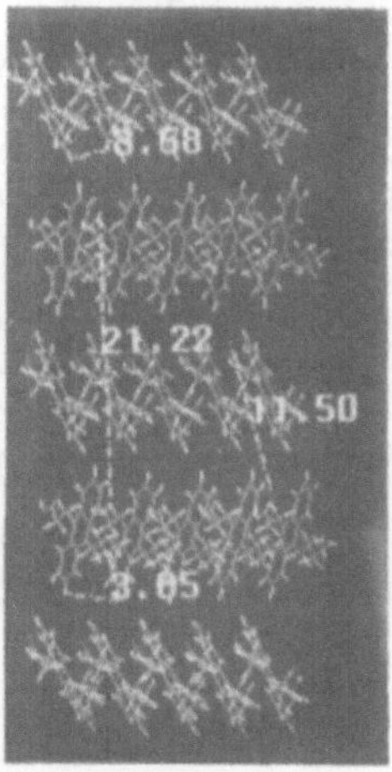

f.416kcals

Fig.8. Modelled arrays of C$_4$anthracene.The energy of the array is
below the diagram.

DISCUSSION

Tetra-t-butyl copper phthalocyanine.

The absence of cross reflections in the diffraction patterns of the monolayer films and the distance of 1.9nm shows that there is no molecular interleaving or fixed relationship between rows,beyond their being parallel columns of face to face stacked molecules. Previously[14] we had suggested from computer modelling that the orientation was tilted with respect to the water surface.Further consideration suggests that this is not consistent with the strong and sharp arcs in electron diffraction,corresponding to 0.34nm at right angles to the 1.9nm reflections.These 0.34nm arcs would correspond to the distance between the phthalocyanine molecular planes-i.e.the face to face Van der Waals separation-in a vertical orientation,otherwise they would not be present.In addition there was only one pair of 0.34nm reflections at right angles to the 1.9nm planes,indicating that the small distances were associated with the same molecular moiety responsible for the large lattice distances. However,the distance between the molecular planes would be greater if the t-butyl groups were eclipsed,and hence the molecules in the row must be in a staggered conformation. The revised structure is shown in Fig.9. This structure would also be favoured as the surface contact points are in a near hexagonal symmetry,which matches that of the water surface.

The areas of periodicity visible in the micrograph are small,and the electron diffraction analysis and optical diffractograms confirm that there is no fixed alignment between them.The mapping showed that the 0.1um sampling areas showed much less multiple orientations present than the 0.2um sampling area.This indicated that the area of continuous lattice(the domain) was approximately 0.1um.The lattice fringes at the ends of the domain do not form a straight edge-as if it were a discrete crystal. This suggests that the ordered regions run into areas of disorder, rather than there being holes in film(N.B.the carbon support film obscures any contrast arising from holes in the film).Also it is not consistant with the structure being that of a discotic liquid crystal with domains parallel and normal to the substrate.

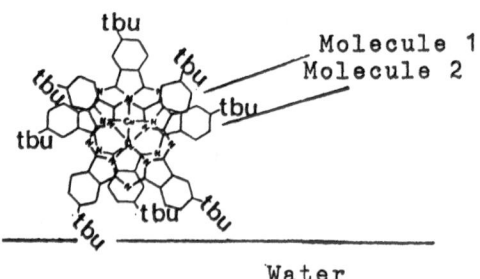

Fig.9. Proposed conformation of tb-CuPc monolayer.

125

Tetra-t-butyl lead phthalocyanine.
 The tetra-t-butyl lead phthalocyanine did not show a diffraction
reflection corresponding to a large lattice spacing but had   well
defined reflections at 0.34nm.Also this 0.34nm set of reflections was
composed of four spots,for a single crystal,with two other spots at
approximately 0.41nm making up the array.Thus the 0.34nm separated
planes were in a two dimensional array,and could not be arranged in one
dimensional rows,as in the copper compound.The limited amount of data
makes interpretation somewhat unreliable, but   such a range   of
reflections would be produced by a random array of discs,separated by
the t-butyl substituents.The two spacings-although reminiscent of
spacings of hk0 oriented paraffins-would arise from the phthalocyanine
nucleii separated by the aliphatic substituents. Even when strong
polycrystalline   patterns-e.g.Fig.4b -were observed, there were no
other reflections which could be associated with the ordering of the
large phthalocyanine moiety.
This structure-or lack of it- would be consistent with the proposal by
Barger et al[19] that the plane of the phthalocyanine is parallel or
at a variable shallow angle to the water surface.The absence of high
lattice spacings in the diffraction pattern shows that is not close
packed.Local areas of ordering occur to either give an   orthorhombic
or hexagonal structure for the film.
The chemical difference between the lead and copper derivatives is
that copper is capable of Jahn-Teller distortion,so that the copper can
adopt a distorted octahedral structure,with d_z2 coordination to a
nitrogen in   an   adjacent   molecule.The staggered   configuration
represented in Fig.9 does not have coincident centres and each copper
has nitrogen atoms in appropriate positions for coordination in
adjacent molecules.This octahedral coordination would stabilise this
structure,but would not be possible in the lead compound.This ability
of copper to achieve partial octahedral coordination in phthalocyanines
of the appropriate crystal structure,has been shown to be responsible
for the increased stability of β phase copper phthalocyanine,compared
the the α phase compound.

C anthracene
 The differences between the modelled molecular structure and that
deduced by previous authors lies mainly in the angle made by the
aromatic nucleus to the tilted aliphatic chain. In Fig 8d-i.e.looking
down on the modelled array,the molecular plane of the aromatic
nucleus lies in the tilt angle, whereas the structure proposed by the
earlier workers had the aromatic plane at 90° to that shown in Fig.8d.
The modelled molecular structure determines the minimum energy
configuration in three dimensions,and since the water surface does not
pose any constraint for an individual molecule,it would be expected to
be the correct structure.
 The diffraction patterns from multilayers and the lattice images of
the   bulk   crystal agree with those of Peterson[10],but this is
reasonable because the modelled structure has a similar tilt angle to
that deduced by him.It is also significant that he found a structural

126

fit at 13° for the aromatic major axis–but rejected it because of the absence of a reasonable Kitaigorodskii fit–and the modelled value is approximately 10°with the adjusted molecular structure which permits a homogeneous array on the substrate.

The monolayer has periodic regions in which the predominant lattice distance is approximately 1.1nm This would correspond to the inter- row separation as shown in Fig.8e. It is noticeable that the inter-row separation is significantly less than the 1.25nm distance observed in the bulk crystal,showing that the monolayer adopts the two dimensional structure as used in the modelling.

The very large drop in energy on tilting the molecules is a result of reducing the interactions between the aliphatic chains.Initially it was thought possible that the aromatic interactions would dominate the packing,but in the energy minimisation the aliphatic chain interactions are more significant.The subsequent reductions in energy with a herring-bone stacking are small,such that it could not be stated that this is the preferred configuration.

However, the modelling shows the optimised molecular packing of a two dimensional film with the hydrophilic groups constrained to one surface,which is similar to the environment of the experimental monolayer.In consequence it is not unexpected that the three dimensional bulk crystal,or multilayers, would have different lattice parameters to both the monolayer and the model.The experimental and modelled structures of the monolayer show that it is more closely packed in two dimensions than a corresponding slice of the bulk crystal.

ACKNOWLEDGEMENT

We wish to thank Dr I.Peterson for discussion on this work.

REFERENCES

1.Bowden,M.J.and Turner,S.R.Eds.1988,Electronic and Photonic
  applications of Polymers.Advances in Chemistry Series 218,ACS.
2.Langmuir,I.,1917,J.Am.Chem.Soc.,39,1848.
3.Blodgett,K.B.,1935,J.Am.Chem.Soc.,57,1007.
4.Blodgett,K.B.and Langmuir,I.,1937,Phys.Rev.,51,964.
5.Pomerantz,M.,1987,Thin Solid Films,152,165.
6.Rabolt,J.R.,Burns,F.C.,Schlotter,N.E.and Swalen,J.D.,1983,J.Chem.
  Phys.,78,946.
7.Fryer,J.R.and Holland,F.M.,1984,Proc.Roy.Soc.(Lond.),A393,353.
8.Peterson,I.R.,Russell,G.J.and Roberts,G.G.1983,Thin Solid Films,
  109,371.
9.Peterson,I.R.and Russell,G.J.1984,Phil.Mag.A.,49,463.
10.Peterson,I.R.,Russell,G.J.,Neal,D.B.,Petty,M.C.Roberts,G.G.,
  Ginnai,T.and Hann,R.A 1986,Phil.Mag.B.,54,71.
11.Dorset,D.L.,1980,Acta Cryst.,A36,592.
12.Dorset,D.L.,1985,J.Elec.Microsc.Tech.,2,89.
13.Baumeister,W.and Hahn,M.,1973,Cytobiologie,7,244.

14.Fryer,J.R.,Hann,R.A.and Eyres,B.L.,1985,Nature,313,382.
15.Uyeda,N.,Takenaka,T.,Aoyama,K.,Matsumoto,M.and Fujiyoshi,Y.1987,
   Nature,327,319.
16.Baker,S.,Petty,M.C.,Roberts,G.G.and Twigg,M.V.1983,Thin Solid
   Films,99,53.
17.Pitt,C.W.and Walpita,L.M.1980,Thin Solid Films,68,101.
18.Kovacs,G.L.,Vincett,P.S.and Sharp,J.H.1985,Can.J.Phys.,63,346.
19.Barger,W.R.,Snow,A.W.,Wohltjen,H.,and Jarvis,N.L.1985,Thin Solid
   Films,133,197.
20.Roberts,G.G.,McGinnity,T.M.,Barlow,W.A.and Vincent,P.S.1980,
   Thin Solid Films,68,223.
21.Giles,C.H.,and Neustadter,E.L.1952,J.Chem.Soc.918,3806.
22.Stewart,F.H.C.1961,Aust.J.Chem.14,57.
23. Steven,J.H.,Hann,R.A.,Barlow,W.A.,and Laird,T.1983,Thin Solid
   Films.99,71.
24.Vincett,P.S.and Barlow,W.A.1980,Thin Solid Films,71, 305.
25.Fryer,J.R.1987,Ultramicrosc.23,321.
26.QUANTA 2.0,Copyright Polygen Corporation,Waltham MA.1988
27.Brooks,B.R.,Bruccoleri,R.,Rhapson,B.,Slantes,D.,Swaninathan,S.and
   Karplus,M.1983,J.Comp.Chem.,4,187.

Maximum Entropy Reconstruction of Low Dose, High Resolution Electron Microscope Images

DAVID C. MARTIN*, KEVIN R. SCHAFFER**, and EDWIN L. THOMAS***

*The University of Michigan
Materials Science and Engineering
H. H. Dow Building
Ann Arbor, MI 48109-2136

**The University of Massachusetts at Amherst
Polymer Science and Engineering
701 Graduate Research Tower
Amherst, MA 01003

***Massachusetts Institute of Technology
Materials Science and Engineering
Room 13-5066
Cambridge, MA 02139

ABSTRACT. Low dose, High Resolution Electron Microscope (HREM) images have provided new information about the ultrastructure of polymers and organic solids. Images of grain boundaries, dislocations, and the structure in thin films have all been successfully obtained. However, low dose HREM images are typically quite noisy and problematic to interpret in detail. Therefore, there is a strong interest in image processing techniques which will take full advantage of the limited information content available in low dose HREM images. At the same time, there is a concern about the possibility of introducing artifacts by image processing methods. Here, we critically examine the use of Maximum Entropy (ME) techniques for improving low dose HREM images of polymers. ME approaches are purported to facilitate the reconstruction of incomplete, noisy data in a rigorous manner which is free from user bias and allows for a minimum amount of adjustable parameters. Our success demonstrates that this approach is indeed an area for further investigation, yet we also find that artifacts can be induced and therefore urge caution. Fourier transform analyses of ME reconstructions in which the entropy of the image is defined only in real space reveal frequency artifacts corresponding to harmonics of strong periodicities in the original data. These artifacts apparently relate to the selective enhancement of "position" information with a corresponding loss in the certainty of "frequency" information. ME reconstructions performed solely in the Fourier space domain show an improvement in periodic information, but a loss in the quality of position information, evidenced by "bleeding" of periodicities over previously sharp boundaries. We suggest the use of a more complete entropy definition which takes into account the known spatial or temporal relationship between data points.

129

J. R. Fryer and D. L. Dorset (eds.), Electron Crystallography of Organic Molecules, 129–145.
© 1990 Kluwer Academic Publishers.

1. Introduction

Maximum Entropy (ME) techniques have found widespread applicability in the reconstruction of incomplete or noisy data. These techniques have been applied in many areas of data analysis including imaging, spectroscopy, and scattering [Gull and Skilling, 1984]. The techniques have proven particularly useful in astronomy [Narayan and Nityanada, 1984]. In many of these applications the goal of the reconstruction is the detection of point objects against a noisy background.

In this work we investigate the applicability of ME techniques to data sets which have strong components which are periodic in space or time. The specific interest in our laboratory is High Resolution Electron Micrographs of beam sensitive materials. However, ME techniques are of general interest for all types of data. These data may or may not have a spatial or temporal character.

Figure 1 shows an HREM image of the rigid-rod polymer poly(paraphenylene benzobisoxazole) (PBZO). The 0.55 nm spacings in the image correspond to the lateral close-packing between the extended polymer molecules. Near the center of this crystallite there is evidence for an edge dislocation. In HREM images both the frequency and position of the information is important for a proper interpretation. Therefore, it is necessary to consider how image processing affects the fidelity of this information in both real and Fourier space.

Our general goal is to establish clearly and in detail the structure of polymers near defects. We anticipate that this will enable us to clarify the role of defects in determining the macroscopic properties of these materials. Often, however, it is problematic to establish the detailed nature of defects from low dose HREM images of polymers. Therefore, it is of interest to try and "improve" a noisy image with numerical image processing algorithms. Given an image like Figure 1 as our input $\{d_i\}$, we are interested in finding that reconstruction $\{f_i\}$ which is the least-biased estimate and incorporates in a well-defined, rigorous way the information at hand. The goal is to find an estimate which is as free from user bias as possible.

The mathematical foundations of the ME methodology have been treated in detail elsewhere [Shore and Johnson, 1982]. The goal is to find that reconstruction $\{f_i\}$ of the data $\{d_i\}$ which maximizes the entropy functional Q:

$$Q = -\Sigma p_i \ln p_i - \lambda \Sigma (d_i - f_i)^2 / \sigma^2$$

(1)

where p_i are "probabilities" associated which each measurement point i. For images with flux f_i in each pixel, p_i has been conveniently defined as $p_i = f_i/\Sigma(f_i)$, which represents the fraction of total flux of the image present in pixel i. In the case of a uniformly flat image, all p_i's are the same. In this case the first term in Q, which has been taken as the "entropy" of the distribution $\{p_i\}$, is a maximum. The second term in Q is a constraint representing the deviation of the reconstruction from the original data. This constraint is in the form of a χ^2 term $C = \Sigma (d_i - f_i)^2 / \sigma^2$. If the standard error in the data σ is known, then C should

131

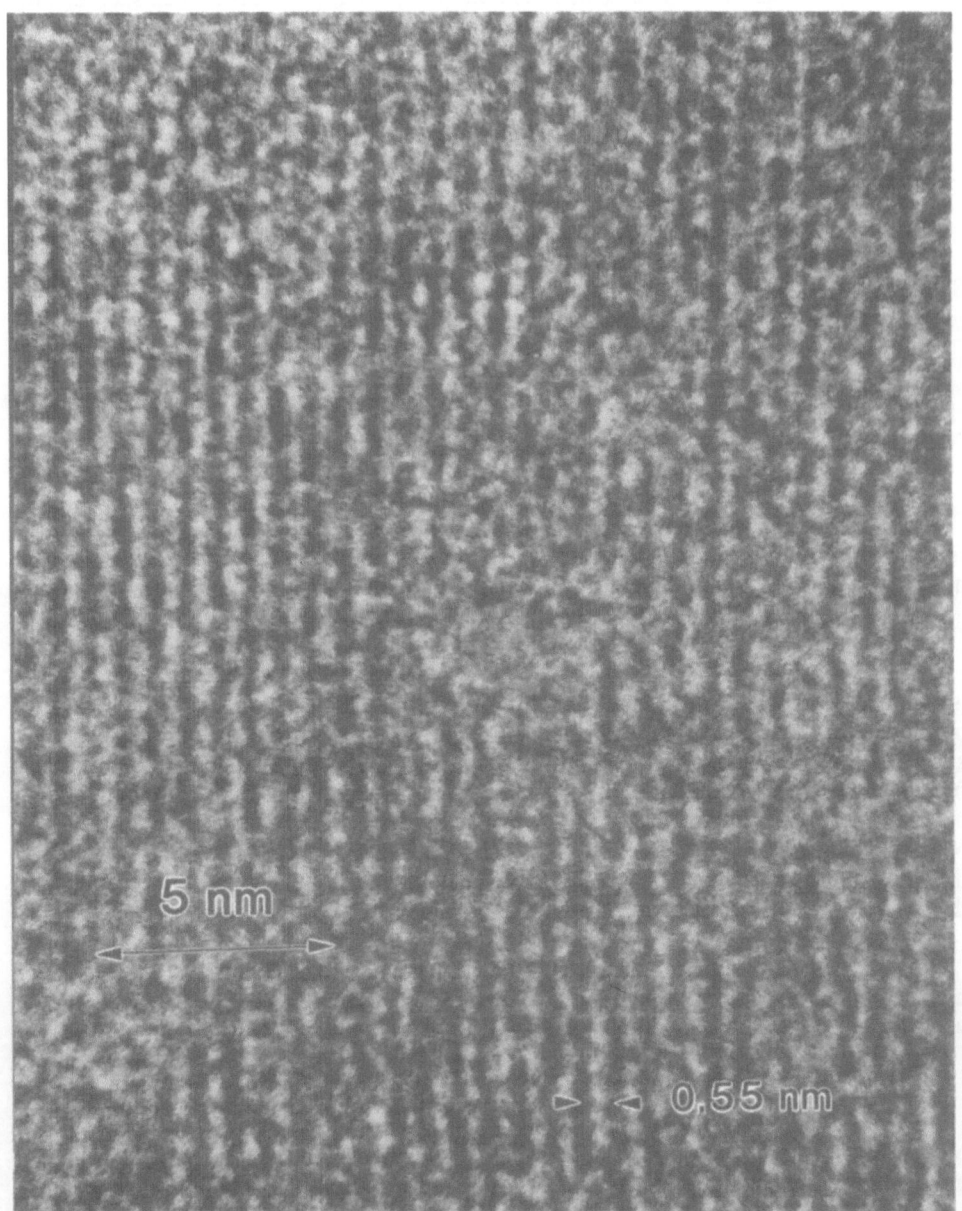

Figure 1: Low dose, High Resolution Electron Micrograph of the rigid-rod polymer poly(paraphenylene benzobisoxazole) (PBZO). The 0.55 nm spacing corresponds to the lateral close packing between the rigid, extended PBZO molecules. In the center of the image is evidence for an edge dislocation with a Burger's vector perpendicular to the axes of the molecules. The position *and* frequency of the information in this noisy image are both important for a proper interpretation.

equal N, the number of measurements i. The Lagrange multiplier λ is chosen such that this is the case. Hence, if σ is known, the problem is specified and the solution $\{f_i\}$ which maximizes Q will be unique. In practice, σ is not well known and is used to estimate the solution $\{f_i\}$.

The use of the form $p_i = f_i/\Sigma(f_i)$ as a "probability" in equation (1) has been rationalized because it represents a fractional flux of the total intensity in the image. It was apparently first derived by Daniell and Gull (1980) who used an analogy between the intensity in a given pixel and the traditional Bayesian statistics problem of arranging balls into boxes. The form $-\Sigma\ p_i \ln p_i$ for entropy which results is similar to that used by Shannon in the development of Information theory [1948], where the p_i represents the probability of a certain signal to occur.

Unfortunately, this definition for the entropy of a signal sampled at positions i in space or time *ignores the spatial or temporal relationship between the events i*. For a given realization of $\{f_i\}$, this definition of entropy will be the same no matter how the individual components of $\{f_i\}$ are arranged.

Spatial relationships between events i have been incorporated into ME reconstructions by considering that there is a point spread function b which relates the ideal undistorted object f'_i to the actual data f_i. Often, the relationship

$$f_i = f'_i * b$$

(2)

is assumed, where * represents the convolution operator. The convolution is applied as a multiplication of the Fourier transforms of f'_i and b. In many physical situations b is known or at least knowable. In HREM imaging, b should incorporate the contrast transfer function (CTF) if the goal of the restoration is the exit wave at the sample. In the weak phase object approximation, this exit wave can be linearly related to the projected electron potential of the object. The CTF relates the final intensity in the image to the exit wave and the experimental conditions such as electron wavelength, spherical aberration coefficient of the objective lens, and objective lens defocus [Cowley, 1988].

Anderson, Martin, and Thomas (1989) have suggested an iterative algorithm which uses ME principles both in real and in Fourier space. Essentially, these authors suggest using an additional entropy definition S_k'

$$S_k' = -\Sigma\ P_k \ln P_k$$

(3)

where the P_k are defined as $/F_k/ \ /\Sigma/F_k/$, with $/F_k/$ the moduli of the (complex) Fourier transform F_k of the image f_i. The rationalization for this approach was based on recent efforts in quantum mechanics which defined entropies simultaneously in both real and Fourier space [Gadre and Bendale, 1985]. For positive, normalized probability distributions the entropies in real and Fourier space are *not* the same. On the other hand, the sum of the real and Fourier space entropies has been shown to be related to the Heisenberg uncertainty principle [Bialynycki-Birula & Mycielski, 1975].

In this work, we critically assess the results obtained when using ME techniques for reconstructing images in which both position and frequency information is important, such as a periodic crystal with a local defect. We first examine the effect of ME reconstructions in which the entropy is defined only in real space. We also investigate the behavior of the ME approach when it is applied in the Fourier space domain.

2. Experimental

Images were taken on a JEOL 2000 FX Transmission Electron Microscope operating at 200 kV. HREM images were obtained using low dose procedures on Kodak Electron Image film SO-163. The micrographs were digitized directly or from intermediate negatives with an Optronics P-1000 microdensitometer.

The algorithms used in this work are based on the original analysis by Willingdale (1981). The algorithms were first written and used by Moriarity-Schieven et al. for the analysis of gas flows from M87 (1988). The speed and convergence of the procedure was improved using a first-order continuation scheme [Anderson, Martin, and Thomas, 1989]. The algorithms were written in FORTRAN 77 and run on a MicroVax II/GPX workstation using the VMS 4.5 operating system. They have been incorporated into the SEMPER image processing system (Synoptics, 1990).

3. Results

Once an image $\{d_i\}$ is available, an estimate of the error σ is necessary to determine the resulting reconstruction $\{f_i\}$. In practice, σ is not well known and therefore the Lagrange multiplier λ is varied instead. For different choices of λ, a maximum for Q is determined and the error σ calculated *a posteriori*. Figure 2 shows the relationship between σ and λ for a particular image. When $\lambda=0$, σ is $\sigma_d = (d_i-f_{avg})^2$, the amount of inherent deviation in the image data. In other words, the error in measuring the image is as large as the experimentally observed fluctuations. In this case the data is completely unreliable and the reconstruction given by this method is a uniformly flat image.

The other extreme is $\lambda => $ infinity; in which case $\sigma=0$. Here, the measuring scheme is perfect and the best estimate of $\{d_i\}$ is $\{d_i\}$ itself. For intermediate values of σ, the maximum entropy approach gives a well-defined means for choosing the "least biased" estimate $\{f_i\}$ for the data $\{d_i\}$.

Some insight into the nature of ME reconstruction can be obtained by the following analysis. A maximum in Q is found when

$$\partial Q/\partial f = 0 \quad \text{for all f.}$$

(4)

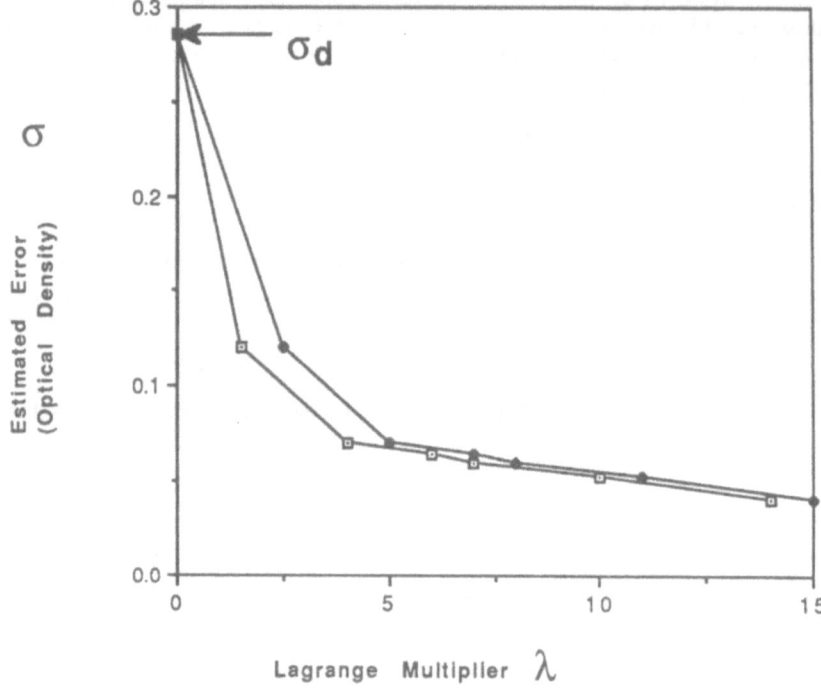

Figure 2: Estimated error in the image σ as a function of the Lagrange multiplier λ. When λ goes to 0, the estimated error goes to σ_d, the standard deviation of the experimental data. In this case the data $\{d_i\}$ are unreliable and the best estimate $\{f_i\}$ is the flat image. When λ goes to infinity the data are perfect and the best estimate is the data itself. Maximum Entropy provides a well-defined means of choosing a reconstructed image $\{f_i\}$ for an estimated value of $\sigma < \sigma_d$.

The boundary conditions are such that $f_i = f_{avg} = \Sigma(f_i)/N$ when $\lambda = 0$, and $f_i = d_i$ when $\lambda =>$ infinity. By taking the derivative of equation 1 with respect to f_i and inserting the appropriate boundary conditions, we obtain:

$$f_i \exp (\lambda\, f_i) = f_{avg} \exp (\lambda\, d_i)$$

$$(5)$$

which, after taking the natural log and dividing by λ may also be written as:

$$\ln (f_i) /\lambda + f_i = \ln (f_{avg}) /\lambda + d_i$$

$$(6)$$

Figure 3 shows the behavior of f_i versus lambda λ for different choices of d_i. The curves shown are solutions f_i for values of d_i with equivalent deviations above and below the mean as a function of λ. When λ is zero, all f_i approach the mean f_{avg}. When λ is large, the solutions f_i approach the data d_i.

Figure 4 shows the relative percentage difference between solutions f_i corresponding to data d_i with equivalent deviations above and below the mean versus λ. Note that the equation causes the rate of conversion of estimates f_i for data points d_i lying below the mean to approach the mean faster than those points for which d_i is above the mean. Hence, the ME approach might be thought of as a non-linear stretch of the uniformly flat image toward the actual data as a function of λ. Starting with a perfectly flat estimate $\{f_i\}$ at the limit $\lambda = 0$, the approach enhances estimates of intensity f_i for points d_i above the mean faster than it suppresses estimates f_i for points d_i below the mean. The disparity in the relative enhancement of points above the mean with respect to those below the mean is a maximum at an intermediate value of λ.

This analysis reveals several problems which might be anticipated when applying ME to periodic objects. Figure 5a shows an image of a perfectly periodic sine wave. Superimposed on this wave is a background of random noise varying linearly in intensity from 0 to 10% of the peak of the sine wave. Figure 5b shows the log of the power spectrum of Figure 5a. Note the strong components in frequency space at positions corresponding to the spacing between peaks in the sine wave. The random background gives rise to a uniform distribution in frequency space.

Figure 5c shows Figure 5b after real-space ME reconstruction. The contrast is somewhat improved over Figure 5a. Figure 5d shows the log of the power spectrum of Figure 5c. Evident in the power spectrum are anomalous frequencies which have been enhanced well above the previously uniform background. These anomalous frequencies correspond to harmonics of the original strong periodicity.

The origin of these anomalous peaks can be understood from the previous analysis. The ME scheme preferentially enhances the "peaks" of the sine wave as compared to the "valleys". This means that the power spectrum of the resulting estimate approaches a series of delta functions corresponding to a set of lines at a spacing corresponding to the "peaks" of the wave form. This results in "oversharpening", producing frequency artifacts.

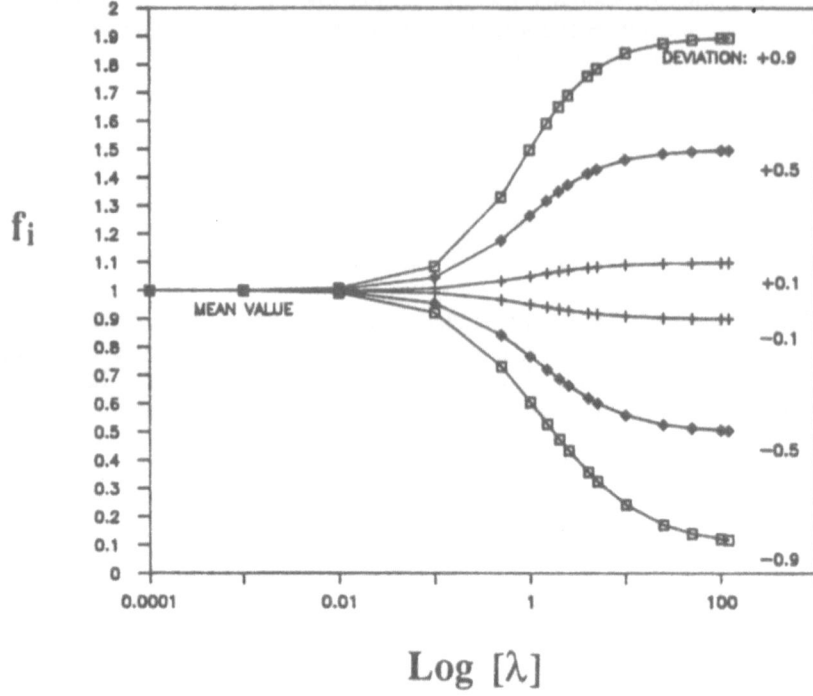

Figure 3: Plots of the estimates f_i for data points d_i both above and below the mean as a function of the Lagrange multiplier λ. The f_i are equal to the corresponding d_i when λ tends to infinity, and all approach the mean value f_{avg} at $\lambda=0$. These results were determined using equation (5) in the text .

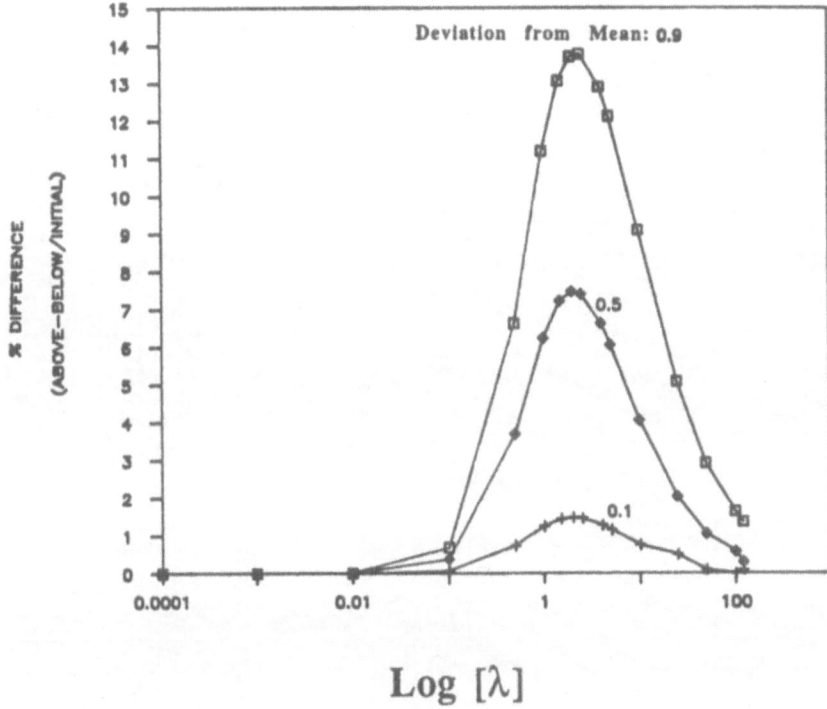

Figure 4: Comparison of the difference between reconstructions f_i for data points with equivalent deviations above and below the mean as a function of the Lagrange multiplier λ. As the value of λ is decreased, reconstructions f_i for points where the d_i is above the mean approach the mean value more slowly than for points where the d_i is below the mean.

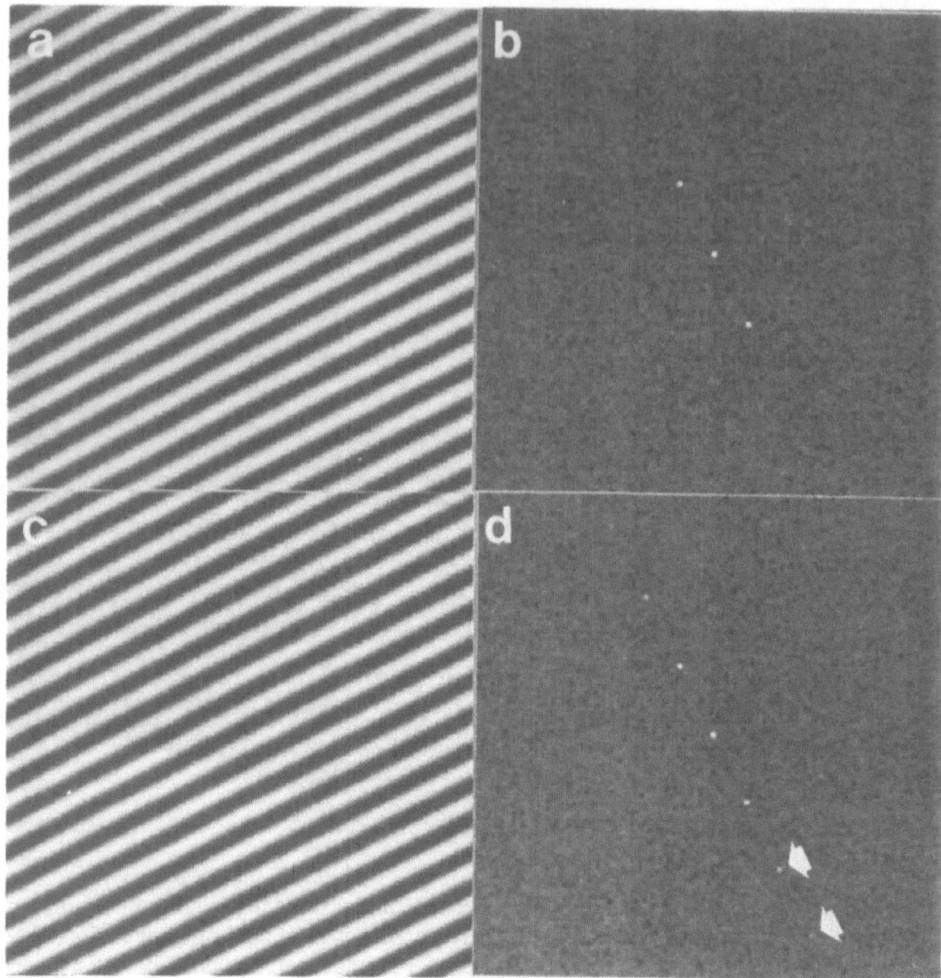

Figure 5: Maximum Entropy reconstructions of a image with a periodic, two dimensional sine wave superimposed on 10% uncorrelated noise. Figure 5a: Original image showing the strong sine wave. Figure 5b: Power spectrum of Figure 5a showing the single peak in frequency space. Figure 5c: ME reconstruction of the image using traditional real space analysis. Figure 5d: Power Spectrum of Figure 5c showing the presence of strong periodicities well above the originally uniform background. These anomalous frequencies correspond to harmonics of the original strong periodicity.

Figure 6 shows a series of images in which the Lagrange multiplier λ was varied to find different reconstructions $\{f_i\}$. These images are displayed so that the grey scale mapping between intensity in the image and that on the graphics monitor is the same for each. As λ decreases (σ increases), the images approach the flat, featureless image (f_i->f_{avg}). It is important to note, however, that the general appearance of the images are not dramatically changed. In particular, if the images are redisplayed on the graphics monitor so that the minimum value is black and the maximum is white, they appear indistinguishable.

Again. equation (5) provides insight into what is occurring. There is nothing in the ME formalism (except the point spread function b) which would cause a reconstruction to have $f_1 < f_2$ for an image in which $d_1 > d_2$. Hence, the relative ranking of intensities will not be disturbed by the ME approach except through the point spread function b. In other words, if two neighboring pixels are of relative brightness $d_1 > d_2$ in the original data , then $f_1 > f_2$ for all real space ME reconstructions $\{f_i\}$. This is consistent with the observation that the traditional entropy definition for an image explicity ignores the precisely known spatial relationship between the pixels.

The effect of the point spread function b is investigated in Figure 7. Here, b was assumed to be a Gaussian function. The characteristic width of b was varied from 0 to 9 pixels. Note from Figure 7 that the appearance of the image is clarified "best" with a p.s.f. of 5 pixels. We conclude that it is the choice of b, not σ, which is most important in determining the appearance of the real space ME reconstructed image.

We now turn our attention to the effect of varying the Lagrange multiplier λ in Fourier space. In other words, we will determine Fourier moduli F_k of the image f_i by using ME. In this case, the data $\{di\}$ are Fourier transformed to give a set of Fourier moduli D_k and phases ϕ_k. The images are examined in real space by recombining the reconstructed moduli F_k with ϕ_k and inverse Fourier transforming.

From our previous analysis, it is possible predict roughly what should happen in this case. In the limit of large λ, F_k will equal D_k and the image will not be altered. However, as λ goes to zero, Fourier components with small moduli will approach the mean faster than those with large moduli. Unlike before (ME analysis in real space only), there is now a chance that pixels with $d_1 > d_2$ in the original data will have $f_1 < f_2$ in the reconstruction. Thus, the reconstruction in this case will not be a simple non-linear scaling operation.

Figure 8 shows the effect of varying λ to find a new distribution of moduli F_k for the Fourier transform of f_i. These images were determined by a numerical inverse Fourier transform using moduli determined by the ME method. Following the example of Anderson, Martin, and Thomas (1989), the phases were determined from the Fourier transform of the input image and were not altered.

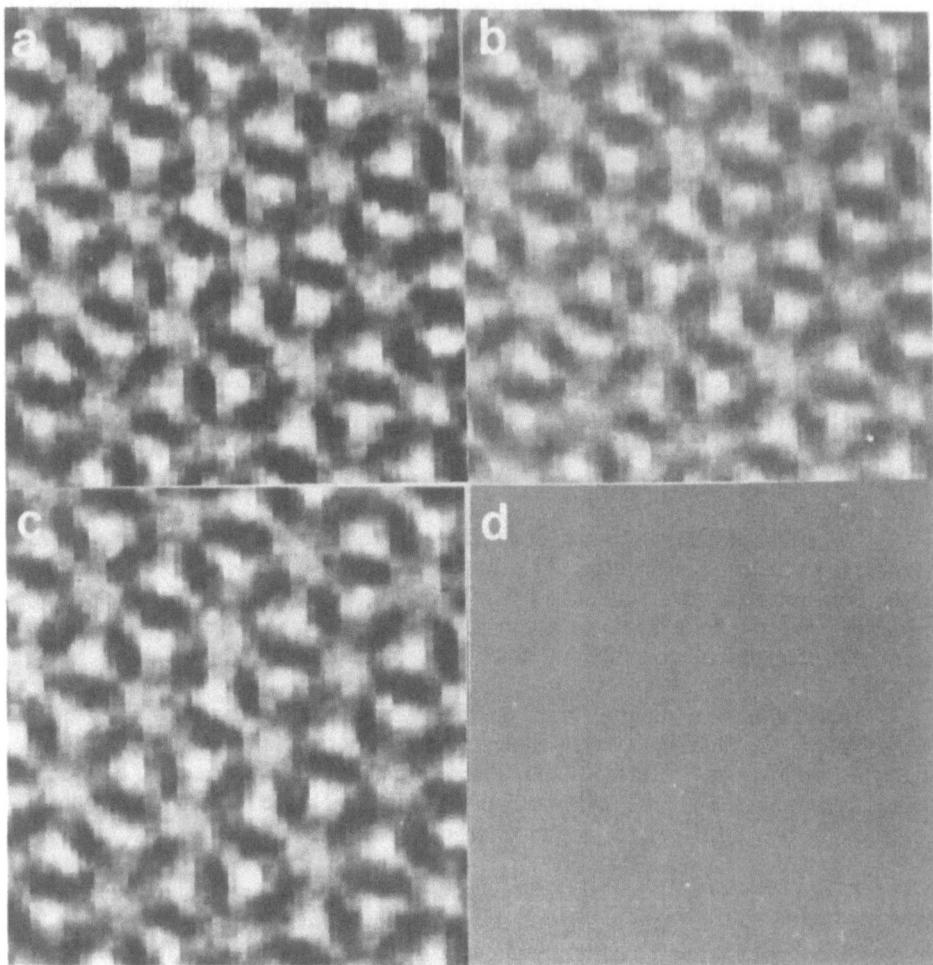

Figure 6: Effect of the Lagrange multiplier on ME reconstructions of a TEM image of the Ordered Bicontinuous Double Diamond (OBDD) structure in microphase separated block copolymers [Thomas et al, 1986]. Figure 6a: Original image, sampled at 1.5 nm/pixel. Figure 6b: $\lambda=10$. Figure 6c: $\lambda=2$. Figure 6d: $\lambda=0.01$. As λ is decreased to zero, the reconstruction approaches the flat image. However, when Figures 6b-d are rescaled such that the minimum value is black and the maximum value is white, they appear virtually the same as the original data (Figure 6a).

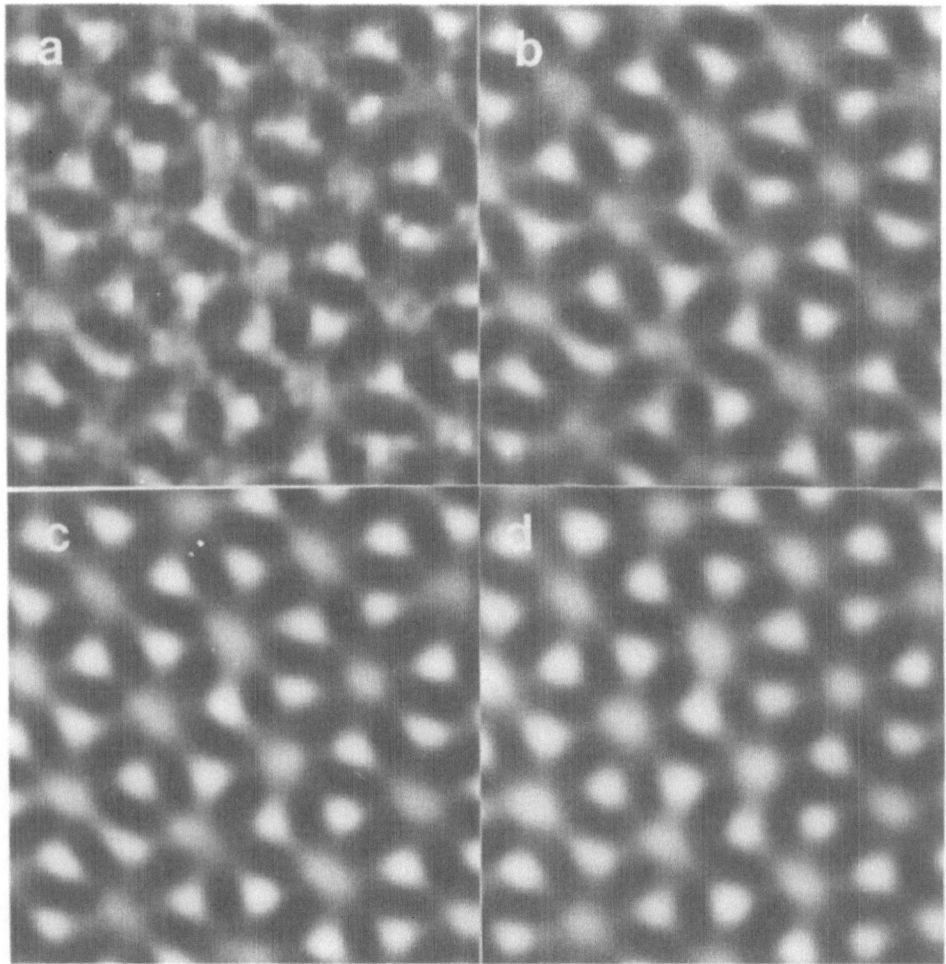

Figure 7: Effect of a Gaussian point spread function b on ME reconstructions of the TEM image of the OBDD microstructure (Figure 6a). Figure 7a: b=three pixels. Figure 7b: b=five pixels. Figure 7c: b=seven pixels. Figure 7d: b=nine pixels. As b is increased, the appearance of the image is changed. Subjectively, image 7b using a point spread function of five pixels appears the most improved by the reconstruction.

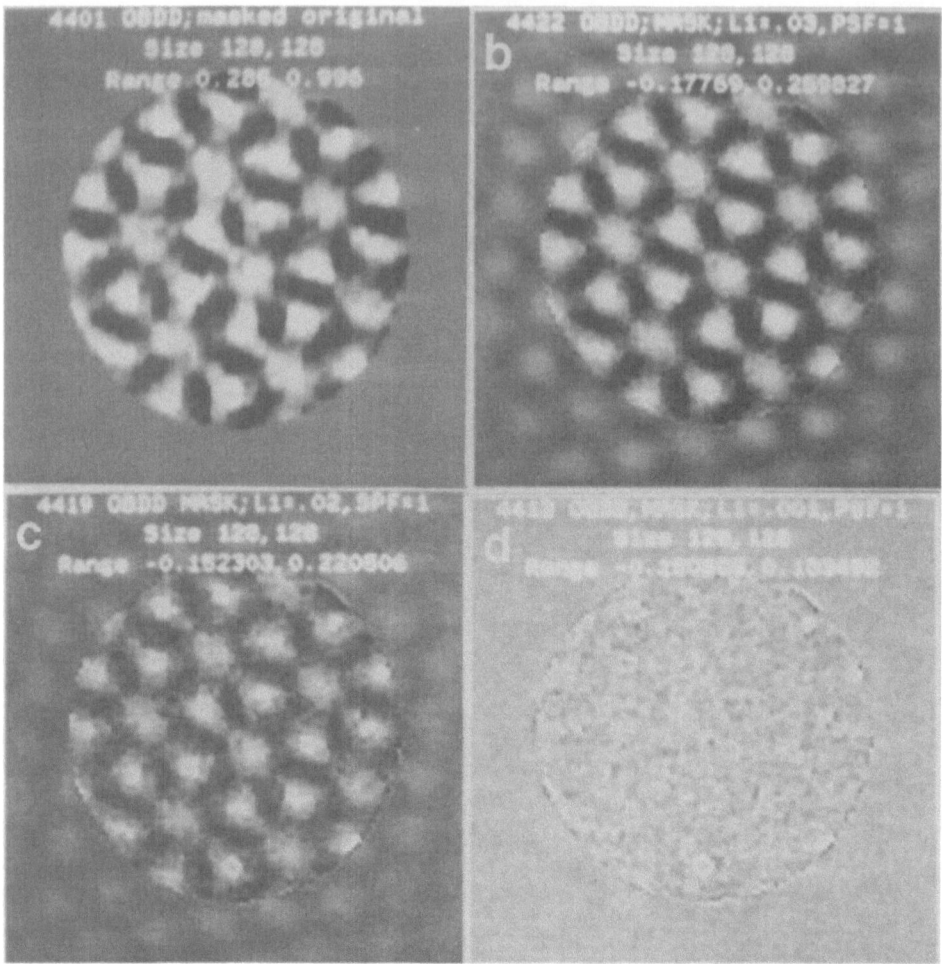

Figure 8: Effect of the Lagrange multiplier on the appearance of an image when ME reconstruction is applied in the Fourier space domain. Figure 8a: Original image masked with a sharp circular boundary. Figure 8b: $\lambda=0.03$. Figure 8c: $\lambda=0.02$. Figure 8d: $\lambda=0.001$. Now, the images do vary significantly as a function of λ as the ME algorithm supresses the weaker frequencies but retains the strongest ones. However, this causes "bleeding" of the strong periodicities into the masked region.

In the limit of small λ, the image becomes noisy and less interpretable. As λ increases, the strong periodicities become more distinct. Eventually, the original image is exactly recovered. Note, however, that for intermediate values of λ there is "bleeding" evident in the area which was originally flat. Hence, while the strong periodicities have been enhanced, the information about where they came from in the image has been compromised. This is apparently because Fourier space ME has suppressed low intensity frequencies which are nevertheless important in determining the precise boundaries of an object.

4. Discussion

We have found that the parameter which is most important in determining the appearance of an image reconstructed in real space by ME is the point spread operator b, not the estimated error σ. Changing σ in real space (through the Lagrange multiplier λ) simply results in non-linear scaling of the reconstruction, and does not alter the relative ranking of the intensity in the pixels. Images subjected to ME reconstructions in Fourier space *can* be effected by the estimate of σ, but this leads to "bleeding" artifacts.

Applied only in real space, ME reconstructions contain frequency artifacts corresponding to harmonics of strong periodicities in the original data. Applied only in Fourier space, ME causes bleeding of frequencies into regions where they previously did not exist.

These findings have important implications for the reliability of information in low dose HREM images reconstructed by ME methods. Despite these lmitations, ME methods have at least two significant advantages over traditional Fourier filtering methods: (1) there is only a limited number of adjustable parameters: the estimated error σ (adjusted through λ) and the point spread function b, and (2) these parameters can both be estimated from additional experimental evidence. This means that ME methods provide at the very least a more systematic approach for reconstructing images in a manner which minimizes user bias.

Future efforts involving the iterative scheme in which ME reconstructions are performed sequentially in real and then Fourier space may provide a compromise for minimizing artifacts. An even more complicated approach in which the entropy is defined simultaneously in both spaces may eventually prove to be the most useful and rigorous, but this would seem to require the Fourier transformation to be a part of the iteration step and therefore would be computationally intensive. Perhaps it will some day be possible to use ME to say whether features in an HREM image are representative of the true sample structure. In the meantime, efforts are probably best expended by trying to obtain better original experimental data $\{d_i\}$ in the first place.

5. Acknowledgments

The algorithms used in this work were written in FORTRAN 77 and are available upon request from the authors. DCM thanks the Shell Company for financial assistance in the form of a Doctoral Research Fellowship, and the E. I. du Pont de Nemours Company. ELT thanks the AFOSR for finanical assistance through grant F 49620-89-C-0073.

6. References

Anderson, D. M., Martin, D. C., and Thomas, E. L., (1989) "Maximum-Entropy Data Restoration Using Both Real- and Fourier-Space Analysis", Acta Crystallographica, A45, 686-698.

Bialynicki-Birula, I., and Mycielski, J. (1975) "Uncertainty Relations for Information Entropy in Wave Mechanics", Communications in Mathematical Physics, 44, 129-132.

Cowley, J. M. (1988) "Imaging Theory", in High Resolution Transmission Electron Microscopy and Associated Techniques, Buseck, P., Cowley, J., and Eyring, L., eds., Oxford University Press.

Daniel, G. J., and Gull, S. F. (1980) "Maximum Entropy Algorithm Applied to Image Enhancement", IEE Proceedings, 127, E(5), 170-172.

Gadre, S. R., Bendale, R. D. (1985) "Information Entropies in Quantum Chemistry", Current Science, 54(19), 970.

Gull, S. F., and Skilling, J. (1984) "Maximum Entropy Method in Image Processing", IEE Proceedings, 131, Pt. F., No. 6, 646-657.

Martin, D. C., Direct Imaging of Deformation and Disorder in Extended-Chain Polymers, Ph.D. Dissertation, The University of Massachusetts at Amherst, Amherst, MA, (1990).

Moriarty-Schieven, G. H., Snell, R. L., Strom, S. E., Schloerb, F. P., Strom, K. M., and Grasdelen, G. L. (1987) "High Resolution Images of the L1551 Bipolar Outflow: Evidence for and Expanding, Accelerated Shell", The Astrophysical Journal, 319, 742-753.

Narayan, R. and Nityananda, R. (1986) "Maximum Entropy Image Restoration in Astronomy", Annual Reviews of Astronomy and Astrophysics, 24, 127-170.

Shannon, C. E., (1948) "A Mathematical Theory of Communication", Bell System Technical Journal, 27, 379, 623.

Shore, J. E., and Johnson, R. W. (1980), "Axiomatic Derivation of the Principle of Maximum Entropy and the Principle of Minimum Cross-Entropy", IEEE Transactions, 26, 26-37.

Synoptics, Ltd. (1990) SEMPER Image Processing Software, Cambridge, England.

Thomas, E. L., Alward, D. B., Kinning, D. J., Martin, D. C., Handlin, D. L., Jr., and Fetters, L. J. (1986) "The Ordered Bicontinuous Double Diamond Structure of Star Block Copolymers--A New Equilibrium Microdomain Morphology", Macromolecules, **19**, 2197.

Willingdale, R. (1981) Monthly Notes of the Royal Astronomical Society, **194**, 359.

THE ACHIEVEMENT AND LIMITATION OF HV-HREM

N. UYEDA
Institute for Chemical Research, Kyoto University
Uji, Kyoto-Fu 611, Japan

ABSTRACT. Since the adoption of 500KV electrons was found to be effective for the high resolution of atomic level at Kyoto University, a 1000kV electron microscope was constructed to obtain a resolution better than 0.12nm especially for organic specimens by adopting a highly sensitive imaging recording system which is also used for the digital image processing.

1. INTRODUCTION

It is well accepted by now that the high resolution molecular and structure imaging has to be based on the multibeam synthesis as demonstrated by Uyeda et al. (1970, 1972) and Iijima (1971). The difficulty in obtaining images at atomic level is due to the lens defects such as spherical and chromatic aberations, which limit the use of higher order reflections necessary for the molecular and crystal structure imaging.

If the higher order reflections representing shorter spacings, d, must be taken into the proper range of the objective lens, one of the reasonable solutions is to reduce the scattering angle, α, by elevating the beam potential to decrease the wavelength λ, according to the common relationship $d = \lambda / \alpha$. This is the basic reason why high voltage electrons are required in general for the high resolution imaging and the principle was well described by Scherzer (1949) on the basis of the electron wave optics.

2. CONSTRUCTION OF HV-HREMS

The first HVEM (500KV) for high resolution work was constructed by Kobayashi et al. (1974) to give 1.5Å resolution and the molecular images of atomic resolution was obtained by Uyeda et al. (1979) to verify the merit of HVEM. While other HR instruments with accelerating voltages up to 1000KV were made since then at several institutions, no constructions of HV-HREM's followed for some 6 years until a new 1000KV HREM was installed at Kyoto University in 1989. The general feature of the tall microscope is shown in Fig.1.

2.1. Specification of the New 1MeV HREM

J. R. Fryer and D. L. Dorset (eds.), Electron Crystallography of Organic Molecules, 147–151.

148

The general principle to construct the new instrument is the upgrading of the former 500kV HREM. However, improved devices were newly added as itemized below.

a. Split type HV supply: The idea to separate the high voltage DC generator and the beam accelerator has been adopted at Kyoto Lab. since 1950 to increase the stability of accelerating potential by avoiding the effect of RF component inherent in the generator, for which a symmetric type of Kockcroft-Walton circuit was adopted in the present case.

As the result, the stability was found to be better than 1×10^{-6}/min and the ripple ratio was limited to 5×10^{-7}. The electron beam source is a precentred LaB_6 single crystal.

Fig. 1. The new 1000KV HREM at Kyoto University, Uji

b. Reduction of spherical aberration: The excitation power of the objective lens has been increased as high as $J/\sqrt{U^*} \gtrsim 28$ holding the current stability better than 1×10^{-6} to reduce the spherical aberration constant (C_s). Together with the computer-designed pole pieces, the spherical and chromatic aberration constants have been reduced to 1.7mm and 3.4mm, respectively, promising the theoretical resolution of 1.2 Å. The pole pieces of a smaller C_s (1.17mm) is being prepared to give a resolution of ~ 1 Å.

c. Observation and recording systems: As the specimen stage, a top-entry goniometer is used so that the specimen plane can be tilted by$\pm40°$ in all directions. The specimen z-control of 2.5mm stroke is also possible so that the optimum specimen position is selected to match the electron beam voltage variable from 400 to 1000KV at the maximum.

The final images are recorded on the imaging plate consisting of photo-stimulative phosphor, $BaFBr(Eu^{2+})$, developed by Mori et al.(1985).
The latent images are read out by scanning the exposed plate with a laser beam which excites phptoluminescence from the colour centres produced by the electron irradiation to be recorded on MT through a photomultiplier circuit.

While the photographic hard copies are reproduced from the output signals, the image data can be used directly for the digital image processing and for the feedback to the objective stigmator facilitating the adjustment of the optical conditions. Since the dynamic range of the phosphor for electrons is so wide as 1×10^{-11} through 1×10^{-14} C/cm², the electron dosage on the specimen can be reduced as low as nearly 1/1000 of the normal photoemulsion. The capability of damage protection is enhanced more effectively by the combined use of an automated minimum dose system (MDS).

d. Electron Energy Loss Spectrometer: A magnetic sector type energy loss spectrometer was attached in view of its efficiency in detecting the lighter atoms. The smaller background noise is another virtue of the high voltage electrons. It is also aimed to observeve the individual atoms in the specimen molecules by recording high resolution filtered images.

2.2. Comparison of Image Features

Although the installation of the total
instrument was completed, it is still
subject to sophisticated adjustments to
produce satisfactory images of expected
resolution. Since the image simulation
was found to reflect the actual features
as reported by Uyeda et al. (1979), the
expected images of polychlorinated Cu-
phthalocyanine molecules were compared
with those based on simulation regarding the former 500KV microscope as shown
in Fig. 2. A comparative examination of
the real images in Fig. 2(a) actually
taken with the 500KV instrument, the
features of benzene rings are observed
being surrounded by four large dots corresponding to Cl atoms at the top and
bottom of each molecule. However, the
five-membered ring, pyrrole, is not well
resolved at all.

The situation is also reproduced in
the simulated images in (b) indicating
the resolution limit of the former case.
The theoretical images for the new case
reproduced in (c) show that the pyrrole
rings are clearly resolved whereas the
sharpness is enhanced in every segment
of macroring as well as heavy atoms such
as central copper and chlorine atoms ar-
ranaged at the molecular margin.

3. THE CONTRAST TRANSFER FUNCTIONS

The contrast transfer functions play the
fundamental role for the high resolution
image simulation. In addition to the
optimum focus, other wide pass bands appear at different defocus positions.

Fig. 2. Cmparison of images for
500 and 1000KV electrons

According to Eisenhandler and Siegel(1966), these defocus values, Δf
is given in a form as $\Delta f = [(2K-0.59) C_s \cdot \lambda]^{1/2}$ with K integer. A set of
transfer functions up to K=5 are shown in Fig. 3. While the basic transfer
function by Scherzer vibrates with a spatial frequency $k(d^{-1}$between +
and -1, modifying the amplitude of the reflections. The actual contrast tra
nsfer function, however, is enveloped, as pointed out by Frank (1973), so
that the vibration damped off failing to reach unities as usual.

The damping effect caused by the chromatic focus spread and partial coherence of electrons is also apparent with the present 1000KV microscope as
indicated by the inside curve for each K in Fig. 3, where the envelope functions were estimated assuming the defocus spread and beam parallellity re-

lated to the beam coherence, to be 0.5nm
and 7mrad. respectively. Though the transfer function K=1 for the optimum focus is
enveloped, the damping cutoff takes place
beyond the first zero of the function, supporting the expected resolution of 0.12nm.

As for the other transfer functions of
different defocus positions, wide pass bands
appear at area for the K-th maximum. The
cutoff point of the envelope usually extends
toward higher k area so that the transfer
function survives to some extent in the pass
band area. Since these areas are overlapped
in K sequence, one can collect information
on reflections by Fourier transforming image
intensities of the defocus series.

By the digital processing with the pass
band data, the image resolution can be enhanced as proposed by Kirkland et al. (1980).
In the present case, the image resolution is
expected to reach about 0.098nm with through
focus images to K=5. The imaging plate recording system is expected to facilitate in
both ways, the through focus observation and
the digital processing.

4. THE LIMITATION OF HV-HREM

The role of high voltage acceleration has
been widely accepted by now as one of the
straightforward approaches to reach the resolution of atomic level. Theoretically, the
adoption of shorter wavelengths extends the
upper limit of the optimum transfer function
which restricts the highest angular range of
reflections. However, the endless elevation
of potential may not necessarily be of special virtue for the resolution improvement.
In order to keep the spherical aberration as
small as possible, the excitation power of
the objective lens must match the accelerating voltage. Owing to the saturation of

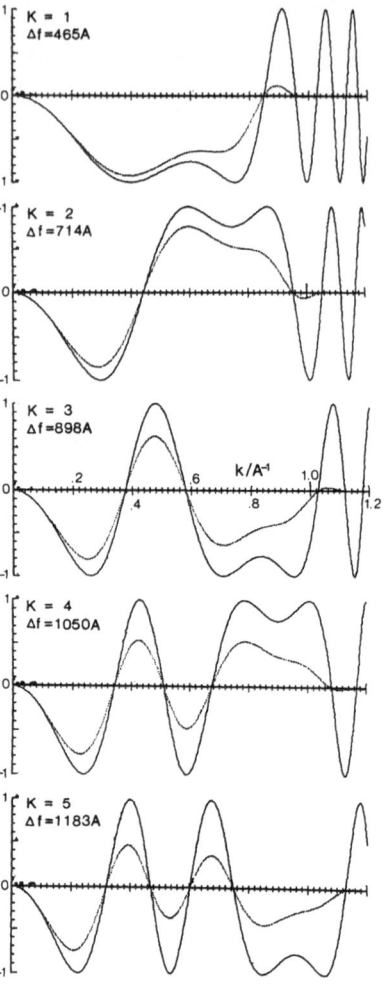

Fig. 3. The contrast
transfer functions
and pass bands of
new 1000KV HREM

magnetic field, it is rather hard to follow up unless new magnetic materials
are found or superconduction lenses gather more popularity.

Usually, the effect of field saturation has been averted by maneuvering
the design of the pole pieces so that $C_s \cdot \lambda$ has been almost kept constant
empirically to be about $1 \times 10^3 nm^2$ up to 1250kV. If it is assumed that this
situation stands even at much higher beam voltages, the resolution may be
improved to 0.5 Å at 10MeV. However, the contrast transfer function will
be quickly damped off at higher space frequencies owing to the increasing
instability in the beam potential as well as in the lens current. Moreover,

the effect of knock-on damage to the specimen will be more serious as the beam potential is elevated so extensively. From a practical view point, it is considered that the limit of a realistic beam potential will be less than 3MeV, which will possibly give the resolution of ca. 0.8Å, theoretically.

It is also known that the contrast of atomic image will be enhanced with the beam potential. However, this effect works only for heavier atoms. For light atoms, it is estimated from simulated images-assuming an imaginary 5MeV HREM-that the contrast may not be improved sufficiently for individual carbon or nitrogen atoms to be clearly resolved.

ACKNOWLEDGEMENTS

The author wishes to thank Dr. S. Isoda, Kyoto University for his kind help in the image simulation. He is also indebted to Mr. T. Honda, JEOL for his useful information. This work is supported by the Grant for Special Equipment from the Ministry of Education, Science and Culture, Japan.

REFERENCES

Eisenhandler, C. B., and Siegel, B. M. (1966) 'A zone-plate aperture for enhancing resolution in phase contrast electron microscopy', Appl. Phys. Letter, 8 , 258-260.

Frank, J. (1977) 'The envelope of electron microscope transfer functions for partially coherent illumination', Optik, 49, 81-92.

Iijima, S.(1971) 'High-resolution electron microscopy of crystal lattice of titanium-niobium oxide', J. Appl. Phys., 42, 5891-5593.

Kobayashi,K., Suito, E., Uyeda, N., Watanabe, M., Yanaka, T., Etoh, T., Watanabe, H. and Moriguchi, M. (1974) 'A new high resolution electron microscope for molecular structure observation' in J. W. Sanders and D. J. Goodchild (ed.), Electron Microscopy 1974, Australian Academy of Science, Canberra, Vol.1, pp. 30-31,

Kirkland, E. J., Siegel, B. M., Uyeda, N. and Fujiyoshi, Y.(1980) 'Digital reconstruction of bright field phase contrast images from high resolution electron micrographs', Ultramicrosc., 5, 479-503.

Mori., N., Oikawa, T., Kato, T. and Harada, Y.(1988) 'Application of the "imaging plate"to TEM image recording', Ultramicrosc, 25, 195-201.

Scherzer, O. (1949) 'The theoretical resolution limit of the electron microscopy', J. Appl. Phys., 20, 20-29.

Uyeda, N., Kobayashi, K., Suito, E., Watanabe, M., and Harada, Y.(1970) 'Direct observation of phthalocyanine molecules in epitaxial films' in P. Favard (ed), Microscopie electronique 1970, Societe Francaise de Microscopie Electronique, Paris, Vol. 1. 1, pp. 23-24.

Uyeda, N., Kobayashi, T., Suito, E., Watanabe, M., and Harada, Y.(1970) 'Molecular image resolution in electron microscope', J. Appl.Phys., 43, 5181- 5189.

Uyeda,N., Kobayashi, T., Ishizuka, K. and Fujiyoshi, Y. (1979), 'High voltage electron microscopy for image discrimination of constituent atoms in crystals and molecules', Chemical Scripta, 14, 47-61.

CRYSTAL STRUCTURES FROM HIGH-RESOLUTION ELECTRON MICROSCOPY

F.H.Li
Institute of Physics
Chinese Academy of Sciences
Beijing 100080
P.R.China

ABSTRACT. This paper gives a review of some research results related to crystal structure determination by high-resolution electron microscopy (HREM) and contains three parts. The first part is about the dependence of image contrast upon crystal thickness. The second part gives some structure images to demonstrate different appearance of atoms. The last part is about image deconvolution and resolution enhancement and shows the possibility of determining crystal structures by using an approach other than the traditional method of trial and error.

1. INTRODUCTION

It is well known that the development of high-resolution electron microscopy(HREM) affords an approach to crystal structure determination[1.2]. Comparing with diffraction methods HREM is appropriate for studying minute crystals and crystals with defects, and has the advantage of intuition. The structure image taken with high-resolution electron microscope mirrors the projected structure. However, the contrast of high-resolution electron microscope images is strongly affected by various factors, especially the defocus amount and the crystal thickness. Even for weak-phase objects the image is not always similar to the structure. The image intensity is the convolution of projected potential distribution function(PPDF) with the Fourier transform(FT) of contrast transfer function(CTF) rather then the PPDF itself. Only under the optimum defocus condition, usually it is near the Scherzer defocus[3], the image can be faithful to the structure. But it is not easy to set the Scherzer defocus exactly during the observation. In addition, the optimum defocus amount may deviate from the Scherzer defocus, and the contrast of optimum defocus image may change enormously with crystal thickness. Furthermore, the resolution of structure image is limited by the electron microscope.

In most cases the structure determination in HREM is carried out by method of trial and error. It is based usually on a series of through focus images so that some prerequisite conditions are indispensable. Firstly, In order to obtain the series of images, the crystal should be strong enough under the electron beam irradiation. Secondly, in order to determine the structure image from series of images some structure features of the crystal under investigation must be known prior. Then structure models are proposed and images are simulated under various electron-optical parameters for different models. The right model would

153

J. R. Fryer and D. L. Dorset (eds.), Electron Crystallography of Organic Molecules, 153–167.

be determined by image matching between observed images and simulated images for different values of defocus and crystal thickness. Obviously, the work is tedious and sometimes the result might be ambiguous. This is to some extend similar to the early stage of X-ray single crystal structure determination. However, since the development of direct methods the technique of X-ray diffraction analysis has made a great progress[4]. It becomes possible to solve the phase problem and hence the crystal structure determination becomes much more straightforward then before. In this sense nowadays the technique of structure determination by HREM remains to be improved.

This paper concerns some research results related to crystal structure determination by HREM carried out in Institute of Physics, the Chinese Academy of Sciences and contains three parts. The first part is about the dependence of image contrast upon crystal thickness. It is illustrated that an optimum crystal thickness is always important for obtaining expected structure information from images. The second part gives three structure images to demonstrate the dependence of appearance of atoms on accelerating voltage and atomic number. The third part is about the image deconvolution and resolution enhancement based on a single image of any defocus condition together with the corresponding electron diffraction pattern(EDP) and shows the possibility of determining crystal structures by an approach other than the traditional method of trial and error.

All observed images given in this paper were taken with a JEM-200CX electron microscope except when a special description is given. Most of them are from inorganic crystals, but they can be used for references in studying organic crystal structures.

2. CHANGE OF IMAGE CONTRAST WITH CRYSTAL THICKNESS

2.1. Pseudo-weak-phase-object approximation

It is well known that for weak phase objects the image intensity under optimum defocus condition is linear to the PPDF $\phi(\mathbf{r})$:

$$I(\mathbf{r}) = 1 - 2\sigma\phi(\mathbf{r}), \qquad (1)$$

here $\sigma = \pi/\lambda U$, λ denotes the electron wave length, U the accelerating voltage. Such image is similar to the projected crystal structure and called structure image. In fact the image contrast also depends upon the crystal thickness. According to the pseudo-weak-phase-object approximation which was derived on the basis of the multiple dynamical electron diffraction theory given by Cowley and Moodie[5] the structure image intensity is roughly proportional to a modified PPDF $\phi'(\mathbf{r})$[6]:

$$I(\mathbf{r}) = 1 - 2\sigma\phi'(\mathbf{r}). \qquad (2)$$

The modified PPDF $\phi'(\mathbf{r})$ was defined as a function of PPDF and crystal thickness[6].

Fig.1 is a schematic diagram showing the variation of the modified PPDF with crystal thickness for a light atom and a heavy atom. It

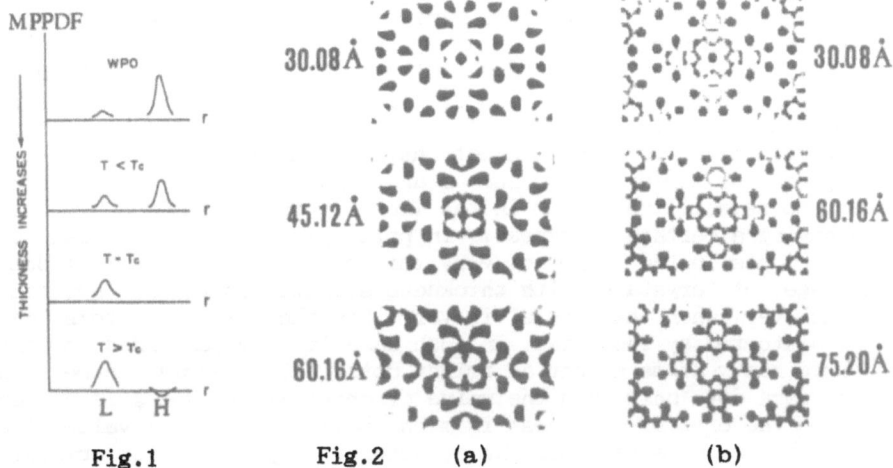

155

Fig.1. Schematic diagram showing the variation of modified PPDF with crystal thickness for a light atom(L) and a heavy atom(H). T denotes crystal thickness and T_c critical crystal thickness.

Fig.2. Simulated optimum defocus images of chlorinated-Cu phthalocyanine with different thickness for (a)accelerating voltage 500KV, C_s = 1.0 mm and (b)accelerating voltage 1MV, C_s = 1.7 mm.

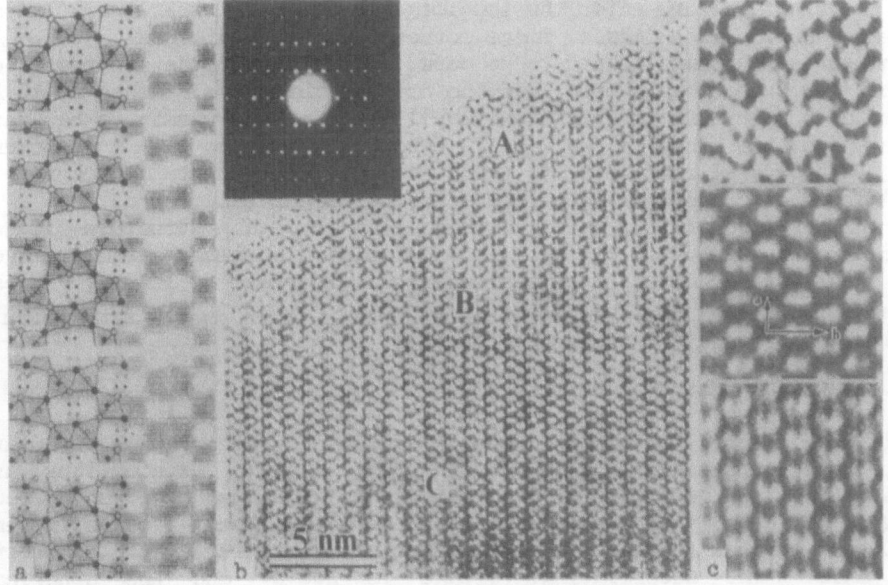

Fig.3. Optimum defocus images of $Li_2Ti_3O_7$. (a)Simulated for crystal thickness from top to bottom 3, 21, 45, 68 and 89 Å, accelerating voltage 200KV, C_s = 1.2 mm. (b)Observed with a wedge shaped crystal. (c)Enlarged photo from regions (top)A, (middle)B and (bottom)C.

illustrates that with the increase of crystal thickness the image contrast of light atoms increases more rapidly than that of heavy atoms.
One can always find a critical thickness above which heavy atoms will
turn from dark to bright in the image while light atoms may still remain
dark. The critical thickness is different for different crystals
and depends upon the atomic number of the heaviest atoms in crystals and
upon the accelerating voltage of electrons. The value of critical thickness is larger for larger atomic number and higher voltage. Therefore,
in order to reveal light atoms in the image it is recommended to use
crystals having a certain thickness. In principle, heavy atoms can be
seen as dark dots in the image of very thin crystals or as bright dots
in the image of crystals with thickness slightly above the critical
value. However, when the crystal is very thin the image of atoms may
seriously distorted by the noise. In such case it is recommended to take
images where heavy atoms appear as bright dots. Heavy atoms can be seen
clearly as dark dots only when the value of critical thickness is rather
large. When the crystal thickness is much above the critical value, the
image would be no more resemble the structure, especially for crystals
with complex structure and large unit cell. Some examples demonstrating
the above argument are give in the following.

2.2. Image contrast of chlorinated copper phthalocyanine

Fig.2a shows simulated optimum defocus images of chlorinated copper
phthalocyanine for different crystal thickness. The accelerating
voltage is 500 KV and the spherical aberration coefficient is 1.0 mm.
The crystal thickness for the top image is 30.08 Å, where all atoms
appear dark. The middle image corresponds to thickness 45.12 A.
Although all atoms remain to be dark in the image, the contrast of
copper atoms becomes lower. When the crystal thickness reaches 60.16 A,
atoms of copper turn to bright while all other atoms remain to be dark
(see the bottom image in Fig.2(a)). Hence, in order to take a structure
image the crystal must be very thin, because for crystal thicker than
45 Å atoms of copper are submerged into the bright background.
    A higher accelerating voltage would allow to obtain structure images
from thicker crystals. Fig.2(b) shows simulated optimum defocus images
for accelerating voltage 1 MV and spherical aberration 1.7 mm. It can
be seen that the appearance of atoms of copper does not turn bright
until the crystal is thicker than 75.20Å.

2.3. Image contrast of R-Li$_2$Ti$_3$O$_7$[7]

R-Li$_2$Ti$_3$O$_7$ belongs to the orthorhombic system. In the crystal structure
there are large channels along the shortest axis[8]. Lithium ions are
located inside the channels. Fig.3(a) shows simulated optimum defocus
images of R-Li$_2$Ti$_3$O$_7$ superimposed with structure model for different
values of crystal thickness. The incident beam is parallel to the shortest axis. The accelerating voltage is 200 KV and the spherical
aberration coefficient is 1.2 mm. It can be seen that lithium ions have
no contrast until the crystal is thicker than 68 A. When the crystal
thickness increases to about 89 A titanium ions turn from dark to

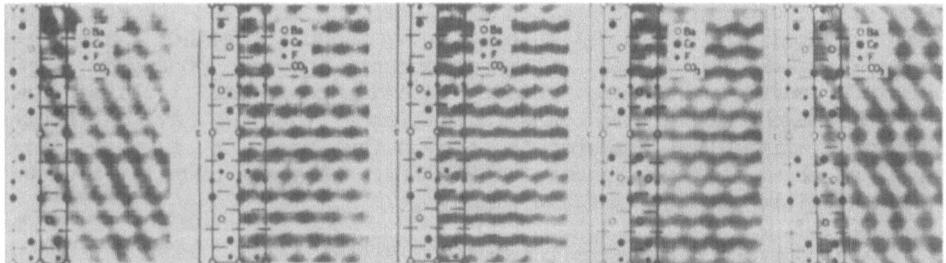

Fig.4. Simulated optimum defocus images of haughoite superimposed with structure model for crystal thickness(from left to right) 20, 30, 40, 45 and 63 A. Accelerating voltage 100 KV, C_s = 0.7mm.

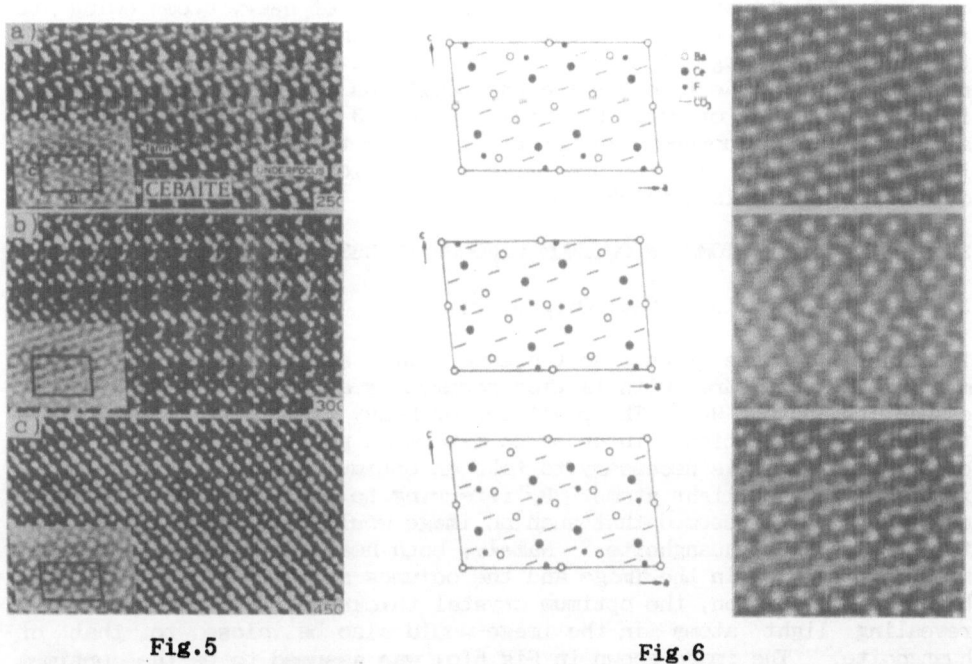

Fig.5   Fig.6

Fig.5. Through focus series of images of cebaite projected along the b axis. (c) is the optimum defocus image. All atoms apear dark.
Fig.6. Three structure models proposed for cebaite and the corresponding simulated optimum defocus images for crystal thickness 50 A. The image contrast is sensitive to positions of light atoms. The simulated image for the bottom model is in agreement with the observed image.

bright while lithium ions remain to be dark. Fig.3(b) shows the observed image of a wedge shaped crystal and Fig.3(c) the enlarged photos from three regions of different thickness A, B and C together with the inserted simulated images. The agreement between observed and simulated images can be seen. Hence, in order to reveal lithium ions in the image the crystal must be thicker than 70 Å. Although principally titanium ions can be seen in the image of a very thin crystal as shown in the top image of Fig.3(a), they are usually blurred by the noise(see the top image in Fig.3(c)).

2.4. Image contrast of huanghoite($BaCe(CO_3)F$)[19]

Huanghoite is a mineral discovered in China. It belongs to the hexagonal system. The lattice parameters are a = 5.06 and c = 38.08 Å[9]. Fig.4 shows a series of simulated optimum defocus images projected along the b axis superimposed with structure model for different crystal thickness (from left to right 20, 30, 40, 45 and 63Å). The accelerating voltage is 100 KV and spherical aberration coefficient 0.7 mm. When the crystal is very thin, heavy atoms of Ba and Ce appear dark and light atoms of C, O and F have no contrast. When the crystal thickness is 45-50 Å, the contrast of light atoms is comparable with that of heavy atoms owing to the dynamical scattering of electrons. This implies that under such a thickness the image contrast would be sensitive to position of light atoms. It is possible to determine the positions of light atoms from the image. When the crystal thickness reaches 63 Å the image is no more similar to the structure. Therefore, in order to determine positions of light atoms from a structure image, the image must be taken with a crystal of appropriate thickness.

3. APPEARANCE OF ATOMS IN OPTIMUM DEFOCUS IMAGES

3.1. Mineral cebaite($Ba_3Ce_2(CO_3)_5F_2$)[9].

The mineral cebaite belongs to the same family as haunghoite. It has a monoclinic structure with lattice parameters a = 21.2 Å, b = 5.06 Å, c = 13.1 Å and β = 90°. The positions of heavy atoms were known before the HREM investigation. In order to determine positions of light atoms of C, F and O, it is necessary to take an optimum defocus image which is favorable to show light atoms. By referring to the result of HREM study on huanghoite we assumed that such an image would have some similar features as that of huanghoite. Namely, both heavy atoms and light atoms will appear dark in the image and the columns among them would appear bright. In addition, the optimum crystal thickness for preferentially revealing light atoms in the image would also be close to that of huanghoite. The image shown in Fig.5(c) was assumed to be the optimum defocus image carrying preferentially the structure information of light atoms. Fig.6 shows three proposed structure models. Their difference is only in the positions of light atoms. The calculated images for the three models indicate that the image contrast is sensitive to position of light atoms and hence it is easy to determine the right model. The contrast matching between observed and simulated images with different

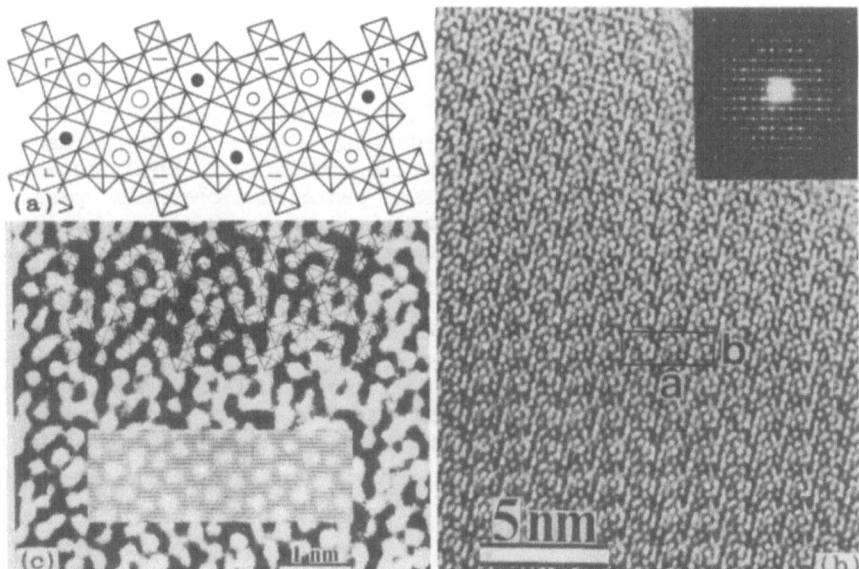

Fig.7. (a)Structure model and (b)image of $K_2Nb_8O_{21}$ projected along the c axis.　　　　　(b) is the enlarged photo for one unit cell. Cations appear as white dots.

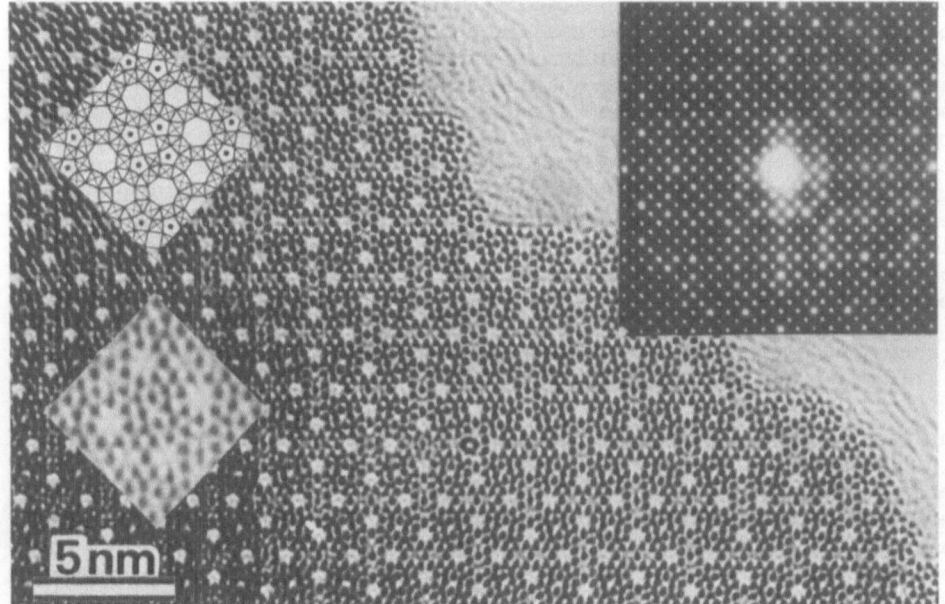

Fig.8. Optimum defocus image of $K_6Nb_{44}O_{113}$. Insets are structure model (top left) and enlarged photo(bottom left) for one unit cell. Nb appear as dark dots.

defocus amounts(Fig.5) confirms further the obtained model.

3.2. $K_2Nb_8O_{21}$

The crystal structure of $K_2Nb_8O_{21}$ was determined by HREM[10]. It belongs to the orthorhombic tungsten bronze type with lattice parameters a = 37.5 A, b = 12.5 A and c = 3.96 A, and consists of Nb–O octahedra which share vertices to form triangle and pentagonal channels parallel to the c axis. The triangle channels are empty and the pentagonal channels are occupied by Nb and K orderly(Fig.7(a)). The crystal was observed under accelerating voltage 200 KV with incident beam parallel to the c axis. It is in agreement with the above argument that all cations appear as white dots in the optimum defocus image which mirrors the projected structure(Fig.7(b)). Fig.7(c) is the enlarged photo from Fig.7(b) superimposed with structure model and simulated image. Hence, the optimum crystal thickness for revealing heavy atoms in the image is slightly above the critical value. The image with cations appearing as dark dots can hardly be found.

3.3. $K_6Nb_{44}O_{113}$[11]

The crystal of $K_6Nb_{44}O_{113}$ belongs to tatragonal system with lattice parameters a = 27.5 A and c = 3.96 A. HREM study[11] indicates that the structure is similar to that of $Rb_3Nb_{54}O_{146}$[12] and of $Cs_xNb_{54}(O,F)_{146}$ [13]. It also consists of Nb–O octahedra which share vertices to form five kinds of channels along the c axis(Fig.6, top left). Triangle and square channels are empty, pentagonal channels are occupied by atoms of Nb, and hexagonal and heptagonal channels are partially occupied by atom of K. Although the constituent atoms in $K_6Nb_{44}O_{113}$ are the same as those in $K_2Nb_8O_{21}$, in the optimum defocus image of $K_6Nb_{44}O_{113}$ taken under 300 KV with electron microscope H-9000 atoms of Nb can clearly appear as dark dots(Fig.6(b)). This is due to two reasons. Firstly, the accelerating voltage of 300 KV leads to a larger value of critical crystal thickness than 200 KV so that the thickness range for seeing heavy atoms as dark becomes larger. Secondly, the resolution of electron microscope becomes better with the increase of accelerating voltage. Both of them enhance the image contrast and hence the noise is reduced relatively.

4. IMAGE DECONVOLUTION AND RESOLUTION ENHANCEMENT BASED ON A SINGLE IMAGE AND THE CORRESPONDING EDP

In this section the image deconvolution is confined to the transformation of a single image taken at any defocus conditions into the optimum defocus image. This is especially important for crystals which are weak under electron irradiation. The resolution of optimum defocus images taken directly under the electron microscope or obtained by image deconvolution, is limited by the electron microscope. It can be improved by combining the structure information included in both image and EDP and using the direct methods developed in X-ray crystallography.

4.1. Image deconvolution for weak phase objects

In the case of weak phase objects the FT of the image intensity is linear to the structure factor F(H) multiplied by CTF:

$$T(H) = \delta(H) - 2\sigma F(H)A(H)W(H), \tag{3}$$

here $\delta(H)$ denotes delta function, A(H) the objective aperture function, W(H) the CTF:

$$W(H) = A(H)Sin\Gamma_1 exp(-\Gamma_2(H)),$$

$$\Gamma_1(H) = \pi \triangle f \lambda H^2 + \pi C_s \lambda^3 H^4/2,$$

$$\Gamma_2(H) = \pi^2 \lambda^2 H^4 D^4/2.$$

$\triangle f$ is the defocus amount (negative for under focus), C_s is the spherical aberration coefficient and D the standard deviation of Gaussian distribution of defocus due to the chromatic aberration[14]. All factors contained in the CTF except the defocus amount can be determined experimentally without much difficulty. Hence, once the defocus amount is determined the image can be deconvoluted and the projected structure is determined straightforward. Among the four proposed methods there is something in common for three methods. Firstly, a series of trial value of $\triangle f$ are assigned in a wide range with a small interval, say 10 A. Then the true $\triangle f$ is determined by using different criterions.

4.1.1. Method using Sayre equation[15,16]. For each given value of trial $\triangle f$ a set of F(H) can be calculated from equation (3). If this value is correct the corresponding set of F(H) should obey the Sayre equation developed in X-ray crystallography[17]:

$$F(H) = (\theta/V)\underset{H'}{\Sigma}F(H)F(H - H'), \tag{4}$$

where θ is the atomic form factor, and V the volume of unit cell. Hence, the true $\triangle f$ can be found by a systematic change of the trial $\triangle f$. Once the true $\triangle f$ is determined, images of any defocus can be transformed into the optimum defocus images without necessity to have any preliminary knowledge about the structure of examined crystal.

4.1.2. Method of maximum entropy[18]. The principle of maximum entropy is applied to solve the phase problem in X-ray crystallography. It gives the real-space counterpart of the classical formulation in reciprocal space and enhances the direct methods[19]. Hence, it is reasonable to use the principle of maximum entropy to determine the value of $\triangle f$.

Since the PPDF is the inverse FT of F(H), for each assigned value of $\triangle f$ we can obtain a PPDF. The entropy of PPDF is defined as

$$S = - \overset{N}{\underset{i=1}{\Sigma}} P_i lnP_i, \tag{5}$$

and

$$P_i = \phi_i / \sum_{i=1}^{N} \phi_i \qquad (6)$$

The maximum entropy will correspond to the true Δf.

Both the method using Sayer equation and the method of maximum entropy are effective except for images taken under the defocus condition close to Scherzer defocus, because near the Scherzer defocus condition the CTF will not be sensitive to small change of Δf. This can be remedied by combining the information included in the corresponding EDP, namely, by deriving the phase of F(H) from image and the magnitude from EDP. The result of image deconvolution by combining the image and EDP will always be more reliable.

4.1.3. Method of square CTF approximation[20]. According to equation (3) the main part of CTF, $\sin\Gamma_1(H)$ which oscillates enormously with the change of H, can be obtained from the image. Its square is expressed as

$$\sin^2 \Gamma_1(H) = |T(H)|^2 / 4\sigma^2 |F(H)|^2 \exp[-2\Gamma_2(H)]. \qquad (7)$$

A set of $\sin^2 \Gamma_1(H)_c$ can be calculated from equation (7) with an assigned value of Δf. If $|F(H)|^2$ in equation (7) is replaced by electron diffraction intensity, another set of $\sin^2 \Gamma_1(H)_0$ can be obtained. The true value Δf would make

$$\sum_H |\sin^2 \Gamma_1(H)_0 - \sin^2 \Gamma_1(H)_c|$$

have a minimum value.

4.1.4 Method based on Wilson statistics[21]. This method is specially proposed for the case that the image is taken near to the Scherzer defocus condition. Considering the average of $|F(H)|^2$ within a narrow ring on the reciprocal plane with a mean radius equal to H, we have from equation (3):

$$\langle |F(H)|^2 \rangle_H \approx \langle |T(H)|^2 \rangle_H / 4\sigma^2 W^2(H). \qquad (8)$$

On the other hand the Wilson statistics (see for instance[4]) gives

$$\langle |F(H)|^2 \rangle_H \approx \sum_{j=1}^{N} f_j^2(H), \qquad (9)$$

where $f_j(H)$ is the atomic scattering factor for electrons. The $W^2(H)$ can be obtained by combing (8) and (9):

$$W^2(H) \approx \langle |T(H)|^2 \rangle_H / 4\sigma^2 \sum_{j=1}^{N} f_j^2(H). \qquad (10)$$

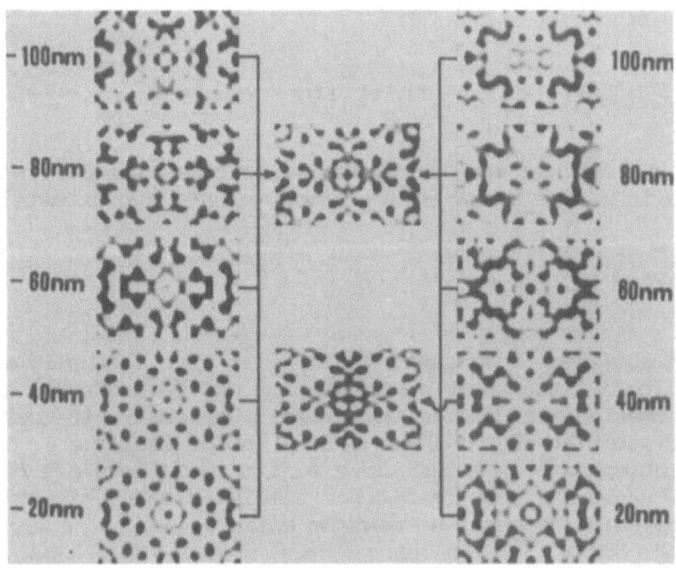

Fig.9. Results of image deconvolution for simulated images of chlorinated copper phthalocyanine treated as weak-phase objects by using method of maximum entropy. Accelerating voltage 200KV. C_s = 1.0mm. Underfocus and overfofus images of different defocus amounts are in left and right columns respectively. Resulted images are in middle column.

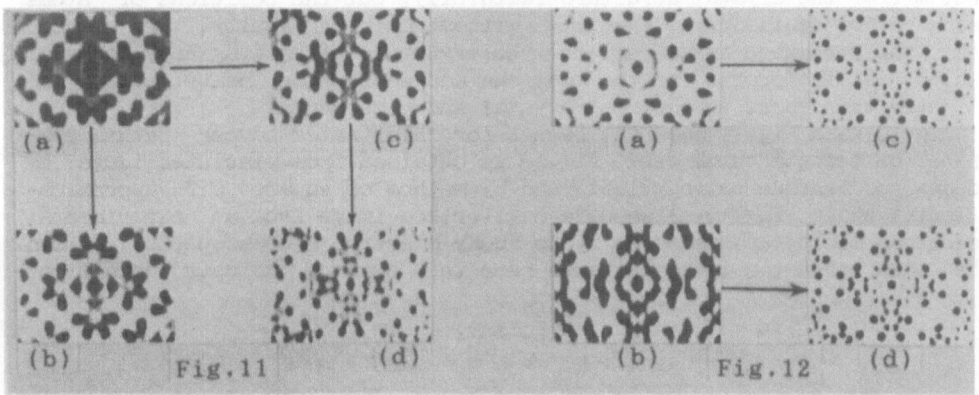

Fig.11. (a)Digiterized observed image of chlorinated copper phthalo-cyanine, (b)deconvoluted image by method based on Wilson ststistics without using EDP, (c)same as (b) but using EDP, and d)image obtain by phase extension from (c).

Fig.12. Results of phase extension for simulated images of chlorinated copper phthalocyanine treated as weak-phase objects. Resolution of starting image is (a)2 A, and (b)2.5 A.

Substituting (10) into (3) and ignoring δ(H) in (3) we have

$$|F(H)| = |T(H)| [\sum_{j=1}^{N} f_j^2(H) / \langle |T(H)|^2 \rangle_H]^{\frac{1}{2}}. \tag{11}$$

When the image is taken near the Scherzer defocus, the CTF will have negative value in a wide range of H. Hence we have approximately

$$F(H) = -T(H) [\sum_{j=1}^{N} f_j^2(H) / \langle |T(H)|^2 \rangle_H]^{\frac{1}{2}}. \tag{12}$$

The FT of F(H) obtained from equation (12) will give the optimum defocus image. Principally, the image deconvolution can be performed by using only the information included in the image. Obviously, to combine the image together with the EDP will leads to a better result.

The results obtained by all the above mentioned methods are of similar level. Fig.9 shows the test result for simulated image of chlorinated copper phthalocyanine by means of maximum entropy.

4.2. Test of image deconvolution for crystals of certain thickness.

In principle all the above mentioned method of image deconvolution are available for crystals thicker than weak-phase objects, but the crystal thickness must be below the critical value. Thus in the modified PPDF, the peaks of heavy atoms become lower and those of light atoms become relatively higher than in PPDF. In this sense the structure image does not mirror the crystal structure faithfully, but the positions of atoms can be recognized from the image without much difficulty. Or we can say, the structure image reflects the structure of an imaginary crystal which is isomorphic to the real one and where the heavy atoms are lighter than those in the real crystal and vise versa.

Fig.10 and Fig.11 show the result for chlorinated copper phthalocyanine of certain thickness. Fig.10 is obtained from simulated image by means of maximum entropy(left) and by method of square CTF approximation(right). Fig.11(a) shows a digiterized image from an experimental electron micrograph taken on Kyoto 500KV electron microscope[22]. It can be seen that the image is taken near the optimum defocus condition.

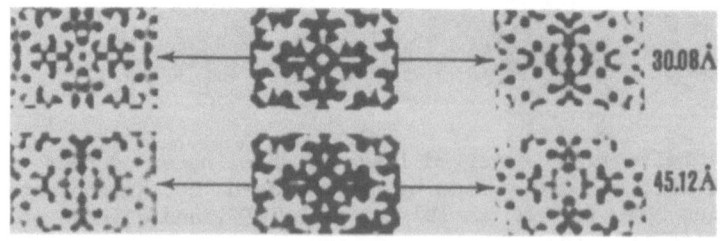

Fig.10. Results of image deconvolution for simulated image of chlorinated copper phthalocyanine by means of maximum entropy(left) and square CTF approximation(right). Unferfocus 800 A. EDP is used.

Although the positions of Cu and Cl atoms are more or less correctly revealed in the image, the position of other lighter atoms are obviously distorted. After performing the image deconvolution by method based on Wilson statistics the image mirrors the structure better than before(Fig.11(b)). If the information included in the corresponding EDP is combined together with the image the result of image deconvolution becomes better(Fig.11(c)).

4.3. Phase extension.

Phase extensions techniques are well developed in X-ray crystallography. The principle of phase extension applied in HREM is as follows. An EDP usually contains observable reflections within a limited sphere of 1 Å radius. This implies the possibility of obtaining a structure image of about 1 Å resolution from the electron diffraction data. Such a resolution is considerable higher than the resolution can be reached by an image. In addition, the intensities of the EDP are independent of defocus and spherical aberration. Accordingly, a set of high-resolution structure amplitudes of good quality can be obtained from an EDP. However, the structure analysis by electron diffraction alone is subject to the well known difficulty of the 'phase problem'. On the other hand, an image can provide phase informations corresponding to resolution about 2 Å. This can greatly reduce the complexity of the solution of the phase problem. Hence, an improved high resolution image may be obtained by a phase interpolation and extrapolation procedure using the amplitudes of the structure factors from EDP and starting phases from image.

4.3.1. Test on weak phase objects. Fig.12(a) shows the simulated optimum defocus image of chlorinated copper phthalocyanine of 2 Å resolution. By Fourier transforming this image and excluding those reflections satisfying the relation $|\sin\Gamma_1(H)\exp[-\Gamma_2(H)]| < 0.2$, a set of phases for 44 independent reflections within 0.5 Å^{-1}. With these starting phases, 116 unknown phase were derived by the MULTAN program[23]. The originally unknown phase belong to two categories. One corresponds to those reflections within 0.5 Å^{-1} and satisfying the relation $|\sin\Gamma_1(H)\exp[\Gamma_2(H)]| < 0.2$, the other corresponds to those reflections between 0.5 and 1 Å^{-1}. There are three unknown phases belonging to the first category and all of them had the correct phase after the MULTAN run. On the other hand, there are 113 unknown phases belongs to the second category and 14 of them have their phase wrongly determined. However, most of the wrong phases are associated with a relatively low weight. The image obtained by inverse Fourier transforming the set of 160 structure factors with phases obtained by phase extension and amplitudes from EDP is shown in Fig.12(b). Its resolution is much improved. Fig.12(c) and (d) show the starting and resulting images respectively for phase extension from 2.5 Å resolution.

4.3.2. Test on crystals of certain thickness[24]. The possibility of applying the phase extension techniques to crystals thicker than weak-phase-objects is as described in 4.2. Fig.11(d) shows the resulting image which is obtained from Fig.11(c) after performing the phase extension.

5. CONCLUDING REMARK

In order to determine crystal structures from high resolution electron microscope images by method of trial and error it is important to have crystals of appropriate thickness. When the structure of examined crystal is completely unknown prior, we need firstly to catch the heavy atoms from the image. In case the accelerating voltage of electron is low and the atomic number of the heaviest atoms contained in the examined crystal is rather large, it is recommended to observe crystals of thickness above the critical value so that heavy atoms may appear clearly in the structure image as white dots. When the voltage is enough high structure images with heavy atoms appearing clearly as dark dots can be obtained. In case the positions of heavy atoms are known, for determining the positions of light atoms it is essential to have structure images where light atoms appear dark and the darkness is comparable with that of heavy atoms. Usually the crystal thickness is slightly below the critical value. In such case although light atoms may not be distinguishable individually in the image, the image contrast is sensitive to position of light atoms so that the structure model can be determined by image matching between experimental and calculated images.

The techniques of image deconvolution together with phase extension afford a new approach to crystal structure determination from high-resolution electron microscope images. The structure analysis techniques developed in X-ray crystallography play an essential role in developing the new approach which is different from the traditional method of trial and error. It needs not to take a through focus series of images, and such determined structure model would include structure details beyond the resolution limit of electron microscope. But it is only suitable for crystals of thickness below the critical value. Therefore, it can be expected to be more effective for high accelerating voltage. The technique applied to thick crystal remains to be improved.

REFERENCES

[1] Cowley,J.M. & Iijima,S.(1972)'Electron microscope image contrast for thin crystals', Z. Naturforsch. A27, 445-451.
[2] Uyeda,N., Kobayashi,T., Suito,E., Harada,Y. & Watanabe,M.(1972) 'Molecular image resolution in electron microscopy', J. Appl. Phys., 43, 5181-5188.
[3] Scherzer,O.(1949) 'The theoretical resolution limit of electron microscope', J. Appl. Phys., 20, 20-29.
[4] Woolfson,M.M.(1970) An Introduction to X-ray Crystallogrphy, Cambridge University Press.
[5] Cowley,J.M. & Moodie,A.F.(1957)'The scattering by atoms and crystals. 1. A new theoretical approach', Acta Cryst., 10, 609-619.
[6] Li,F.H. & Tang,D.(1985)'Pseudo-weak-phase-object approximation in

high-resolution electron microscopy. 1.Theory', Acta Cryst., **A41**, 376-382.

[7] Tang,D., Teng,C.M., Jou,J. & Li, F.H.(1986)'Pseudo-weak-phase-object approximation in high resolution electron microscopy. 2.Feasibility of directly observing Li ions', Acta Cryst., **B42**, 340-342.

[8] Morosin,B. & Mikkelsèn,J.C.(1979)'Crystal structure of the Li ion conductor dilithium trititanate, Acta.Cryst., **B35**, 798-800.

[9] Li,F.H. & Hashimoto,H.(1984)'Use of dynamical scattering in the structure determination of a minute fluorocarbonate mineral cebaite by high-resolution electron microscopy', Acta. Cryst., **B40**, 454-461.

[10] Teng,C.M., Tang,D. & Li,F.H.(1986)'Crystal structure investigation of $K_2Nb_8O_{21}$ by HREM', Proc. 11th Intern. Congr. on Electron Microscopy, Tokyo, 815-816.

[11] Li,F.H., Teng,C.M., Hu,J.J., Nagata,F. & Tsuruta,C. (1990)'Crystal structure of $K_6Nb_{44}O_{113}$', Proc. 12th Intern Congr. on Electron Microscopy, Seattle, in press.

[12] Gatehouse,B.M., Lloyd,D.J. & Miskin,B.K.(1972)Nat. Bur. Stand. (U.S.A.)Spec. Publ., 364, 15.

[13] Wang,D.N. Hovmöller,S., Kihlborg,L. & Sundberg,M.(1988) 'Structure determination and correction for distortions in HREM by crystallographic image processing', Ultramicroscopy, **25**, 303-316.

[14] Fejes,P.L.(1977)'Approximation for the calculation of high-resolution electron microscope images of thin films', Acta Cryst., **A33**, 109-113.

[15] Li,F.H. and Fan,H.F.(1979)'Image deconvolution by use of Sayre equation', Acta Physica Sinica, **28**, 276-278.

[16] Han,F.S., Fan,H.F. & Li,F.H.(1986)'Image processing in high resolution electron microscopy using the direct method. 2. Image deconvolution', Acta Cryst.,**A42**, 353-356.

[17] Sayre,D.(1952)'The squaring method: A new method for phase determination', Acta Cryst., 5, 60-65.

[18] Hu,J.J. & Li,F.H.(1990)'Image deconvolution in high resolution electron microscopy by means of maximum entropy', unpublished.

[19] Bricongne,G.(1984)'Maximum entropy and the foundation of direct methods', Acta Cryst., **A40**, 410-445.

[20] Tang,D. & Li,F.H.(1988)'A method of image restoration for pseudo-weak-phase objects', Ultramicroscopy, **25**, 61-68.

[21] Liu,Y.W., Xiang,S.B., Fan,H.F., Tang,D., Li,F.H., Pan,Q., Uyeda,N. & Fujiyoshi,Y.(1990)'Image deconvolution of a single high-resolution electron micrograph', Acta Cryst.,**A46**, in press.

[22] Uyeda,N., Kobayashi,T., Ishizaka,K. & Fujiyoshi,Y.(1978-1979)'High voltage electron microscopy for image discrimination of constituent atoms in crystals and moleculas', Chem.Scr., 14, 47-61.

[23] Main,P, Fiske,S.J., Hull,S.E., lessinger,L., Germain,G., Declercq, J.P. & Woolfson,M.M (1990) MULTAN80. A System of Computer Programs for the Automatics Solution of Crystal Structures from X-ray diffraction Data. University of York, England, And Louvain, Belgium.

[24] Fan,H.F., Xiang,S.B., Liu,Y.W., Tang,D., Li,F.H., Pan,Q., Uyeda,N. Fujiyoshi,Y.(1989)'Image processing in high resolution electron microscopy using X-ray crystallographic methods', Proc. 3rd Beijing Conference and Exhibition on International Analysis.',A109-110.

ELECTRON CRYSTALLOGRAPHY OF RADIATION-SENSITIVE POLYMER CRYSTALS

J.-F. Revol
Paprican and Department of Chemistry,
Pulp and Paper Research Centre, McGill University,
3420 University St., Montreal, Quebec, Canada H3A 2A7

ABSTRACT. Low-dose techniques allow high resolution electron microscopy (HREM) of polymers extremely susceptible to irradiation damage to be performed with a standard microscope operating at 120 keV and at room temperature. After a brief discussion on irradiation damage of polymers, with wood polymers being given as example, it is shown that electron microdiffraction patterns, lattice images and even molecular images can be obtained with many natural and synthetic polymer crystals. The observation of structural defects within the crystals of radiation-sensitive polymers remains however very difficult. The prospects for the improvement of such techniques by the use of higher accelerating voltages or by cryoprotecting the specimen are also discussed.

INTRODUCTION

In the last few years the resolution of electron microscopes has been improved considerably. It is now possible to observe the crystal structure on a molecular or atomic level. However, in practice, it is difficult to study electron-beam sensitive materials at these levels. Polymer crystals, for instance, are damaged by electron irradiation and are changed to an amorphous state by a relatively small irradiation dose. For this reason, lattice imaging of polymer crystals was restricted until recently to ones that are relatively resistant to irradiation [1-10].

At the present time high-resolution can be maintained at very low magnifications, thus allowing the use of lower radiation doses. In addition, the use of a low-dose unit system makes it possible to record pictures of areas without prior irradiation; the focusing being carried out on an adjacent area [11]. As a result, crystalline polymers extremely sensitive to damage by electron irradiation have been studied by "low-dose imaging" in high-resolution [12-14]. Such low-dose imaging was mentioned by R.M. Glaeser as early as 1971 [15] as a possible means to overcome radiation damage. He suggested the recording of images at a very low

J. R. Fryer and D. L. Dorset (eds.), Electron Crystallography of Organic Molecules, 169–187.

electron-exposure followed by averaging techniques to produce statistically well defined information. This pioneering technique was then first applied by Unwin and Henderson who obtained a clear map featuring the molecular structure of the purple membrane protein, bacteriorhodopsin, at 0.7 nm resolution [16] from low-dose images containing no apparent structural information. This heralded a new era in electron microscopy at molecular dimension for electron beam sensitive material.

In the following, results already presented in recent literature and some unpublished work will show that by using low-dose techniques, high-resolution electron microscopy (HREM) of polymers extremely susceptible to irradiation damage can be performed with a standard microscope operating at 120 keV and at room temperature. In particular, it will be seen after a brief discussion on irradiation damage of polymers that electron microdiffraction patterns, lattice images and even molecular images can be obtained with many natural and synthetic polymer crystals. The prospects for improvement of such techniques by the use of higher accelerating voltages or by cryoprotecting the specimen will be discussed briefly.

IRRADIATION DAMAGE

When a sample of organic material is under observation in a transmission electron microscope (TEM), it is severely damaged by the ionizing effects of the electron radiation, leading to drastic changes in specimen structure [17]. The most spectacular result of this damage is a rapid loss of mass, which ranges from 10 to 90% depending on the chemical composition of the specimen [18-20]. A recent study done on wood and its major components by measuring the decrease in the continuum X-ray intensity for various doses of irradiation showed that for doses higher than 5×10^{-8}C/μm^2, with the microscope operating at 100 keV, about 30% of mass is lost for lignin while close to 70% is lost for cellulose [21]; this confirms the chemical dependence of this phenomenon. In the same paper it was also shown that the crystallinity of the cellulose is totally destroyed at a much lower irradiation dose of about 2×10^{-11} C/μm^2, however the mass loss is still negligible. Since less spectacular, this effect is the most important. Indeed, the dose at which the crystallinity of a polymer disappears corresponds to the total end-point dose (TEPD) that should not be exceeded when imaging by diffraction contrast. Figure 1 illustrates the effect of the disappearance of the crystallinity while imaging cellulose by diffraction contrast. The images were obtained from an ultrathin section of the cell wall of <u>Valonia-ventricosa</u> at an accelerating voltage of 120 keV. The objective aperture was 20 μm allowing all the diffracted beams to be stopped, which produced a high diffraction contrast in the bright field mode. Figure 1a shows an image obtained at a very low dose condition i.e. well below the critical dose. The cellulose microfibrils in cross section are seen in very high contrast and with sharp contours. In such an image, the cellulose microfibrils present clearly an approximate square cross-sectional shape with about a 20 nm side [22]. In Figure

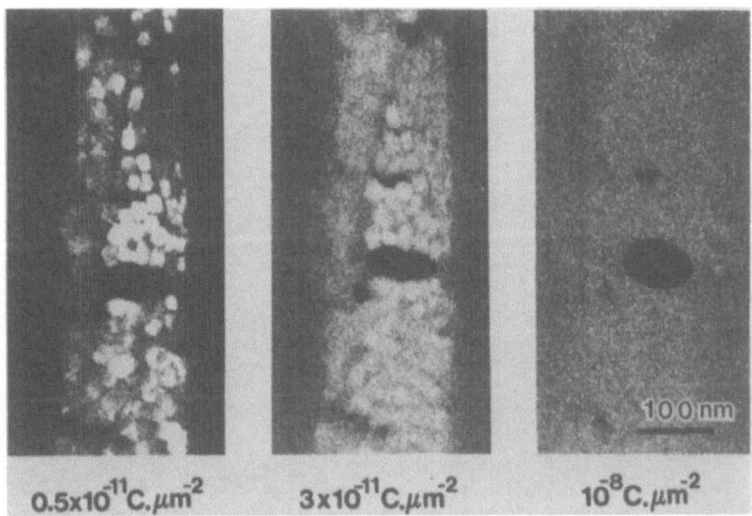

0.5x10^{-11}C.μm^{-2} 3x10^{-11}C.μm^{-2} 10^{-8}C.μm^{-2}

Figure 1:   Series of diffraction contrast electron micrographs of a section of the cell wall of <u>Valonia-ventricosa</u> (Printed in reverse contrast). The cellulose microfibrils are mostly perpendicular to the plane of observation and are seen as white objects in Fig. 1A. They are totally invisible in Figure 1C.

1b, the total accumulated dose used to obtain the image was 3×10^{-11}C/μm^2 which exceeds the critical dose by about 50%. In this image, the cellulose microfibrils in cross section are still visible, but their contrast is drastically reduced and their contours blurred and poorly defined. Finally in Figure 1c where the sample received a very high dose of 10^{-8}C/μm^2, substantial mass was already lost and no cellulose microfibrils can be recognized. In addition, it is to be noted that the gross morphology of the whole section is changed as seen by the deformation of the hole at the center of the image.

To obtain images of crystalline organic materials at the much higher resolution required for lattice images, not only should the critical dose that destroys the crystallinity not be exceeded, but attention should be also paid to the damage which is always present before the TEPD is reached as revealed by the continuous decrease in diffraction intensity. Also, changes in the spacing of the crystallographic planes is often observed for polymers. For example, some lattice spacings of polyethylene (PE) increase gradually with increasing irradiation dose, corresponding to a change of the crystal structure from orthorhombic to hexagonal [23,24]. Similarly, cellulose [13] and chitin [25] undergo gradual changes of their equatorial lattice spacings before the crystallinity disappears. While the true mechanism of the electron

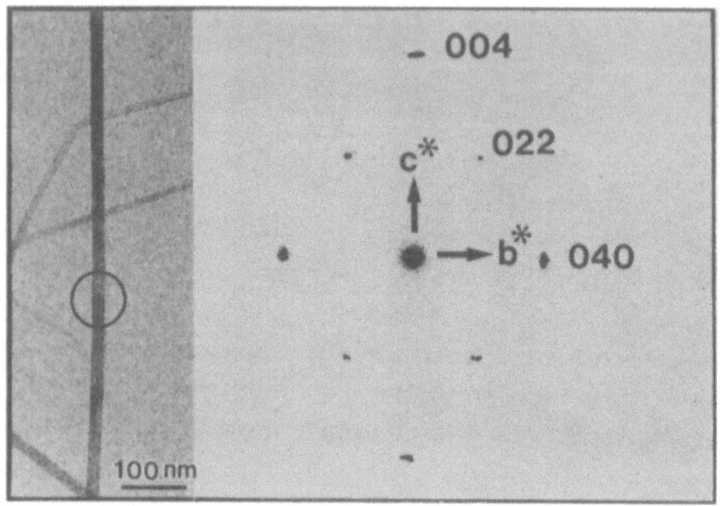

Figure 2:   Cellulose microfibrils of <u>Valonia-ventricosa</u> deposited parallel to the supporting film.
Inset: a corresponding microdiffraction pattern.

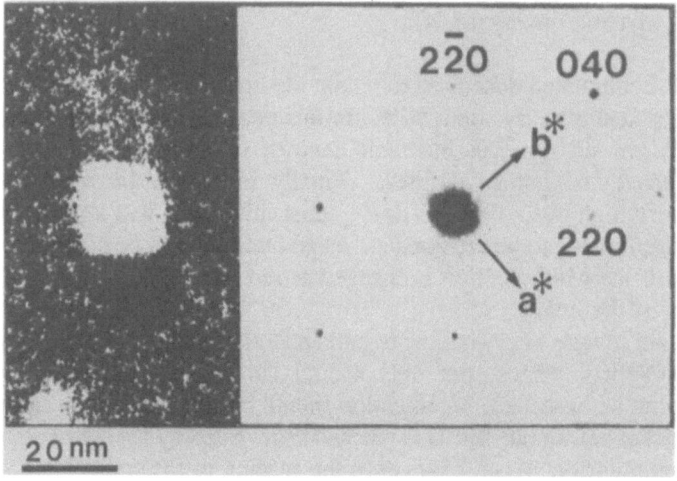

Figure 3:   Cross-sectioned view of a cellulose microfibril from <u>Valonia-ventricosa</u>.
(Picture printed in reverse contrast)
Inset: a corresponding microdiffraction pattern.

irradiation damage of polymers remains still unknown, its effects, as previously described, are known. They should be considered with great care when using a transmission electron microscope for imaging, for any electron diffraction or for quantitative microbeam analysis, and caution should be exercised in the interpretation of the results.

MICRODIFFRACTION

Although microdiffraction is a well known technique for crystalline materials, it has rarely been performed on polymers because the small spot size used in such a technique results in a high number of electrons per square unit. As a consequence, the crystallinity is rapidly destroyed and no information can be obtained from the original crystalline structure. However, if conditions are carefully chosen, and by using a beam diameter as small as 40 nm, microdiffraction can be performed on materials as sensitive as cellulose or polyethylene. This is done at a very low dose condition by overfocusing the first condenser lens and by reducing to a minimum the time for searching a suitable area. In order to reduce the beam damage and the beam convergence during the recording of the pattern, the second condenser aperture was small [5 μm and 2 μm]. The shutter of the low dose unit system was closed as soon as a suitable pattern was seen on the screen, so as not to expose the sample to irradiation longer than needed. The shutter was then reopened for the exposure time only, which was 5 seconds in the following examples.

Figure 2 represents a diffraction contrast image of a cellulose microfibril from Valonia-ventricosa running parallel to the supporting film with its corresponding microdiffraction pattern. In this particular case, the cellulose microfibril is oriented in such a way, with respect to its fibre axis, that it produces a pattern corresponding to the b*c* reciprocal lattice of native cellulose which is the cellulose I polymorph. The single crystal nature of individual microfibrils is clearly revealed by this technique. When this same technique is applied to individual microfibrils seen in cross section, such as the one shown by diffraction contrast in Figure 3, the microdiffraction diagram obtained exhibits sharp reflection spots corresponding to the a*b* reciprocal lattice of native cellulose. This indicates again that a cellulose microfibril, also termed crystallite, is a single crystal of high perfection. Similar work done on individual microfibrils and microfibrils contained in the same lamella of the cell wall revealed that each lamella is composed of parallel microfibrils with two opposite directions for the fibre c-axis [26]. This technique can also be applied to other types of polymers such as single crystals in order to have local information across the whole crystal. A good example of this is work done recently on polyethylene (PE) single crystals [14]. Figure 4 shows a PE single crystal imaged by diffraction contrast and exhibiting the typical succession of bright and dark bands. The corresponding microdiffraction patterns obtained from these different bands indicate clearly that the orientation of the fibre c-axis changes

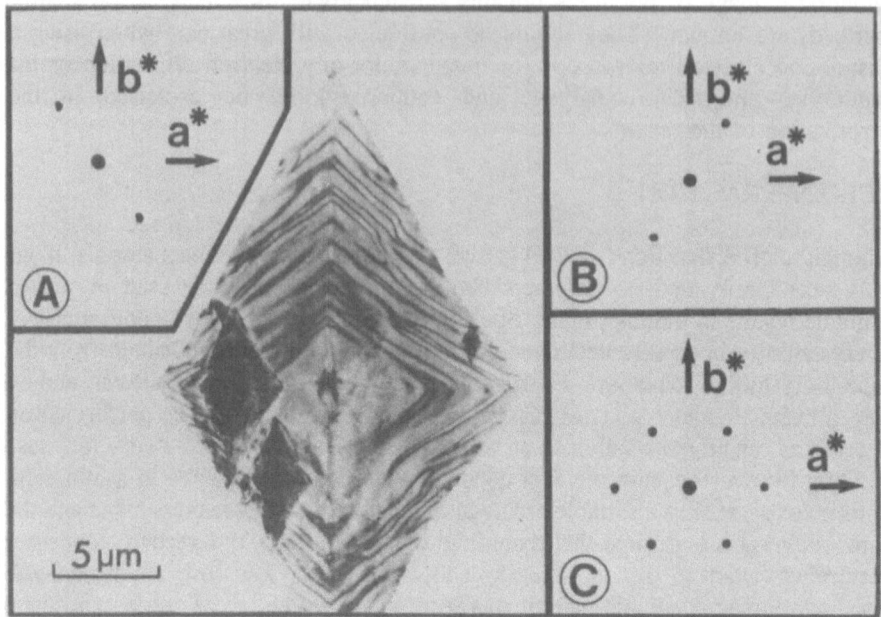

Figure 4:   PE single crystal imaged by diffraction contrast in the bright-field mode
and the electron microdiffraction patterns obtained respectively from:
a) the bright bands in the upper right and lower left sectors; b) the
bright bands in the upper left and lower right sectors; c) the dark
bands.

continuously with respect to the plane of observation and that probably they remain
at the same angle to the surface of the lamellar crystal and accordingly follow the
undulation of the lamella.

These microdiffractions on sections perpendicular to the fiber axis of
fibrous polymers or on polymer single crystals thus give a direct local
crystallographic analysis at a level well below 0.1 μm which cannot be reached by
the conventional selected-area electron diffraction.   Microdiffraction is thus a
powerful and elegant technique that is applicable to a wealth of fibrous polymers,
polymer single crystals and other crystalline polymers.

HIGH RESOLUTION IMAGING

Until recently, lattice imaging of polymer crystals has been successfully performed
only on polymers with a high resistance to radiation damage (greater than 2 x 10^{-10}
C/μm^2), most of them containing aromatic rings [1-10]. However, lattice imaging

is not limited to such polymers, but can be applied also to most of the natural and synthetic polymers having a TEPD lower than $2 \times 10^{-11} C/\mu m^2$, even though the feasibility of this imaging technique has been seriously questioned when using a standard microscope [24, 27]. The following is a brief description of investigations by HREM of the two most abundant naturally occurring polymers in the world: cellulose and chitin, and of two common synthetic polymers: polyethylene (PE) and polytetrafluoroethylene (PTFE). In order to obtain lattice images of these polymers, certain critical experimental conditions were carefully chosen which are as follows:

1.   The electron microscope, a Philips 400T instrument operating at 120 keV, was carefully adjusted for high resolution (perfect alignment, no astigmatism, etc).

2.   The microscope was equipped with a μ-metal shield, which allowed us to obtain high resolution at the lowest magnification compatible with the size of the photographic emulsion grain. A resolution of 0.53 nm was thus obtained at a magnification as low as x 13 000 [25].

3.   The size of the objective aperture should be large enough to allow the desired diffracted beams to be used.

4.   Two films were selected. Ilfoset SP425 from Ilford and Mitsubishi MEM. They were developed at 20°C for 5 min. in Kodak developer D-19 full strength.

5.   The final optical density obtained on the film was chosen to be approximately 0.5.

6.   With these films, images were recorded at magnifications of x 22 000 to x 46 000 which allowed the desired resolution to be obtained.

7.   Under the above conditions, the total irradiation dose for one image ranged from $0.6 \times 10^{-11} C/\mu m^2$ to $2.5 \times 10^{-11} C/\mu m^2$.

8.   Finally the low-dose unit [11] allowed the focusing on an area adjacent to the area of interest which was thus irradiated at a minimum dose before the image was recorded. The selection of suit able specimen areas were made as quickly as possible in a very low dose condition.

All the micrographs were analysed with a Polaron electron micrograph optical diffractometer. Regions that displayed optical diffraction patterns were enlarged 10 to 15 times on Kodak electron microscope film 4489 for further printing or digitizing.

Figure 5 represents a lattice image obtained on a cellulose microfibril from Valonia-ventricosa running parallel to the plane of observation. The corresponding optical diffraction pattern indicates that the lattice lines have a spacing of 0.53 nm. The lines can be seen across the width of the microfibril from one edge to the other and along the fiber axis over distances exceeding 0.1 μm. This lattice imaging confirms, without ambiguity, the single crystal nature of the cellulose microfibrils. In a recent study on the same substrate [13] 3 successive lattice images of the

Figure 5:   Lattice image of one cellulose microfibril from <u>Valonia-ventricosa</u>
             running parallel to the supporting film.
             Inset: Corresponding optical diffraction pattern.

equatorial plane from the same area of the microfibril were reported showing an
increase of the d-spacing from 0.53nm to 0.54nm with increasing doses.
This series of lattice images of the same area, the first to be published for polymers,
opens the way for further investigations on the detection of eventual defects resulting
from electron irradiation.

   β-chitin microfibrils, which bear a close resemblance to cellulose microfibrils
have also been investigated recently by HREM [28]. This work reported the
imaging of a set of crossed meridional and equatorial lattice lines. Figure 6 shows
one β-chitin microfibril, parallel to the supporting film, where only meridional lines
are visible. The presence of the 2nd order at 0.26 nm spacing on the corresponding
optical diffraction pattern indicates the improved resolution for β-chitin due to its
higher resistance to electron-beam irradiation. A dose of $2.5 \times 10^{-11} C/\mu m^2$ was
used in this particular case instead of $1.8 \times 10^{-11} C/\mu m^2$ which was used for
cellulose. When deposited on the carbon supporting film, the β-chitin microfibrils
present a selective uniplanar orientation, with the b-axis mostly perpendicular to the
plane of observation. As a result, it is possible to record the lattice lines
corresponding to the a*c* projection (along the b-axis) of the crystal structure.
Figure 7a represents a general view of a β-chitin microfibril preparation. Figure 7b
shows an electron diffraction pattern of a single microfibril having its b-axis

177

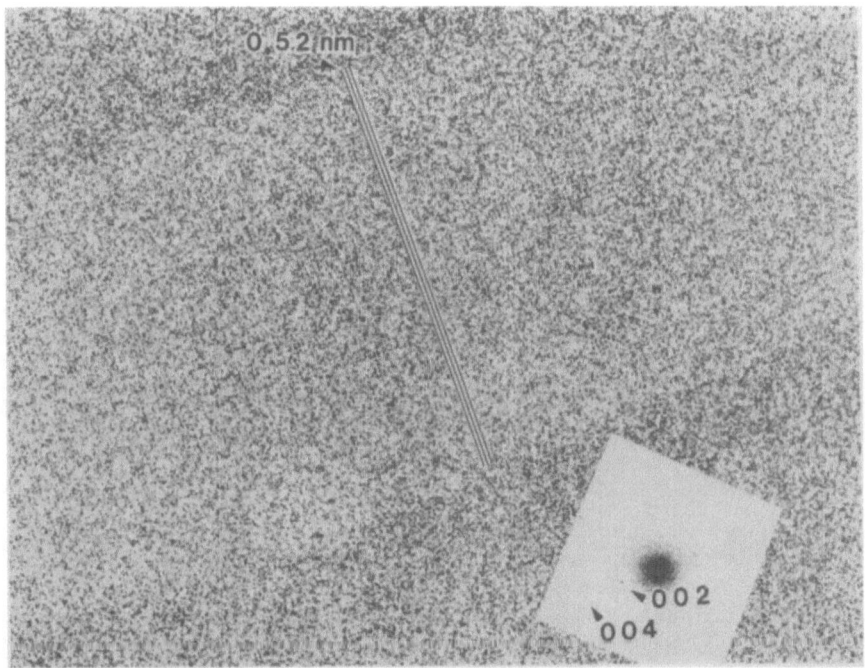

Figure 6:   Lattice image of a β-chitin microfibril from <u>Thalassiosira-fluviatilis</u>.
Only the meridional planes are imaged.
Inset: Corresponding optical diffraction pattern.

perpendicular to the plane of observation and Figure 7c an optical diffraction pattern
of a high-resolution electron image of such a microfibril. By subsequent image
processing of appropriate areas, a clear molecular image of the β-chitin crystal was
obtained at 0.35 nm resolution [29] and is shown in Figure 8 with the known crystal
model for comparison.

Figure 9 shows a high resolution image of a typical corrugated PE single
crystal [14]. On careful inspection, intersecting lattice lines are visible in three
directions, as confirmed by the corresponding optical diffraction pattern. The
difficulty of seeing the intersecting lattice lines is due to the very low signal-to-noise
ratio which results from the low exposure used to record the image, and also from
the granularity of the photographic emulsion. By spacial filtering of the image
transform, the image shown in Figure 10 was obtained [14]. The straight chain
segments are seen in c-axis projection as white dots arranged according to the
orthorhombic unit cell of polyethylene, which is indicated on the inset.

Figure 11 shows a lattice image of a thin film of PE prepared by ultradrawing
gels of ultrahigh molecular weight [30]. Lattice imaging with these specimens
allows the identification of each crystalline block, together with its relative

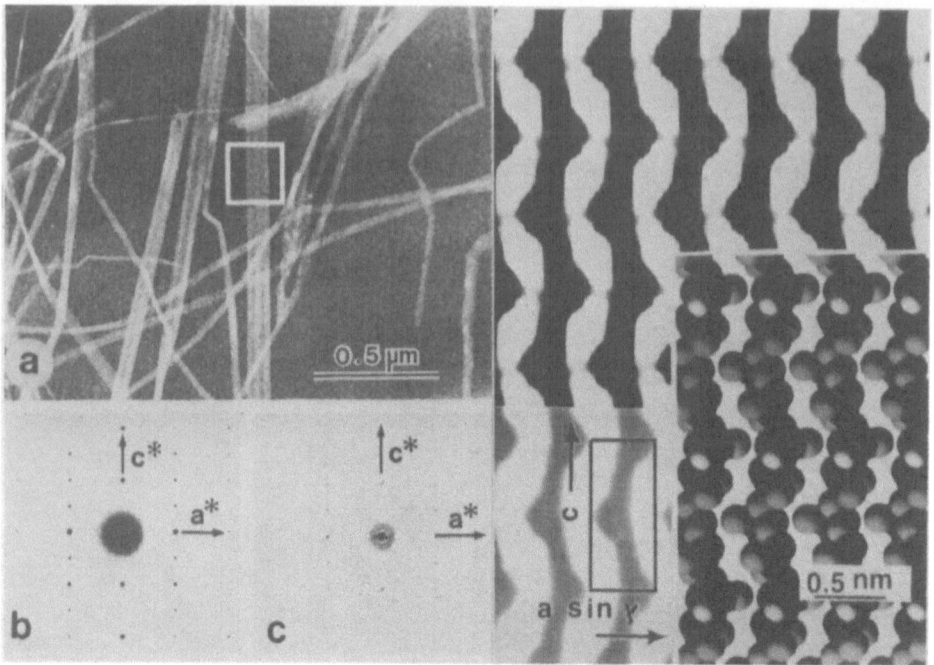

Figure 7: a) Diffraction contrast electron micrograph of a dispersion of β-chitin microfibrils from the diatom Thalassiosira-fluviatilis. The contrast has been reversed during printing.

b) Electron diffraction diagram representing the a*c* reciprocal lattice of one isolated B-chitin microfibril. Diffraction spots corresponding to spacings as low as 0.1 nm are visible on the original negative.

c) Optical diffractogram from a high resolution low-dose image of an area similar to the one shown in Figure 1a. Diffraction spots corresponding to the 100 (0.48 nm), 101 (0.44 nm), 102 (0.35 nm) and 002 (0.52 nm) are clearly visible.

Figure 8: Computer-reconstructed image of a high resolution low-dose image showing the projection of the -chitin crystal along the b-axis.The black rectangle is an outline of the unit cell.Inset:known structure based on atomic coordinates.

orientation with respect to the drawing direction and to the adjacent blocks. Crystalline blocks of the triclinic form have also been identified and localized in similar samples [30].

Hexagonal single crystals of PTFE are obtained in the very early stage of a conventional tetrafluoroethylene emulsion polymerization. A low-dose high-resolution image of such thin crystals is presented in Figure 12 [31]. The inset represents a computer-filtered image showing the chain molecules in projection along their axis, with the expected circular shape. When the polymerization is more advanced rod-like particles are formed, together with spherical particles. Figure 13 represents lattice lines obtained from the rod like particles [32]. These lattice lines reveal that these rods are extended chain crystals with the chain axis parallel to the long axis of the rods. Both images in Figures 12 and 13 illustrate the perfection of the crystalline lattice within the single crystals and the rods of PTFE.

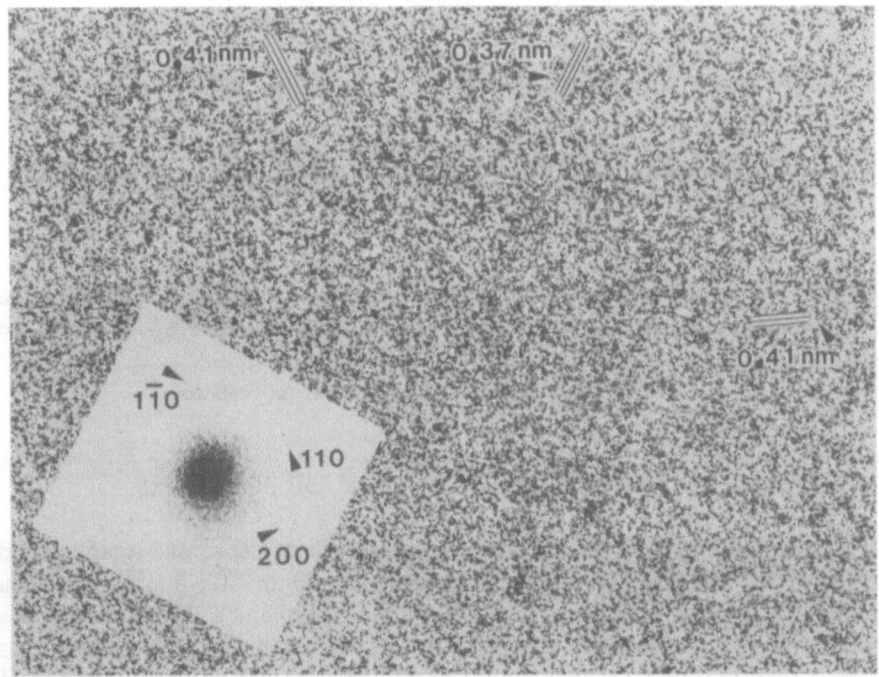

Figure 9:   High resolution low-dose image of a portion of a PE single crystal which on careful inspection shows the 110, 1̄10 and 200 lattice lines. The total irradiation dose accumulated to record such an image was 1.8 x 10^{-11}C/μm^2.
Inset: corresponding optical diffraction pattern.

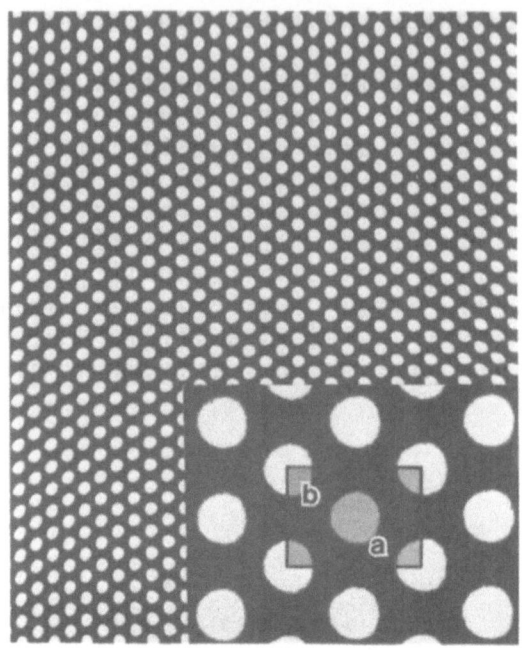

Figure 10:   Computer-reconstructed image (obtained by Fourier filtering) of a
portion of the area in Figure 9. White dots indicate the positions of the
straight chain segments.
Inset: an enlarged area with the unit cell represented.

LOOKING AHEAD AT "LOW-DOSE IMAGING" OF POLYMERS

In the results reported above, the electron dose used to record one image varied
from 0.6 x 10^{-11} to 2.5 x 10^{-11}C/μm^2 (approximately 50 to 150 e$^-$/nm^2). At such a
low level of electron exposure, there is no hope of obtaining direct images at a
molecular or atomic resolution. At best, only lattice lines can be detected in images
having a very low signal-to-noise ratio; more electrons are needed to produce a
statistically well defined image [15]. As was seen before, spatial averaging of such
images can however overcome these low signal-to-noise ratio limitations and
well-defined molecular features can be obtained [16,29].
     If spatial averaging is sufficiently powerful to extract information from the
periodic crystal structure, it will suppress any local and non-periodic information
such as structural defects. Polymer properties are directly affected by such defects;
therefore it is very important to detect them and determine their nature. With

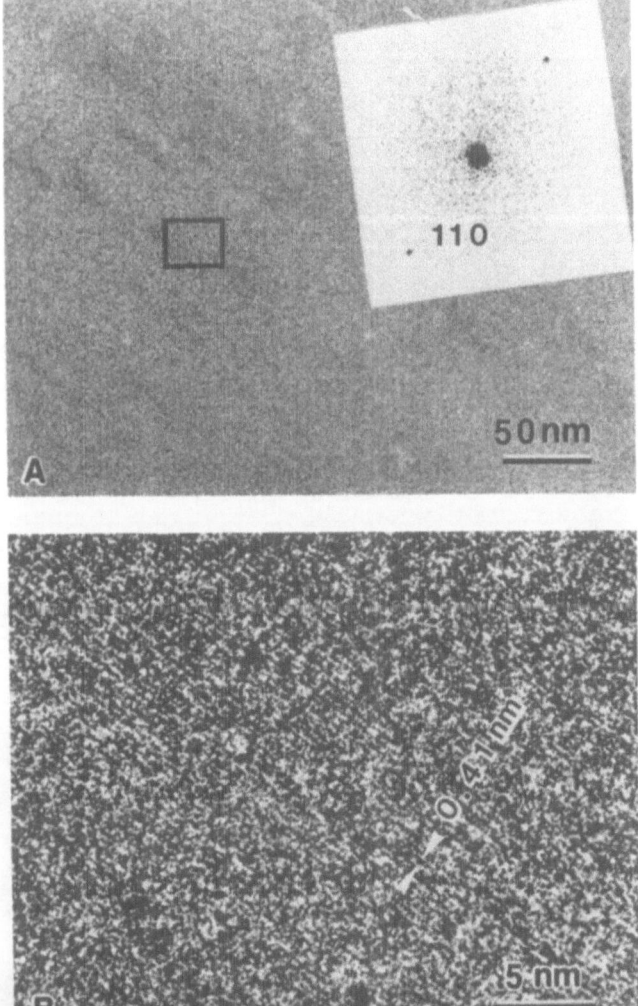

Figure 11: A)High resolution low-dose electron micrograph of a stretched film of
PE.
Inset: optical diffractogram of the framed area. This diffractogram
displays one pair of strong spots that correspond to the (110) planes of
the polyethylene in its orthorhombic form.
B)Enlargement of the framed area of Figure 11a. Lattice lines separated
by 0.41 nm [(110) planes of the orthorhombic phase of polyethylene]
are seen throughout the micrograph.

182

Figure 12:   High resolution low-dose electron image of one portion of a PTFE single crystal. The inset represents a computer-filtered image showing the projection of individual PTFE chain molecules which are packed in a hexagonal lattice.

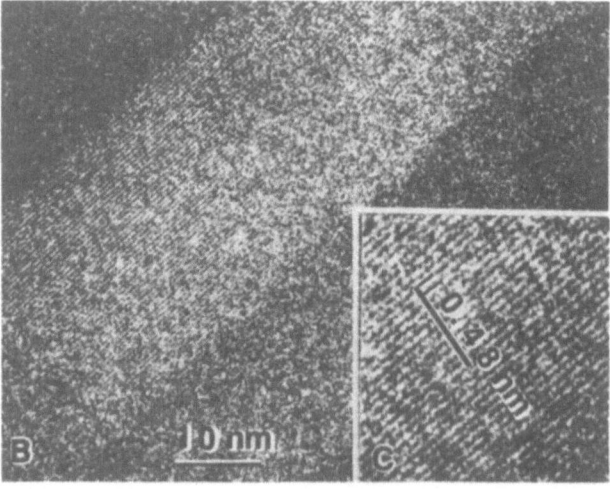

Figure 13:   High resolution low-dose electron image of a rod-like PTFE particle. The inset is an enlargement revealing the high perfection of the lattice fringes.

poly(p-xylylene) which is relatively resistant to electron damage, E.M. images of lattice defects were reported for the first time [8]. However, with other polymer crystals which are much less favorable it has never been possible, so far, to detect such structural defects by direct imaging. In principle, crystal defects such as edge dislocations can be enhanced by the use of large mask windows which have the effect of decreasing the averaging condition. In that case, the amount of noise selected through the mask is higher. Recent results have shown that the selected noise creates a lattice containing defects such as dislocations [33]. Depending on the noise level and the size of the windows, the lattice created by the noise may be dominant and thus generate artifacts. It was concluded that apparent dislocations can only be interpreted with confidence when the polymer is sufficiently resistant to radiation damage, so that the signal-to-noise ratio $S/N \geq 1$ for large windows. This probably means that the defect should be visible on the original micrograph. A prerequisite for that is to use more electrons. As higher electron doses destroy the polymer crystal structure, some method is needed to increase the resistance of the polymers.

One approach is the use of higher accelerating voltage. For instance, it was reported that between 300 and 1000 keV the resistance increases 3 times for cellulose [34] and 5 times for PE [35]. On the other hand, the sensitivity of the photographic emulsion decreases at the same time, and exploratory experiments showed that for PE and paraffin the gain in resistance is about a factor of 2 from 100 to 300 keV while the decrease of the emulsion sensitivity for all the films tested was a little less than 2. Recent developments of new recording systems, which are independant of the voltage used [36], could be of great help in the near future, but the prospect for using high voltages is limited by the fact that, even though the inelastic scattering electrons are reduced, the elastic scattering amplitude is also reduced, resulting again in a need of more electrons. If the main goal is to preserve the crystal structure during imaging then only intermediate voltages up to approximately 400 keV should be considered.

When higher resolution is required, the use of such voltages may be very advantageous since it is only at these voltages that modern instruments are capable of operating at 0.2 nm resolution or better. When thicker specimens are to be investigated intermediate voltages are also needed, since in order to resolve a distance d_{min}, the specimen thickness must not exceed approximately $R_{max} = d^2_{min}/2\lambda$ [37]. For example, a high resolution E.M. study done on 30 nm thick cross-sections of Valonia-ventricosa cellulose microfibrils showed lattice lines at 0.39 nm resolution using an accelerating voltage of 200 keV [38]. At 120 keV this same resolution was obtained only for thinner samples [13]. For such a resolution, according to the above equation, specimen thickness must not indeed exceed about 23 nm at 120 keV and 32 nm at 200 keV. The selection of higher accelerating voltages depends thus on the thickness of the specimen and on the resolution required.

If preservation of the crystal structure is of prime importance then cryoprotection of the specimen is much better. When cooled to a temperature of 170°k or lower, organic materials can withstand electron exposures 5 to 7 times greater than at room temperature [39]. However, the cold stages presently used in the standard E.M. present some limitations which are due mainly to poor mechanical stability. The use of cryomicroscopes equipped with superconducting objective lens [40] are capable of much higher resolution, as was recently demonstrated using crystalline crotoxin [41], and paraffin crystals [42]. Such cryomicroscopes are however special prototypes and the development of better cryostages is perhaps a more urgent need for the success of high resolution imaging of polymer crystals with standard microscopes.

CONCLUSION

The results presented in this paper demonstrate that many polymer crystals can be analyzed by HREM, in spite of the well known sensitivity of these materials to damage by electron irradiation. More specifically, electron microdiffraction and lattice imaging can be performed when conditions are carefully chosen. These techniques yield a wealth of structural information on, for example, the local orientation of microcrystals or crystallites of fibrous polymers, and the detection of crystalline blocks with their respective orientation and their crystalline form in polymer films. By spatial averaging techniques applied to the original images, in which no molecular details can be seen, direct views of the shape of projected macromolecules can also be obtained. Although direct observation of dislocations has been reported in poly(paraxylylene) single crystals [8], a material unusually resistant to electron beam damage, the detection of structural defects within the crystals of radiation-sensitive polymers remains in general difficult. Research methodology is required to preserve the molecular structure while the specimen is under electron irradiation. At present, the most promising of the existing methods is cryoprotection, which can increase the electron beam resistance of the polymers by a factor of better than 5. Accordingly a corresponding higher dose of electrons can be used and a better statistically-defined image obtained. With the development of cryospecimen holders capable of maintaining high resolution at temperatures below 150°k the stage is now set for imaging polymer crystals at molecular and atomic resolution, especially with the new generation of electron microscopes operating at intermediate voltages at a resolution of 0.2 nm or better, and with future recording systems more sensitive to accelerating voltages higher than 100 keV than are the current photographic emulsions. The success of lattice imaging of sensitive materials as shown here with a standard microscope at room temperature, holds great promise for the investigation of polymer crystals at the molecular level as more sophisticated methods are developed.

REFERENCES

1. Keller, A. (1969) 'Solution grown polymer crystals', Kolloid Z. Z. Polymers 231, 386-418.
2. Dobb, M.G., Hindeleh, A.M., Johnson, D.F. and Saville, B.P. (1975) 'Lattice resolution in an electron beam sensitive polymer', Nature 253, 189-190.
3. Read, R.T. and Young, R.J. (1981) 'Direct lattice resolution in polydiacetylene single crystals', J. Mater. Sci. 16, 2922-2924; Young, R.J. and Yeung, P.H. (1985) 'Molecular detail in electron micrographs of polymer Crystals', J. Mater. Lett. 4, 1327-1330.
4. Tsuji, M., Isoda, S., Ohara, M., Kawaguchi, A. and Katayama, K. (1982) 'Direct imaging of molecular chains in a poly(p-xylylene) single crystal', Polymer 23, 1568-1574.
5. Isoda, S., Tsuji, M., Ohara, M., Kawaguchi, A. and Katayama, K. (1983) 'Structural analysis of β- form poly (p-xylylene) starting from high-resolution image', Polymer 24, 1155-1161.
6. Dobb, M.G., Johnson, D.J. and Saville, B.P. (1977) 'Direct observation of structure in high-modulus aromatic fibers', J. Polym. Sci. Polym. Symp. 58, 237-251.
7. Katayama, K., Isoda, S., Tsuji, M., Ohara, M. and Kawaguchi, A. (1984) 'High-resolution electron microscopy of polymers with fiber structure', Bull. Inst. Chem. Res., Kyoto Univ. 62, 198-207.
8. Isoda, S., Tsuji, M., Ohara, M., Kawaguchi, A. and Katayama, K. (1983) 'Direct observation of dislocations in polymer single crystals', Makromol. Chem., Rapid Comm. 4, 141-144.
9. Kawaguchi, A., Isoda, S., Petermann, J. and Katayama, K. (1984) 'High-resolution electron microscopy of $(SN)_x$', Colloid & Polymer Sci. 262, 429-434.
10. Tsuji, M., Roy, S.K. and Manley, R. St. John (1984) 'Direct lattice imaging in single crystals of isotactic polystyrene', Polymer 25, 1573-1576.
11. Handbook of "Low Dose Unit for EM400T", Philips (1979).
12. Sugiyama, J., Harada, H., Fujiyoshi, Y. and Uyeda, N. (1984) 'High resolution observations of cellulose microfibrils', Mokuzai Gakkaishi 30, 98-99.
13. Revol, J.-F. (1985) 'Change of the d-spacing in cellulose crystals during lattice imaging', J. Mater. Sci. Lett. 4, 1347-1349.
14. Revol, J.-F. and Manley, R. St. John (1986) 'Lattice imaging in polyethylene single crystals', J. Mater. Sci. Lett. 5, 249-251.
15. Glaeser, R.M. (1971) 'Limitations to significant information in biological electron miscroscopy as a result of radiation damage', J. Ultrastruct. Res. 36, 466-482.

16. Unwin, P.N.T. and Henderson, R. (1975) 'Molecular structure determination by electron microscopy of unstained crystalline specimens', J. Mol. Biol. 94, 425-440.

17. Cosslett, V.E. (1978) 'Radiation damage in the high resolution microscopy of biological materials: A review', J. Microsc. 113, 113-129.

18. Bahr, G.R., Johnson, F.G. and Zeitler, E. (1965) 'The elementary composition of organic objects after electron irradiation', Lab. Invest. 14, 377-395.

19. Reimer, L. (1965) 'Irradiation changes in organic and inorganic objects', Lab. Invest. 14, 344-358.

20. Egerton, R.F. (1980) 'Measurement of radiation damage by electron energy-loss spectroscopy', J. Microsc. 118, 389-399.

21. Mary, M., Revol, J.-F. and Goring, D.A.I. (1986) 'Mass loss of wood and its components during transmission electron microscopy', J. Appl. Polym. Sci. 31, 957-963.

22. Revol, J.-F. (1982) 'On the cross-sectional shape of cellulose crystallites in Valonia-ventricosa', Carbohydrate Poly. 2, 123-134.

23. Orth, M. and Fisher, E.W. (1965) 'Anderungen der gitterstruktur hochpolymerer einkristalle durch bestrahlung im elektronenmikroskop', Makromol. Chem. 88, 188-214.

24. Tsuji, M., Roy, S.K. and Manley, R. St. John (1985) 'Lattice imaging of radiation-sensitive polymer crystals', J. Polym. Sci. Polym. Phys. Ed. 23, 1127-1137.

25. Revol, J.-F. (1988) 'Microscopie électronique à haute résolution appliquée à la cellulose et à la chitine', Thesis (thèse d'état), Grenoble (France).

26. Revol, J.-F. and Goring, D.A.I. (1983) 'Directionality ofthe fibre c-axis of cellulose crystallites in microfibrils of Valonia-ventricosa', Polymer 24, 1547-1550.

27. Boudet, A. and Kubin, L.P. (1982) 'The limitations to resolution in the observation of radiation-sensitive specimens by electron microscopy', Ultramicroscopy 8, 409-415.

28. Revol, J.-F. and Chanzy, H. (1986) 'High-resolution electron microscopy of -β-chitin microfibrils', Biopolymers 25, 1599-1602.

29. Revol, J.-F., Gardner, K.H. and Chanzy, H. (1988) 'β-Chitin: Molecular imaging at 0.35 nm resolution', Biopolymers 27, 345-350.

30. Chanzy, H., Smith, P., Revol, J.-F. and Manley, R. St. John (1987) 'High-resolution electron microscopy of ultradrawn gels of high-molecular weight polyethylene', Polymer Communications 28, 133-136.

31. Chanzy, H., Folda, T., Smith, P., Gardner, K.H. and Revol, J.-F. (1986) 'Lattice imaging of poly(tetrafluoroethylene) single crystals', J. Mater. Sci. Lett. 5, 1045-1047.

32. Chanzy, H., Smith, P. and Revol, J.-F. (1986) 'High-resolution electron microscopy of virgin poly(tetrafluoroethylene)', J. Polym. Sci., Polym. Lett. 24, 557-564.

33. Chanzy, H. (1975) 'Irradiation de la cellulose de Valonia au microscope à 1 MV', Bull. BIST 207, 55-57.

34. Pradère, P., Revol, J.-F., Nguyen, L. and Manley, R. St.John (1988) 'Lattice imaging of poly-4-methyl-pentene-1 single crystals; Use and misuse of Fourier averaging techniques', Ultramicorscopy 25, 69-80.

35. Boudet, A. and Roucau, C. (1985) 'Degradation of polyethylene single crystals in electron microscopy between 1 and 2.5 MV', J. Physique 46, 1571-1579.

36. Mori, N., Katoh, T., Oikawa, T., Miyahara, J. and Harada, Y. (1986) 'Imaging plate - A new recording material for electron microscopy', Proc. 11th I.C.E.M., Kyoto 1, 29-32.

37. Cowley, J.M. (1975) Diffraction Physics, North Holland Pub. Co., Amsterdam, New York, Chapter 13.

38. Sugiyama, J., Harada, H., Fujiyoshi, Y. and Uyeda, N., Planta 166, 161-168.

39. Hayward, S.B. and Glaeser, R.M. (1979) 'Use of low temperature for electron diffraction and imaging of biological macromolecular arrays', Ultramicroscopy 4, 201-209.

40. Dietrich, I., Fox, F., Knapek, E., Lefranc, G., Nachtrieb, K., Weyl, R. and Zerbst, H. (1977) 'Improvement in electron microscopy by application of superconductivity', Ultramicroscopy 2, 241-249.

41. Jeng, T.W., Chiu, W., Zemlin, F. and Zeitler, E. (1984) 'Electron imaging of crotoxin complex thin crystals at 3.5°', J. Mol. Biol. 175, 93-97.

42. Zemlin, F., Reuber, E., Beckmann, E., Zeitler, E. and Dorset, D.L. (1985) 'Molecular resolution electron micrographs of monolamellar paraffin crystals', Science 229, 461-462.

HIGH RESOLUTION ELECTRON MICROSCOPY ON V_H AMYLOSE CRYSTALS

D. MILLER
Dept. of Physics & Astronomy, Clemson University,
Clemson, SC 29631-1911, USA,

J. SUGIYAMA
Dept. of Forest Products, Faculty of Agriculture,
University of Tokyo, Tokyo 113, JAPAN,

J. BRISSON
CERSIM, Faculté des Sciences et de Génie,
Laval University, Québec, PQ, CANADA G1K 7P4

and H. CHANZY
CERMAV/CNRS, BP 53X, 38041 Grenoble Cedex, FRANCE

Abstract

High resolution TEM lattice images can be recorded from beam sensitive (i.e., cellulose, amylose and chitin) crystalline biopolymers. Images showing information to 0.3-0.4 nm resolution are achieved with electron doses of 100-1000 elec/nm2, at magnifications of 20000-40000 X and accelerating voltages of 100-200 kV. Lattice images are very noisy, but lattice information can be observed by He-Ne interferometry of the micrograph film (even in absence of a visual lattice image). A power spectrum of the photoscanned data array of the same region gives a similar spot pattern. Lattice enhancements (from windowed fast Fourier transforms) show effects of lattice tilt and spot window shape. Diffraction phases are not yet accurately computed from motif-averaged, cell-origin centered lattice images.

Introduction

Most polymers and biopolymers are extremely sensitive to electron beam irradiation. When studied by electron microscopy at high resolution and under standard conditions, they are so damaged as to be totally decrystallized. Starting in 1971, to address this problem, Glaeser developed a technique of low illumination, coupled with periodic averaging [1]. This method has proven quite valuable for looking at periodic biological objects. In particular, in 1975, its use led to the now classical reconstruction of bacteriorhodopsin at 0.7 nm resolution [2].

J. R. Fryer and D. L. Dorset (eds.), Electron Crystallography of Organic Molecules, 189–195.

With many crystalline polymers and biopolymers, the lattice details that one would like to resolve occur at resolution lower than 0.7 nm. The electron microscope, therefore, has to be tuned such that higher resolution is achieved, even at magnifications as low as 20000-40000 X. It was only in 1984 that Sugiyama et al. [3] were able to produce significant lattice images of cellulose microfibrils, showing 0.54 nm interchain spacings. These images were produced at a magnification of 23000 X and with an accumulated dose of 300 elec/nm^2, using an acceleration voltage of 200 kV. The technique of lattice imaging of polymers has now been applied to a wealth of polymers, ranging from synthetic (poly β hydroxybutyrate, polyethylene, poly 4-methyl pentene-1, etc...) to natural (chitin and a variety of celluloses). In this paper, the focus is on V_H amylose single crystals. They yield interesting electron diffraction diagrams, where the symmetry is such as to indicate a statistical packing of the amylose chains within the lattice [4]. Because these crystals are hydrated, they need be handled under frozen hydrated conditions. This brings an extra degree of experimental difficulty into the recording of the images. Images of V_H amylose, showing lattice information, are herein presented. There is also described an attempt to use this information to obtain diffraction phases for the computation of a molecular projection of the structure.

Materials and Methods

Preparation of Samples

Single crystals of amylose were prepared from DP 15 fractions (Nakarai chemicals, amylose A) [5]. The crystals were hexagonal platelets. They yielded electron diffraction patterns which could be indexed to give the hexagonal lattice dimensions a = b = 1.365 nm, c = 0.805 nm. The space group was P6$_5$22. A drop of suspension of these crystals was deposited on a microgrid. The grids were prepared by flotation of a micronet carbon membrane, dried and gold coated. A thin carbon film (<3nm), obtained by evaporation onto a freshly cleaved mica surface, was deposited on each grid, again by flotation. The carbon and gold membrane provides good mechanical stability and good electrical conductivity. The crystals are observed on the region of the thin carbon film covering the holes. This contributes much less background than would the usual carbon film (it allows the sample to rest on a thin surface while minimizing the risk and inconvenience usually associated with such a support).

Electron Microscopy

High resolution lattice images were recorded on a JEM-2000SCM JEOL super-conducting electron cryomicroscope, equipped with a LaB$_6$ filament and operated at an accelerating voltage of 160 kV. The spherical coefficient of the objective was 1.2 mm. The expected specimen temperature was 4.2 K. The lattice images were recorded on Mitsubishi MEM films and developed in Kodak D-19. Defocus and astigmatism were adjusted, prior to recording the high magnification images, at higher magnification on a portion of the grid close to the crystal to be studied. The magnification was reduced to the desired value (typically 37000 to 45000 X). Defocus was adjusted to the Scherzer defocus for this magnification, in accord with prior calibration. The photograph was then taken, using the minimum dose technique. The total dose is estimated to be 1000 elec/nm2.

Image processing

Over 200 electron image negatives were inspected, by interference, on a laser optical interference bench. Only two were found to show diffraction spots to 0.38 nm resolution. One had very weak diffraction spots and was discarded. The other, although it did not exhibit perfect 6mm symmetry, was accepted for photoscanning. This film was exposed at a magnification of 45000 X (Figure 1A). Its optical diffraction pattern is figure 1B. Its electron diffraction pattern is figure 1C.

Prior to digitization, the negatives were enlarged onto Kodak EM film. The positive, so obtained, incurred an additional magnification factor of 3.4 and 5.6. Arrays 1024 by 1024 pixels were digitized on an Optronics rotating drum digitizer, using a 25 μm raster. This corres-

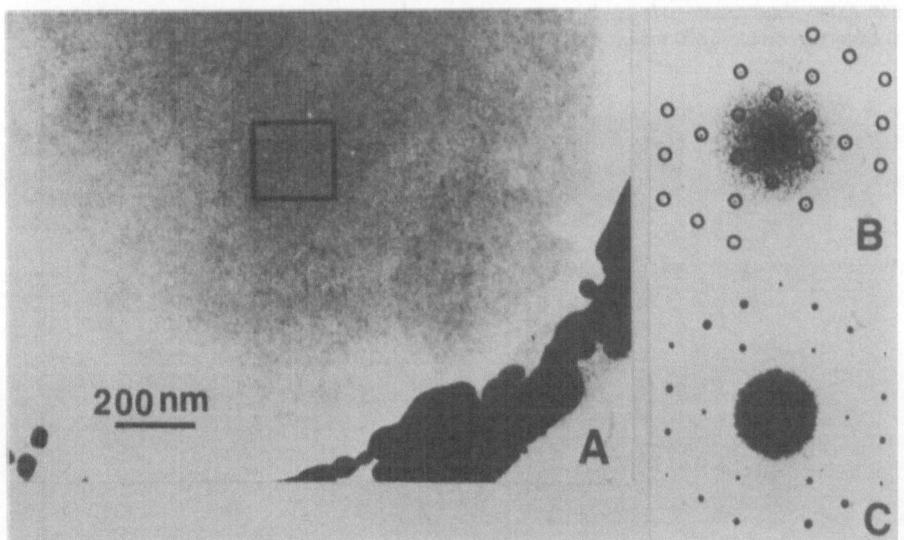

Figure 1. Micrograph of V_H amylose, 1A, its laser optical diffraction pattern, 1B, and its electron interference pattern, 1C.

ponds to a 101 nm by 101 nm (for the 5.6 factor) area on the specimen.

The digitized images were processed on a SUN3 graphics workstation, using the SEMPER V-3 image analysis system. From the large array, 512 by 512 pixel arrays were selected. For each array, a

power spectrum (PS) was computed and examined. That PS which showed
maximum resolution and signal to noise ratio was deemed acceptable.

Lattice Image Analysis

The electron diffraction pattern of the sample normally exhibits
the same spot pattern as that obtained by laser interference of the
micrograph. Figure 1B shows the laser interference pattern for V_H
amylose. Diffraction spots are circled. Not all spots exhibited by
the electron diffraction, Figure 1C, show in the laser interference
pattern. The sample plane was tilted slightly from perpendicular to
the electron beam, causing a small progressive defocus away from the
sample tilt axis. Diffraction spots progressively weaken away from the
tilt axis.

The PS computed from the real and imaginary parts of the Fourier
transform (FFT) of the photoscanned micrograph, usually exhibits the
same spot pattern as that obtained by laser interference, if the
photoscanned area used in computing the PS is at least as large as the
area illuminated by the laser. Effects of the sample tilt are also
seen in the PS. For lattice imaging, the film image must be enlarged,
again, from 3 to 6 X. This brings the lattice dimensions, in the data
array, to a size such that several orders of the diffraction can be
exhibited in the PS. The penalty is that the film grain size is a
reasonable fraction of a lattice dimension in the data array. Figure
2A shows an enlarged area of a V_H micrograph and 2B, its computed PS.

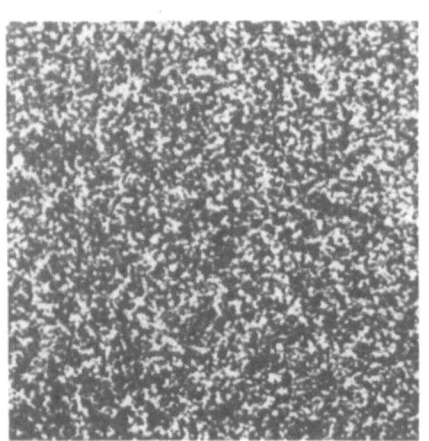 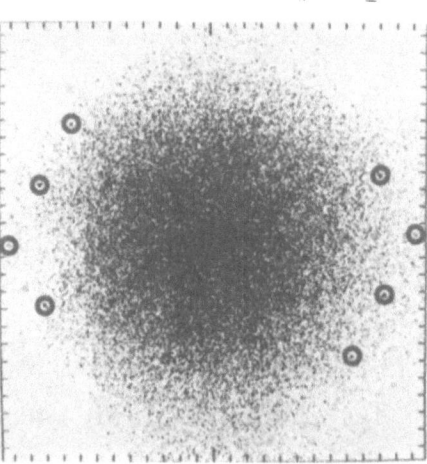

Figure 2. V_H amylose image, A) enlarged 5.6 X from original
micrograph, and B) its computed power spectrum.

If one assumes a grain size of 2 μm in the original micrograph and a 5X
magnification, grain size becomes 10 μm in the enlargement. The
photometer scan step was 25 μm, so grain noise in the enlargement is
severe. This greatly increases the non-periodic contribution to the PS

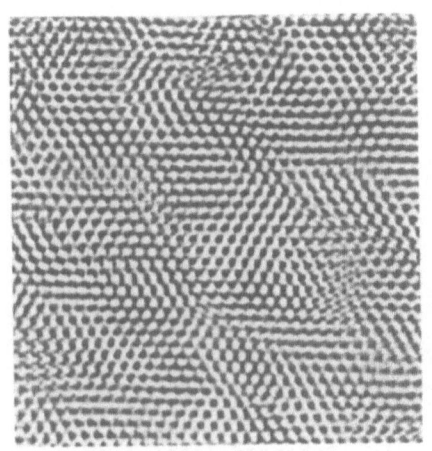

Figure 3 Enhnaced lattice
computed from the FFT of
figure 2A window width
0.014 lattice vector.

so much so that many diffraction
spots in the PS are lost in noise.
The minimal beam dosage, used to
obtain the initial micrograph, also
produced enhanced noise, in the
micrograph and then in its
enlargement. The diffraction spots do
not have symmetrical shape,
indicating that a portion of a spot
may come from periodicity of a
particular portion of the enlarged
image. The spots, therefore, do not
lie upon exact lattice sites, but
near them. The sample tilt and local
thermal curvature and twist cause a
diffraction spot to be computed of
several parts, one or more of which
may be slightly off the average
lattice site. Cell constants,
calculated from the average positions
of the several diffraction spots,
have large standard deviations. Small
window apertures, centered on averaged lattice sites of the FFT, in an
attempt to exclude non-periodicity
from the inverse FFT, often exclude parts of diffraction spots. If
large, to include all diffraction spots, they often include considerable noise as well. Windows were, therefore, individually centered on
the spots (no more than 1 pixel off a lattice site), where spots are
discernible above the noise. Further, windows of lattice shape
introduce artifacts into the enhanced lattice image, Figure 3. The
nonperiodic contributions, appearing in the FFT windows outside the
diffraction spots, inverse transform to produce a lattice image that
appears to contain periodic dislocations and regions of defocus. Motif
averages, even if cross (position) correlated, exhibit severe
distortion and artifacts due to the tilt and thermal curvature in the
original sample and the large amount of noise. Figure 4 shows such an
image, motifed from figure 2A at ten neighboring lattice sites. The
noise largely obscures the lattice, which appears only in an overall
modulation along the unit cell vectors. Individual cells cannot be
distinguished, much less any cell contents. Attempts to use a small
part of the enhanced lattice as a mask for motif cross correlation of
the original lattice image are also unsatisfactory. The diffraction
spots are, after all, but a small fraction of the FFT. It is often
more revealing to use the reverse procedure: inverse transform the
nonperiodic contributions to the FFT and subtract it from the original
image.

Much of the distortion in the enhanced lattice image can be
avoided by use of windows shaped to the diffraction spots of the PS.

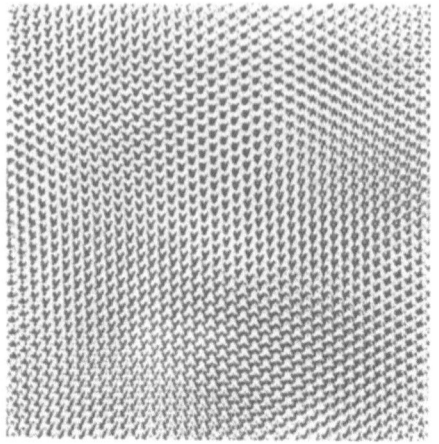

Figure 4. Lattice motif of figure 2A.

Figure 5. Enhanced lattice from FFT, circular windowed, of phase symmetric motif of 2A.

When the FFT is so windowed and inverse transformed, the enhanced lattice appears much clearer. No "dislocations" or defocus is apparent. Lattice dimensions are maintained across the width of the image. Instead, gray-scale reversal varies from side to side in the image, produced by the tilt and curvature of the sample, when irradiated. Some accommodation for variations in lattice geometry, across the image, can be made by phase centering each image subarray before motif

Table 1. Averaging phases of diffraction (FFT) spots of V_H amylose .

Miller Indices h k l	d-spacing (nm)	Phase (degrees)	Phase Error ($\pm$ degrees)
1 0 0	1.19	180	2
1 1 0	0.68	155	8
2 0 0	0.59	171	6
1 2 0, 2 1 0	0.45	129	21
3 0 0	0.40	113	39
2 2 0	0.39	125	31
3 1 0, 1 3 0	0.33	87	33

averaging. The subarray can be chosen at sites where the computed FFT exhibits near symmetrical intensities and phases. The motif average can be transformed, circular windows applied and inverse transformed to give (Figure 5) a much sharper enhanced lattice image. Its appearance suggests that an accurate computation of the phases, by averaging the magnitudes of symmetry equivalent diffraction spots in the real part and imaginary part of the FFT may be possible.

Before the phases were computed, however, the averaged lattice array was again cell (phase) centered. This was achieved by permutating the array, while computing phases of the (1,0,0) equivalent spots in the FFT, until the average phase was 180 degrees, with small error. As can be seen in Table 1, this was achieved. The phases of other diffraction spots were then computed and symmetry-equivalent averaged. The table shows that the error reflects the noise level of the data array. For diffraction spots of d-spacing 0.45 nm, or less, the computational error becomes large and the computed phases become suspect or patently incorrect. This is not surprising, since the grain size of the enlarged image subtends this dimension on the original sample.

Discussion

With reasonable care in sample preparation, electron irradiation and exposure, using the low dose technique and lattice averaging technique, it is certainly possible to obtain information about the lattice perio-dicities and symmetries of a beam-sensitive polymer. At present, howe-ver, phases may only be accurately obtained for diffraction spots of d-spacing larger than the minimal d-spacing spots observed in a laser interference pattern of the micrograph. This is due both to the requirement of low dose, moderate magnification, for preservation of specimen integrity, and to noise from film grain during exposure and subsequent enlargement, before photoscanning. If the initial image can be taken at a small increase in magnification (e.g., 60000 X) and the film can be scanned to smaller raster dimension (e.g., 10 μm) most of the current limitations should be removed. Use of an in-instrument detector array of sufficient resolution to record the enlarged image without the need for photographic processing and of increased sensitivity would also remove the limitations.

Acknowledgments

The authors wish to acknowledge the invaluable assistance of S. Lavaitte and R. Vuong of CERMAV/CNRS and M. Iwatsuki of JEOL Corp.

References

[1] Glaeser, R. M., J. Ultrastruct. Res. 36, 466 (1975).

[2] Henderson, R. and Unwin, P.N.T., Nature 257, 28 (1975).

[3] Sugiyama, J., Harada, H., Fujiyoshi, Y and Uyeda, N., Mokuzai Gakakaishi 31, 61 (1985).

[4] Brisson, J., Chanzy, H. and Vuong, R., Food Hydrocol. 1, 523 (1987).

[5] Brisson, J., Chanzy, H., and Winter, W. T., Int. Jour. of Biol. Macromol., submitted (1990).

HIGH RESOLUTION ELECTRON MICROSCOPY ON LIQUID CRYSTALLINE POLYMERS

I.G.VOIGT-MARTIN
Inst.Physik.Chemie der Univ.Mainz,Germany

ABSTRACT. Two types of liquid crystalline materials with specific
molecular structures were investigated by electron diffraction and high
resolution electron microscopy.The two classes of substances were:
(1) Smectic liquid crystals(main chain/side chain group polymalone
    ester
(2) Discotic liquid crystals(triphenylene derivative)
In both cases the liquid crystalline materials were compared with the
crystalline in order to aid interpretation.The investigation included
four aspects:(1)Electron diffraction,(2)High resolution imaging,
(3)Image restoration,(4)Image simulation.

The diffraction patterns from the crystalline materials show the
expected sharp diffraction patterns with a large number of higher order
maxima.By choosing a suitable frequency window in the contrast transfer
function it is possible to obtain high resolution images.These show
straight,perfectly regular crystallographic planes with occasional edge
dislocations-Fig.1[1].Similarly the columns of discotic mesogenic units
are viewed from above in a perfect hexagonal arrangement-Fig.2[2].

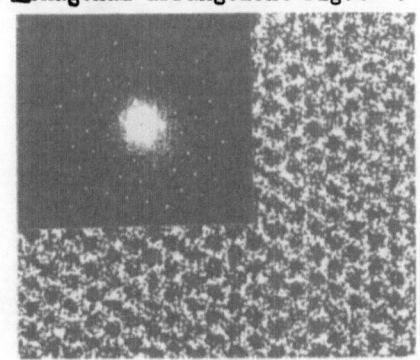

Fig.1,High resolution micrograph of a   Fig.2. Discotic liquid
main chain liquid crystalline   crystal.
polymalonate in the crystalline phase.

J. R. Fryer and D. L. Dorset (eds.), Electron Crystallography of Organic Molecules, 197–201.
© 1990 Kluwer Academic Publishers.

With the aid of microdiffraction,it is possible to detect regions of
tilted columns separated by grain boundaries,and to relate these to the
appropriate microdiffraction pattern-Fig.3². By analysing such
micrographs and microdiffraction patterns,it is shown that the
crystalline structure has a lower symmetry than the hexagonal due to
the triangular shape of the discotic molecule-Fig.4.

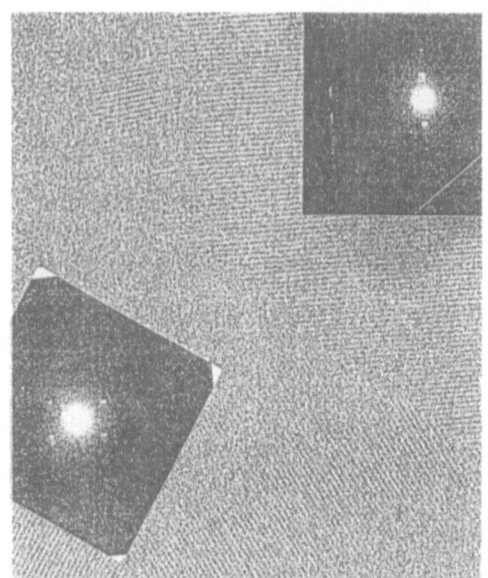
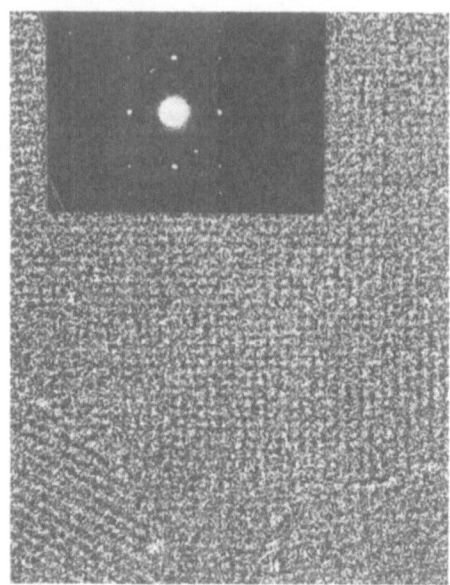

Fig.3.Grain boundaries and   Fig.4.Symmetry related structure
microdiffraction in discotic crystals.   in discotic crystals.

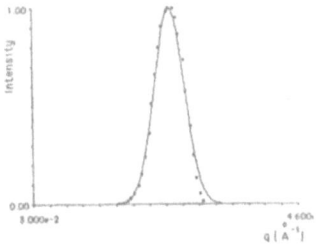

Fig.5.Smectic liquid crystal

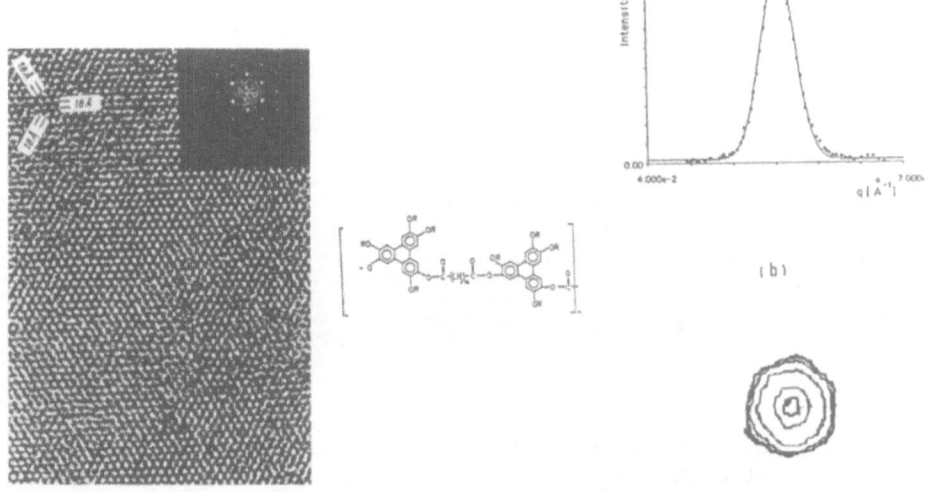

Fig.6.Discotic liquid crystal.

In comparison, these materials are shown in the smectic-Fig.5-and discotic-Fig.6- state with line profile analysis and contour plots of the diffraction maxima.Due to the analysis of the crystalline state, these features can now be better understood.The most dramatic feature in liquid crystals is the loss of higher order diffraction maxima plus the appearance of an oriented amorphous halo in the smectic case.The smectic planes are shown to undulate, a feature which is confirmed by a streak on the diffraction maximum.The long spacings decrease since the molecules are no longer extended.Furthermore many single and multiple dislocations appear.The diffraction patterns from the triphenylene columns generally give rise to hexagonal symmetry and loss of higher orders.These features arise due to rotation of the triphenylene molecules in subsequent layers,giving the projected potential from smeared discs rather than triangles.The individual positions of the columns are slightly displaced from the ideal crystallographic position.

With the aid of various computational techniques both in real and reciprocal space,the signal/noise ratio was reduced.The problem of artefact production was carefully controlled-Fig.7[3].

In order to check whether image interpretation using the weak phase approximation is justified, image simulations,using the full dynamical approach, were performed.Fig.8 shows the model of the triphenylene lattice with the resulting diffraction patterns and Fig.9 shows the simulated images at various defocus values[4].

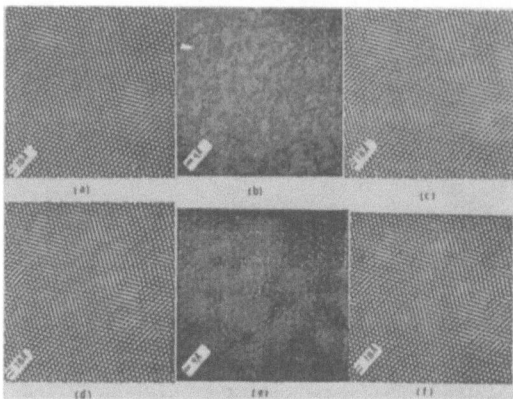

Fig.7.Frequency filtered images from discotic triphenylene columns.
  (a).First order   r=1/p
  (b).Second order(negative phase) r=4/p
  (c).First and second order(negative)
  (d).First order r=2/p
  (e).Second order(negative phase) r=2/p
  (f).First and second order(negative)

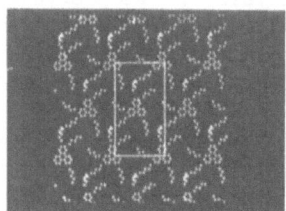 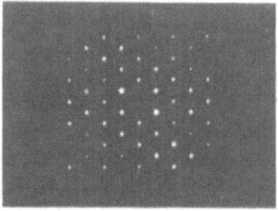

Fig.8.(a)Model of triphenylene single crystal.
      (b)Calculated diffraction pattern.

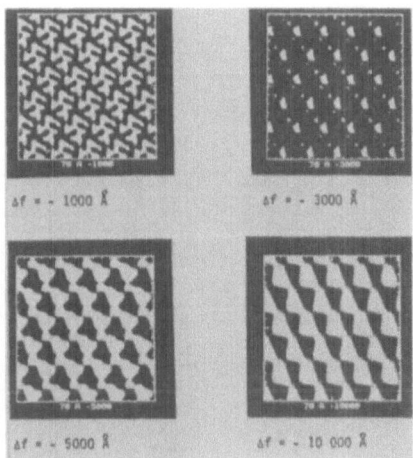

Fig.9.Simulated images of the triphenylene lattice as a function of defocus.

REFERENCES

1.Voigt-Martin,I.G.and Durst,H.1989 Macromol.,22, 168.
2.Schumacher,M.PhD.Thesis,University of Mainz.
3.Voigt-Matin,I.G.,Krug,H.and Durst,H.1989 Macromol.,22. 595.
4.Voigt-Martin,I.G.,Krug,H.and van Dyck,D. J.de Physique,submitted for publication.

Scanning Tunneling Microscopy of Surface Structures

James K. Gimzewski
IBM Research Division
Zurich Research Laboratory
CH-8803 Rüschlikon
Switzerland

ABSTRACT. A discussion of scanning tunneling microscopy for surface-structure determination with an emphasis on molecular and biological systems is presented. Basic considerations involved in imaging and electron transfer between tip and sample are evolved from the ideal case of simple metals up to complex biological systems. The aim of the paper is to illustrate some of the interpretational aspects involved in relating STM images to true structural details of molecules on surfaces and the local form of their associated electron charge density contours.

1. Introduction

In context with the aim of the workshop on electron crystallography, emphasis will be given to the utility of STM in determining the structure of organic and biological systems with a focus on practical and theoretical considerations. Rather than present a detailed discussion, an attempt has been made to guide the reader to relevant key references and review papers, and hopefully this paper will provide a basic framework for fundamental considerations that require appreciation.

The key advantages of STM stem not only from its unsurpassed lateral resolution and vertical sensitivity but also from the variety of environments under which it can perform. These vary from ultra-high-vacuum (UHV) to gases and various polar and nonpolar liquids. Temperatures may range from cryogenic to several hundred °C. In addition, the electron energies involved can be as low as several meV and electron currents currently down to the pA range can be used. These conditions are, to date, generally viewed as nondestructive, even for delicate biological systems.

STM is based on electron tunneling from an atomically sharp metallic tip to a conducting surface. If one considers a metallic surface with a low voltage applied to the tip in molecular proximity, then the current flow in the tunnel mode is sensitive to the local density of electronic states in the region of the Fermi level (E_F). Under these circumstances, STM topographs reflect contours of constant charge density which can be *directly* related to surface structure. Figure 1 shows an example of the surface topography of an Au(110)-(1×2) reconstructed surface in UHV [1]. Rows of atoms separated by $\sim 8\,\text{Å}$ are clearly resolved in the [$1\bar{1}0$] direction with a corrugation amplitude of 0.6 Å as well as a single atomic step $\sim 1.5\,\text{Å}$ in height which crosses the

J. R. Fryer and D. L. Dorset (eds.), Electron Crystallography of Organic Molecules, 203–215.
© 1990 *Kluwer Academic Publishers.*

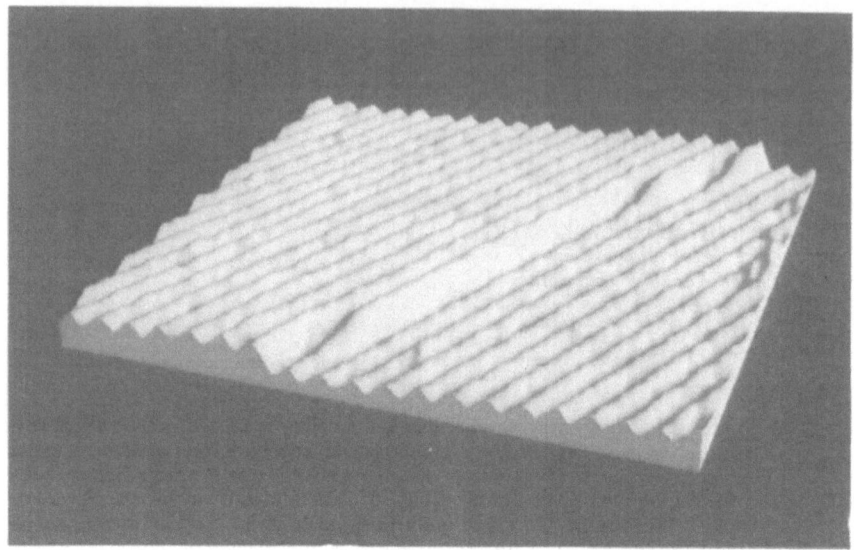

Figure 1.   Pseudo-illuminated   three-dimensional   image   of   an   Au(110)-(1×2) reconstructed surface.   Atomic rows are separated by ~8 Å.   Note single atomic step crossing image almost diagonally. (Gimzewski and Schlittler, unpublished work).

image almost diagonally.   Theoretically, Tersoff and Hamann have shown that the observed corrugation can be modelled by a simple superposition of atom charge densities [2], confirming the *direct* nature of STM.   Their calculations also indicate the the topographs are relatively insensitive to the position of atoms beyond the first atomic layer.

Semiconductor surfaces represent a next higher level of complexity in terms of electron transfer and imaging.   Their nonuniform charge density results from spatial variations in contributions from valence or conduction bands and in certain cases surface states.   This presents a limitation for the *direct* use of STM to measure surface atomic structure [3].   To illustrate this point, Figure 2 shows an STM topograph of the Si(111)-(7×7) reconstruction which was first resolved by Binnig and co-workers [4]. This image, recorded with a negative sample bias shows a slight height asymmetry in both halves of the unit cell.   Imaging at similar bias voltage in the opposite polarity results in the disappearance of this asymmetry.   The dimer-adatom-stacking fault (DAS) model proposed by Takayanagi and co-workers [5] includes a stacking fault in the outermost double layer of one half of the unit cell.   The different onsets for current flow from the two halves in the unit cell have been related to the subtle voltage and polarity dependence of the stacking fault [6].   Detailed electronic structure calculations, other surface physics data and tunneling spectroscopy studies have provided quite a detailed picture of the local electronic structure of the surface [3,4,6,7].   The level of understanding has been increasing for STM, allowing a qualitative understanding of more heterogeneous semiconductor systems.

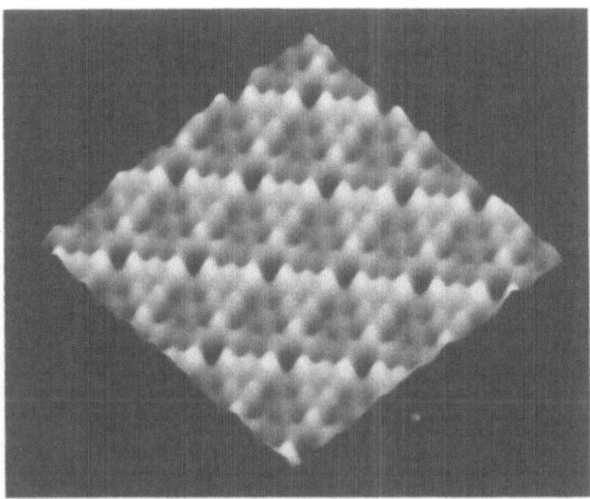

Figure 2. Pseudo-three-dimensional image of the Si(111)-(7×7) reconstruction recorded at a sample bias of − 1.8 V and a tunnel current of 0.16 nA. Approximately 14 unit cells are resolved. Note the asymmetry in both halves of the unit cell. (Gimzewski and Schlittler, unpublished work).

2. Molecular Imaging of Delocalized Electron Systems

If we extend the basic concepts involved in imaging semiconductors to molecular systems, then in place of the valence and conduction bands, as an ersatz one may consider the highest occupied molecular orbital (HOMO) and lowest unoccupied molecular orbital (LUMO), respectively. For states of s or p_z symmetry with nonzero weight at $E = E_F + eV$, tunneling to or from the molecule is expected to occur, whereas in situations with zero weight at $E = E_F + eV$ or for orbitals with a higher azimuthal quantum number $m \neq 0$ (i.e., p_x or p_y states), tunneling to the underlying substrate is expected to dominate [8]. The expected visibility or invisibility of a molecule (or part thereof) in STM defined by these criteria has been discussed by Lang for atomic adsorbates in which calculated curves of the difference in state density between metal-adatom systems and bare metal were predicted theoretically [9].

Many typical molecules with aromatic ring systems have HOMO and LUMO levels consisting of π and π^* molecular orbitals. It should be noted that considerable broadening and energetic shifts in the positions of these states are to be expected owing to interaction with the substrate and/or surrounding neighboring molecules. Examples of STM studies where the role of HOMO and LUMO levels have been invoked in the interpretation of images include phthalocyanines [10-14], protoporphirin [15], benzene [16], phenol [17,18], TTF-TCNQ [19] and (BEDT-TTF) charge-transfer salts [20-22].

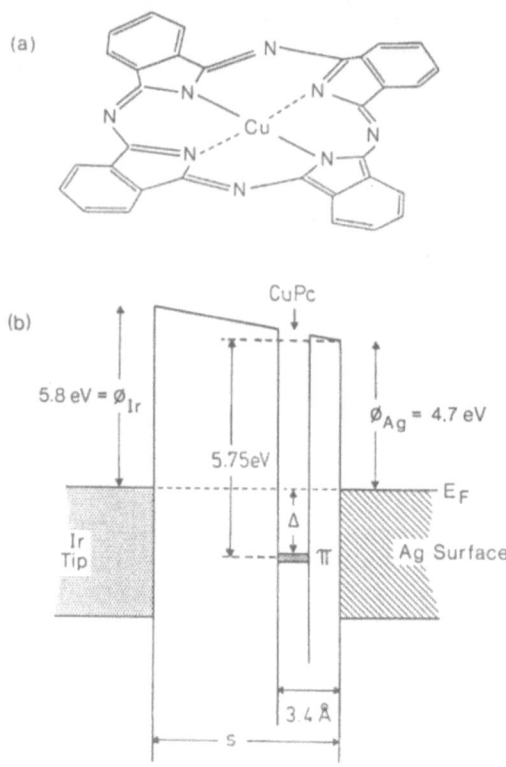

Figure 3. (a) A schematic perspective view of a CuPc molecule. (b) A schematic electric potential diagram for a tunnel barrier consisting of an Ir tip, Ag sample and a CuPc adsorbate bound to the bare Ag surface. Image potential effects are omitted. From [10].

Copper phthalocyanine (CuPc) was one of the first such molecules to be investigated using STM [10]. Figure 3 shows a schematic model of the electric potential energy diagram for the tunnel barrier for an Ir tip, an adsorbed CuPc molecule and an Ag substrate based on a variety of published data. The HOMO is situated on the aromatic ring system, corresponding to a π-electron system, whereas the occupied state of the central copper atom lies significantly deeper in energy and would not be expected to contribute to tunneling from the molecule. Figure 4 shows an early STM topograph of an individual CuPc molecule recorded at a sample bias voltage of − 0.25 eV. The albeit slightly asymmetric, four-lobed structure and central minimum are consistent with the dominant tunneling current arising from the π-electron ring system with no observable contribution from the copper atom. Recent Hückel molecular orbital calculations of the charge densities of the HOMO and LUMO states 2 Å above the molecular plane show a similar pattern [11]. For Cu(100) substrates at near 1 monolayer coverages, high-resolution images of partially ordered layers have been obtained [11], and for GaAs(110) surfaces individual CuPc molecules were resolved as a symmetric structure (see Figure 5) [12,13]. On substrates such as Ag [10,12] or Au [11], it was difficult to obtain stable images, particularly at low voltages.

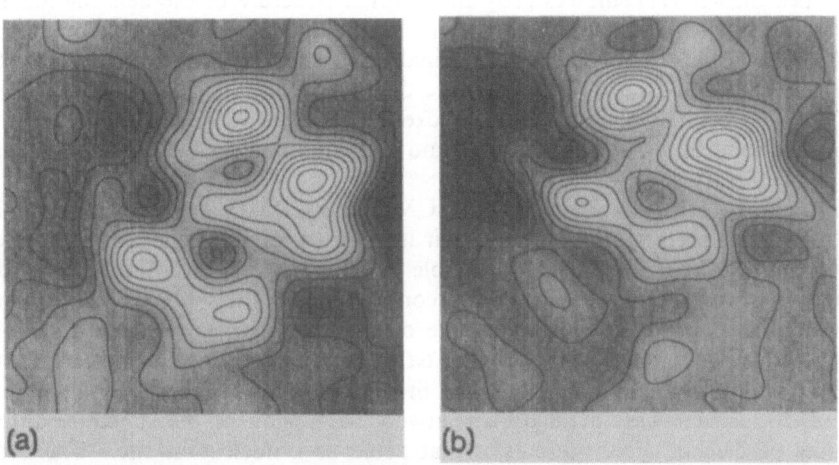

Figure 4.   Grey-scale image with contours of a CuPc molecule adsorbed on polycrystalline Ag recorded at a sample bias of − 0.25 V and a tunnel current of 0.35 nA. The total height of the structure is approx. 3 Å. From [10].

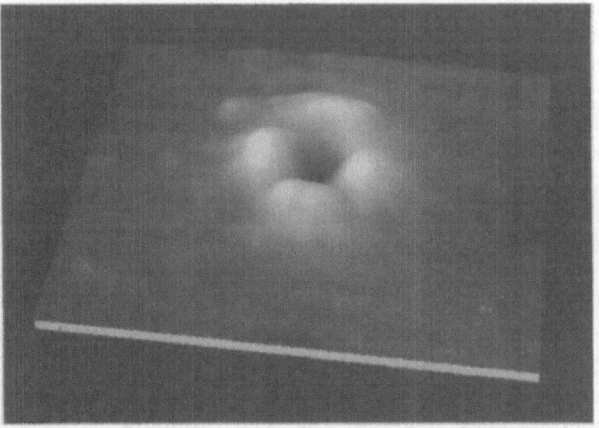

Figure 5.   Pseudo-three-dimensional image of a CuPc molecule deposited onto a GaAs(110) surface using the STM tip, and subsequently imaged at a sample bias of + 0.24 V and a tunnel current of 0.2 nA. From [12].

This effect has been interpreted as evidence of molecular motion, including translation and rotation, induced by the tip, and in certain cases it was possible to actually identify molecular transfer to the tip itself [12,13]. These instabilities appear to be minimized by packing the molecules in an ordered layer, thereby restricting their translational and rotational motion. For (3×3) superlattices of benzene and carbon monoxide co-adsorbed on Rh(111) surfaces [16], the internal structure of the benzene molecular could be resolved, in accord with such steric confinement. Likewise for bulk organic crystals such as TTF-TCNQ, the principal molecular features observed in STM topographs could be simulated from charge density contours of the TTF-HOMO and TCNQ-LUMO orbitals obtained from Hückel calculations [19]. In the last example, the tip was found to dig into the surface, and imaging may involve mechanical contact of tip and surface.

To summarize, Hückel calculations of LUMO and HOMO levels near E_F provide a useful means for interpretation of molecular images. Additional considerations include the symmetry of the orbitals and the role of the tip in perturbing the adsorbed molecule. Also to be included in imaging considerations is the effect of electron traps [23], which can temporarily create positive or negative charges on the surface giving rise to noise in the tunnel current. Figure 6a shows an example of a molecular cluster on polycrystalline Ag exhibiting both well ordered rows and disordered structure. In this case, the cluster was found on a relatively flat region of the substrate although there may be defects under the cluster that acted as a nucleation site. The ordered rows have a lateral periodicity of ~12.7 Å with no strong corrugation along the rows. Other sets of rows with a similar periodicity and a clear corrugation were also observed [12]. The corrugation amplitude is ~2.2 Å between rows. These data were interpreted in terms of the various polymorphic forms that CuPc can assume in the crystalline state. The two basic types, α and β, involve stacking sequences that favor coordination of the central Cu atom with nitrogen atoms on adjacent molecular rings. Coordination in an octahedral geometry with the outer nitrogen atom results in the more stable β-form, with is monoclinic [$a = 19.4$ Å, $b = 4.8$ Å, $c = 14.6$ Å] with a canting angle of the stack of 44.8°. Coordination with an adjacent nitrogen on the inner ring of an adjacent molecule results in the α-phase which is tetragonal [$a = 25.9$ Å, $b = 3.8$ Å, $c = 23.9$ Å] with a canting angle of 26.5°. In the α-form, the molecules are arranged in a series of parallel canting rows, whereas in the β-phase the rows alternate in canting, producing a zigzag set of alternating rows. Our measured corrugation of 12.7 Å is in good agreement with the corrugation periodicity in the a or c-directions of the α-form (12 - 13 Å) but not with the β-form corrugation periodicity (9.5 Å). The results show that a variety of stacking sequences within an individual cluster is observable. For instance, in addition to the packing described above, an apparently amorphous surface topography and a hexagonal pattern with a spacing of ~13 Å are also observed. The latter region is indicated by D in Figure 6b. Figure 6b shows a tunnel current noise map for the cluster of CuPc molecules shown in Figure 6a [12]. The contour lines represent the STM topograph. The white patches, particularly in area D, are of atomic dimensions in a partially ordered three-fold symmetry pattern. These sites are associated with electron trapping in the center of CuPc molecules.

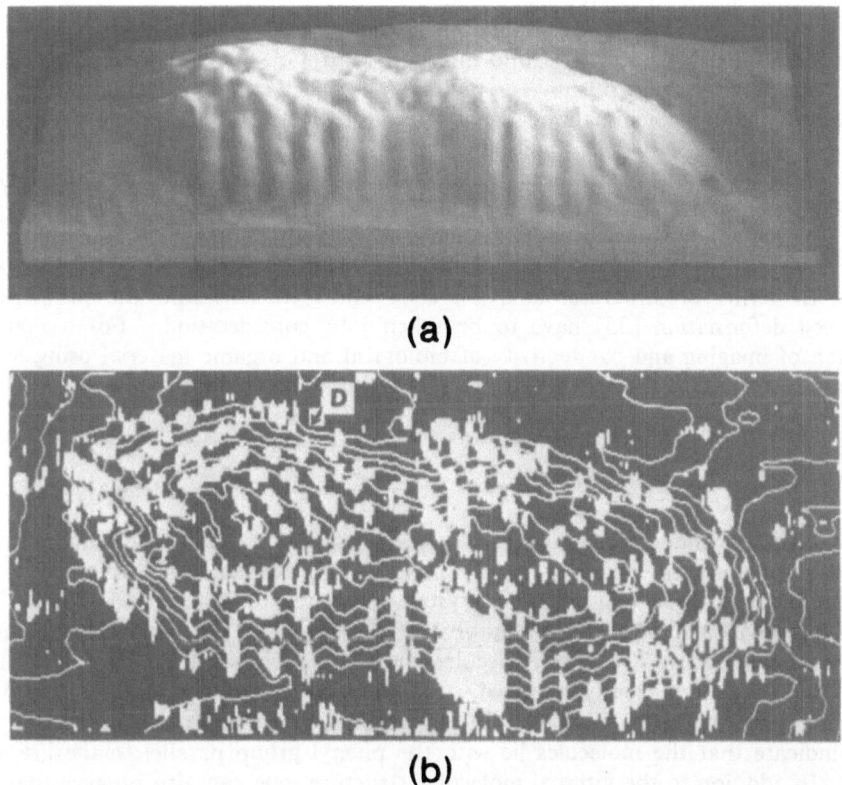

(a)

(b)

Figure 6. (a) Pseudo-illuminated three-dimensional image of a molecular cluster of CuPc on Ag. Area is 315×120 Å^2. (b) Tunnel current noise map for cluster shown above. Contour lines are STM topography. Both images were recorded at a sample bias of $+0.7\,V$. From [12].

An increase in the noise characteristics during molecular imaging has been noted for a wide variety of molecular systems and provokes further questions, especially as regards the role of trapping and hopping electron transfer processes.

3. Molecular Imaging of More Localized Electron Systems and *Insulators*

During the evolution of STM, researchers have investigated molecular and biological systems which, if judged by the predicted availability of bulk conduction states near E_F, would have been deemed unimaginable. Although some STM studies of biological samples used a conductive coating (e.g., recA-DNA [24] and HPI layers [25]) to obtain sufficient surface electrical conductivity, many STM investigations are conducted on apparently uncoated samples with similar results [26-33]. A variety of problems occur in the interpretation of such images, for example, the nature of the electrical conductivity and sufficient fixation of the molecule to the surface. Similar to the considerations described for CuPc molecules, tip-molecule interaction appear to play an important role in pushing or pulling the object under investigation. For a discussion of these problems, the reader is referred to Refs. 31, 32 and 34. A striking example of imaging *insulating* molecules is the recent observation of thick alkane crystals (n-heptadecane, n-octadecane and n-hexatriacontane) adsorbed on HOPG graphite and gold imaged in air [34]. The STM topographs obtained clearly suggest that alternative conduction pathways for electron transfer other than tunneling occur. These processes may include intrinsic, extrinsic, atmospheric and defect-induced surface states. In terms of structural analysis, such considerations and the problem of tip-induced deformation [35] have to be taken into consideration. For a detailed discussion of imaging and conductivity of biological and organic material using STM, the reader is referred to a recent review by Travaglini *et al.* in [36].

4. Liquid-Crystal Molecular Films

A current subject that has stimulated much interest in the STM community is the high-resolution images obtained from ordered liquid-crystal molecules such as alkylcyanobiphenyls [37-41]. Liquid crystal molecules tend to self-assemble into ordered phases. Figure 7 shows a beautiful STM image of a liquid-crystal monolayer structure studied in air [38]. The molecules consist of a biphenyl head group with an alkyl tail. Both are clearly resolved, the biphenyl group appears as a double "doughnut" structure in this picture, with the alkyl tail being of weaker intensity. The results indicate that the molecules lie with the phenyl group parallel to the graphite surface. In addition to the internal molecular structure, one can also observe that the molecules lie head to head and in repetitive groups of four followed by a phase shift. Interestingly, if thick layers of the crystal are deposited onto HOPG graphite, then the tip moves through this phase as if it were electronically invisible and imaging occurs only directly to the first ordered molecular layer on the surface of the substrate [38]. This observation indicates that the substrate plays an important role in both stabilizing the molecule and modifying its electronic structure.

In most of the cases discussed above, additional information on the dependence of the tunnel current on the bias voltage could not be obtained since stable images could only be recorded within a narrow range of tunnel voltages. The dependence of the tunnel current on the gap separation has probably provided the most interesting information. For clean metal surfaces under UHV condition, the apparent tunnel barrier height, ϕ_{app}, defined as d ln I/ds, where I is the tunnel current and s the gap separation, is of the order of 3 - 4 eV, except at very small gap spacing ($s < 3$ Å) [42,43] In many STM studies in air or liquids, ϕ_{app} may be as low as several meV with values of ~ 1 eV in certain experiments. The low values of ϕ_{app} cannot be simply explained in terms of tunneling theory, and give further evidence that additional or alternative mechanisms involving atomic forces [44] and/or electron-transfer mechanisms come into play.

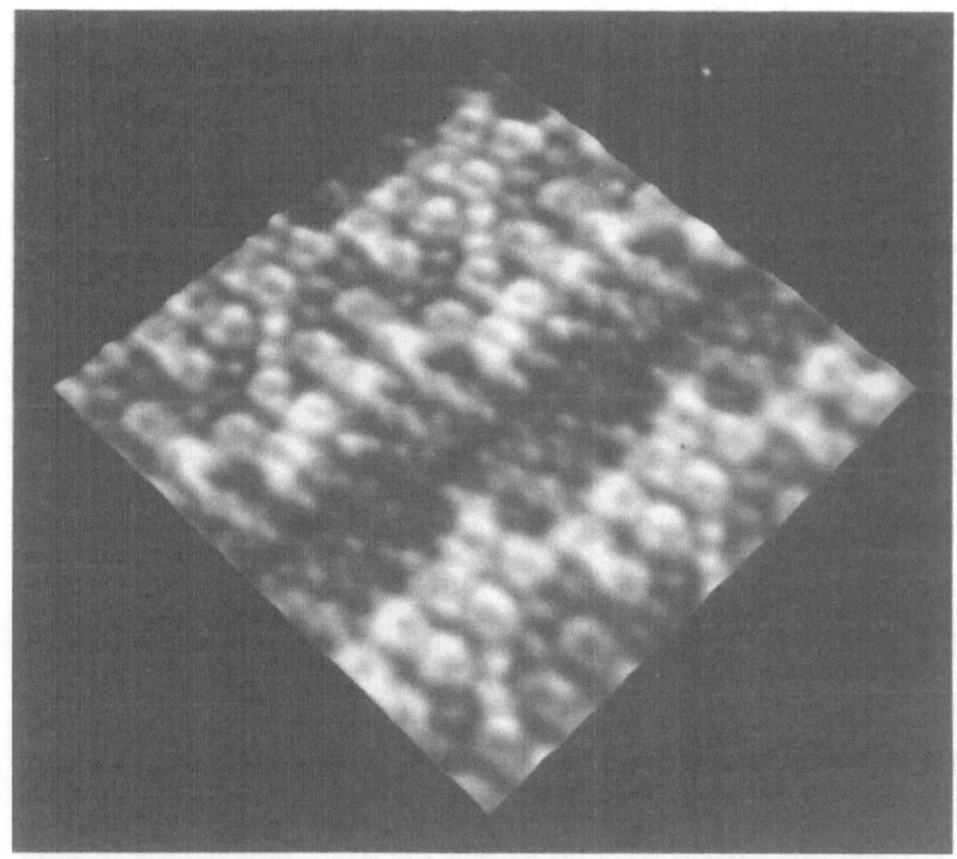

Figure 7. STM image of 8CB-cyanobiphenyl molecules adsorbed on graphite and imaged in air. Images are 114×114 Å^2 recorded at a sample bias of − 0.8 V and a tunnel current of 0.1 nA. Taken from [38], courtesy of Ch. Gerber.

5. Concluding Remarks

Little theoretical progress in our understanding of tunneling through organic material has been achieved to date. However, in complete contrast, the variety of molecular and biological images obtained with STM that bear an obvious and clear similarity to what is known of their structure is evidence of the potential of the technique. A similar state of affairs is now emerging in the related field of atomic force microscopy (AFM) reviewed elsewhere [45]. The understanding of mechanisms for electron-transfer processes in complex molecular systems may result from further experimentation rather than extended theoretical speculation and, in analogy to the development of STM applied to metal and semiconductor surfaces, result in further refinement of the method for surface structural analysis in molecular and biological systems. In addition to our ability to obtain structural information on molecular systems, it is also important to note that the mechanisms for electron transfer under spatially confined conditions of atomic dimensions may ultimately be of greater importance.

Acknowledgments

I wish to thank my colleagues in Rüschlikon for many illuminating discussions, in particular B. Michel, Ch. Gerber and H. Rohrer. R.R. Schlittler is also acknowledged for his expertise in many experiments. J.K. Sass provided a comprehensive insight into alternative methods of electron transfer.

References

[1] Gimzewski, J.K., Berndt, R., and Schlittler, R.R. (1990) 'Observation of mass transport on Au(110)-(1×2) reconstructed surfaces using scanning tunneling microscopy', *Proc. 14th Int'l Seminar on Surface Physics*, Przesieka, Poland, May 1990, in: *Surface Sci.* (in press)

[2] Tersoff, J., and Hamann, D.R. (1985) 'Theory of the scanning tunneling microscope', *Phys. Rev. B* **31**, 805.

[3] Demuth, J.E., Koehler, U., and Hamers, R.J. (1988) 'The STM learning curve and where it may take us', *J. Microsc.* **152**, Pt. 2, 299.

[4] Binnig, G., Rohrer, H., Gerber, Ch., and Weibel, E. (1983) '7×7 reconstruction on Si(111) resolved in real space', *Phys. Rev. Lett.* **50**, 120.

[5] Takayanagi, K., Tanishiro, T., Takahashi M., and Takahashi, S. (1985) 'Structural analysis of Si(111)-7×7: UHV-transmission electron diffraction and microscopy', *J. Vac. Sci. Technol.* A **3**, 1502.

[6] Hamers, R.J., Tromp, R.M., and Demuth, J.E. (1987) 'Electronic and geometric structures of Si(111)-(7×7) and Si(001) surfaces', *Surface Sci.* **181**, 346.

[7] Berghaus, Th., Brodde, A., Neddermeyer, H., and Tosch, St. (1988) 'On the interpretation of current images in scanning tunneling spectroscopy of the Si(111)7×7', *J. Vac. Sci. Technol.* A **6**, 483.

[8] Persson, B.N.J., and Demuth, J.E. (1986) 'Inlastic electron tunneling from a metal tip', *Solid State Commun.* **57**, 769.

[9] Lang, N.D. (1986) 'Spectroscopy of single atoms in the scanning tunneling microscope', *Phys. Rev. B* **34**, 5947.

[10] Gimzewski, J.K., Stoll, E., and Schlittler, R.R. (1987) 'Scanning tunneling microscopy of individual molecules of copper phthalocyanine adsorbed on polycrystalline silver surfaces', *Surface Sci.* **181**, 267.

[11] Lippel, P.H., Wilson, R.J., Miller, M.D., Wöll, Ch., and Chiang, S. (1989) 'High-resolution imaging of copper-phthalocyanine by scanning-tunneling microscopy', *Phys. Rev. Lett.* **62**, 171.

[12] Gimzewski, J.K., Coombs, J.H., Möller, R., and Schlittler, R. (1989) 'Scanning tunneling microscopy of molecular clusters of copper phthalocyanine adsorbed

on silver surfaces', in *Molecular Electronics—Science and Technology*, Engineering Foundation, New York, p. 87.

[13] Möller, R., Coenen, R., Esslinger, A., and Koslowski, B. (1990) 'The topography of isolated molecules of copper-phthalocyanine adsorbed on GaAs(110)', *J. Vac. Sci. Technol.* **A 8**, 659.

[14] Mizutani, W., Sakakibara, Y., Ono, M., Tanishima, S., Ohno, K., and Toshima, N. (1989) 'Measurements of copper phthalocyanine ultrathin films by scanning tunneling microscopy and spectroscopy', *Jpn. J. Appl. Phys.* **28**, L-1460.

[15] Coombs, J.H., Pethica, J.B., and Welland, M.E. (1988) 'Scanning tunneling microscopy of thin organic films', *Thin Solid Films* **159**, 293.

[16] Ohtani, H., Wilson, R.J., Chiang, S., and Mate, C.M. (1988) 'Scanning tunneling microscopy observation of benzene molecules on the Rh(111)-(3×3) (C_6H_6 + 2CO) surface', *Phys. Rev. Lett.* **60**, 2398.

[17] Sakamaki, K., Itoh, K., Fujishima, A., and Gohshi, Y. (1990) 'Surface density of states of TiO_2 (110) single crystal and adsorbed molecular observation by scanning tunneling microscopy and tunneling spectroscopy', *J. Vac. Sci. Technol.* **A 8**, 614.

[18] Sakamaki, K., Matsunaga, S., Itoh, K., Fujishima, A. and Gohshi, Y. (1989) 'Imaging the phenol molecule adsorbed on TiO_2(110) by scanning tunneling microscopy' *Surface Sci.* **219**, L531.

[19] Sleator, T., and Tycko, R. (1988) 'Observation of individual organic molecules at a crystal surface with use of a scanning tunneling microscope' *Phys. Rev. Lett.* **60**, 1418.

[20] Bando, S., Kashiwaya, S., Tokumoto, H., Anzai, H., Kinoshita, N., and Kajimura, K. (1990) 'Tunneling spectroscopy on an organic superconductor $(BEDT-TTF)_2Cu(NCS)_2$', *J. Vac. Sci. Technol.* **A 8**, 479.

[21] Yoshimura, M., Fujita, K., Ara, N., Kageshima, M., Shioda, R., Kawazu, A., Shigekawa, H., and Hyodo, S. (1990) 'Observation of an organic superconductor [bis(ethylenedithio)-tetrahiafulvalene]$_2$[Cu(NCS)$_2$]' by scanning tunneling microscopy', *J. Vac. Sci. Technol.* **A 8**, 488.

[22] Bai, C., Dai, C., Zhu, C., Chen, Z., Huang, G., Wu, X., Zhu, D., and Baldeschwieler, J.D. (1990) 'Scanning tunneling microscopy of silver containing salt of bis(ethylenedithio)tetrahiafulvalene', *J. Vac. Sci. Technol.* **A 8**, 484.

[23] Welland, M.E., and Koch, R.H. (1986) 'Spatial location of electron trapping defects on silicon by scanning tunneling microscopy', *Appl. Phys. Lett.* **48**, 724.

[24] Amrein, M., Stasiak, A., Gross, H., Stoll, E., and Travaglini, G. (1988) 'Scanning tunneling microscopy of recA-DNA complexes coated with a conducting film', *Science* **240**, 514.

[25] Michel, B., and Travaglini, G. (1988) 'An STM for biological applications: bioscope', *J. Microsc.* **152**, Pt. 3, 681.

[26] Amrein, M., Dürr, R., Stasiak, A., Gross, H., and Travaglini, G. (1989) 'Scanning tunneling microscopy of uncoated recA-DNA complexes', *Science* **243**, 1708.

[27] Keller, R.W., Dunlap, D.D., Bustamante, C., Keller, D.J., Garcia, R.G., Gray, C., and Maestre, M.F. (1990) 'Scanning tunneling microscopy images of metal-coated bacteriophages and uncoated, double-stranded DNA', *J. Vac. Sci. Technol.* **A 8**, 706.

[28] Bendixen, C., Besenbacher, F., Lægsgaard, E., Stensgaard, I., Thomsen, B., and Westergaard, O. (1990) 'Deoxyribonucleic acid structures visualized by scanning tunneling microscopy', *J. Vac. Sci. Technol.* **A 8**, 703.

[29] Miles, M.J., McMaster, T., Carr, H.J., Tatham, A.S., Shewry, P.R., Field, J.M., Belton, P.S., Jeenes, D., Hanley, B., Whittam, M., Cairns, P., Morris, V.J., and Lambert, N. (1990) 'Scanning tunneling microscopy of biomolecules', *J. Vac. Sci. Technol.* **A 8**, 698.

[30] Jericho, M.H., Blackford, B.L., Dahn, D.C., Frame, C., and Maclean, D. (1990) 'Scanning tunneling microscopy imaging of uncoated biological material', *J. Vac. Sci. Technol.* **A 8**, 661.

[31] Salmeron, M., Beebe, T., Odriozola, J., Wilson, T., Ogletree, D.F., and Siekhaus, W. (1990) 'Imaging of biomolecules with the scanning tunneling microscope: problems and prospects', *J. Vac. Sci. Technol.* **A 8**, 635.

[32] Lindsay, S.M., and Barris, B. (1988) 'Imaging deoxyribose nucleic acid molecules on a metal surface under water by scanning tunneling microscopy', *J. Vac. Sci. Technol.* **A 6**, 544.

[33] Elings, V.B., Edstrom, R.D., Meinke, M.H., Yang, X., Yang, R., and Evans, D.F. (1990) 'Direct observation of enzymes and their complexes by scanning tunneling microscopy', *J. Vac. Sci. Technol.* **A 8**, 652.

[34] Michel, B., Travaglini, G., Rohrer, H., Joachim, C., and Amrein, M. (1989) 'Images of crystalline alkanes obtained with scanning tunneling microscopy', *Z. Phys. B—Cond. Matter* **76**, 99.

[35] Lindsay, S.M., Thundat, T., and Nagahara, L. (1988) 'Adsorbate deformation as a contrast mechanism in STM images of biopolymers in an aqueous environment: images of the unstained, hydrated DNA double helix', *J. Microsc.* **152**, Pt. 1, 213.

[36] Travaglini, G., Amrein, M., Michel, B., and Gross, H. (1990) 'Imaging and conductivity of biological and organic material', in *Basic Concepts and Applications of Scanning Tunneling Microscopy (STM) and Related Techniques*, Proc. NATO Meeting, Erice, Italy, April 1989, Kluwer Academic Publishers, Dordrecht (in press)

[37] Foster, J.S., and Frommer, J.E. (1988) 'Imaging of liquid crystals using a tunnelling microscope', *Nature* **333**, 542.

[38] Smith, D.P.E., Hörber, H., Gerber, Ch., and Binnig, G. (1989) 'Smectic liquid crystal monolayers on graphite observed by scanning tunneling microscopy', *Science* **245**, 43.

[39] McMaster, T.J., Carr, H., Miles, M.J., Cairns, P., and Morris, V.J. (1990) 'Adsorption of liquid crystals imaged using scanning tunneling microscopy', *J. Vac. Sci. Technol.* **A 8**, 672.

[40] Mizutani, W., Shigeno, M., Sakakibara, Y., Kajimura, K., Ono, M., Tanishima, S., Ohno, K., and Toshima, N. (1990) 'Scanning tunneling spectroscopy study of adsorbed molecules', *J. Vac. Sci. Technol.* **A 8**, 675.

[41] Rabe, J.P., Buchholz, S., and Ritcey, A.M. (1990) 'Reactive graphite etch and the structure of an adsorbed organic monolayer—a scanning tunneling microscopy study', *J. Vac. Sci. Technol.* **A 8**, 679.

[42] Gimzewski, J.K., Möller, R., Pohl, D.W., and Schlittler, R.R. (1987) 'Transition from tunneling to point contact investigated by scanning tunneling microscopy and spectroscopy', *Surface Sci.* **189/190**, 15.
Gimzewski, J.K., and Möller, R. (1987) 'Transition from the tunneling regime to point contact studies using scanning tunneling microscopy', *Phys. Rev. B* **36**, 1284.

[43] Lang, N.D. (1987) 'The resistance of a one-atom contact in the scanning tunneling microscope', *Phys. Rev. B* **36**, 8173.
Ferrer, J., Rodero, A.M., and Flores, F. (1988) 'Contact resistance in the scanning tunneling microscope at very small distances', *Phys. Rev. B* **38**, 10113.

[44] Dürig, U., Gimzewski, J.K., and Pohl, D.W. (1986) 'Experimental observation of forces acting during scanning tunneling microscopy', *Phys. Rev. Lett.* **57**, 2043.

[45] For a review see Drake, B., Prater, C.B., Weisenhorn, A.L., Gould, S.A.C., Albrecht, T.R., Quate, C.F., Cannell, D.S., Hansma, H.G., and Hansma, P.K. (1989) 'Imaging crystals, polymers, and processes in water with the atomic force microscope', *Science* **243**, 1586.

POLYMER DECORATION: RECENT DEVELOPMENTS

B. LOTZ and J.C. WITTMANN
Institut Charles Sadron (CNRS-ULP)
6 rue Boussingault
67083 Strasbourg
France

ABSTRACT. The main characteristics of the "polymer decoration" based on the vaporization and condensation-crystallization of polymers (notably polyethylene) are presented. For a number of crystalline substrates, the decoration rests on epitaxial relationships. The variability in decoration patterns produced on the end (fold) surface of polymer crystals is analyzed. Recent significant contributions of the polymer decoration to the elucidation of the structure of crystals of polyethylene and its low M.w. paraffin analogs, either linear or cyclic, are reviewed.

1. Introduction

Whereas transmission electron microscopy (TEM) has lateral, i.e. two dimensional resolution approaching the angstroem, it has inherent limitations as regards the third dimension, parallel to the electron beam. TEM is not therefore well adapted for surface structure and topography investigations. A most classical means to overcome these limitations is to use shadowing techniques which however result in significant loss of resolution.

The depth perception on surfaces by TEM techniques can only be indirect, i.e. rely on the use of an extra component with specific interactions with the surface. Along this line, a significant contribution has been the introduction of the gold decoration technique by Bassett (1958). It rests on the vaporization under vacuum (ca. 10^{-3} torr) of gold. Upon condensation on the substrate, gold vapors form droplets which can be seen to move on the surface, and become anchored at topographic features, such as cleavage steps. Although indirect, the technique is quite powerful : steps of atomic dimensions have been revealed on cleaved NaCl crystals. In later experiments, (Bassett et al, (1967), Blundell and Keller (1968)), the fold surface of polyethylene (PE) single crystals has been gold decorated. The decoration density was found to vary with the crystallization temperature and subsequent physical treatments (washing with solvents) but the origin of these differences could not be determined.

Gold particles are isotropic, and thus cannot reveal orientational features, except through their prefered alignment. A few years ago however, a decoration technique similar to gold decoration in its principle and experimental set-up has been introduced (Wittmann and Lotz (1985)) which, contrary to gold atoms, uses highly anisometric markers, namely polymer fragments, typically 100 Å long by 5 Å in diameter (PE fragments of Mw ~1300). As a consequence the resulting decoration pattern uses a structural marker which is not the density and alignment of decorating units, but rather the orientation of the units, which is a very sensitive means to analyze the surface structure and, under favourable circomstances, the surface topography.

Polymer decoration has been used first to investigate the structure of polymer crystals, and notably the problem of fold structure and orientation (Wittmann and Lotz, (1985)). It has found however wider uses since decoration patterns may originate from specific polymer-substrate interactions. The present paper, after a brief presentation of the specificities of the technique, deals mainly with two aspects :

217

J. R. Fryer and D. L. Dorset (eds.), Electron Crystallography of Organic Molecules, 217–226.
© 1990 *Kluwer Academic Publishers.*

(a) the analysis of decoration patterns resulting from specific polymer- crystalline substrate epitaxies. These are most vividly illustrated for substrates displaying differences in structure such as (i) phyllosilicates (talc and mica) in which abrupt changes of the decoration orientation reveal the cleavage steps of the layers and (ii) a ferroelectric crystal of triglycine sulphate in which different epitaxial relationships on the plus and minus end faces make it possible to locate differently charged domains on a cleavage surface.

(b) the ability to reveal the fold orientation and, to some extent, differences in fold structure of polymer crystals. This is a significant feature of the polymer decoration technique. It allows insight into the structure of a thin (ca 20 Å) and relatively disordered layer which has a profound influence on the structure and physical properties of the polymer as a whole, but for the investigation of which only few experimental techniques are available. Drawing on recent works from various laboratories, several contributions of the decoration technique will be reviewed. They deal with the sectorization of polymer crystals, the fold structure and organization in paraffins and their blends, in cyclic paraffins, and with the fold structure in relation to chain tilt in polymer lamellae, ultimately leading to lamellar twist in spherulitic crystallization from the bulk.

2. The polymer decoration technique

The vaporization of polymers - typically polyethylene (PE) in our experiments - is achieved by heating to 300-350°C of ~2 mg of material in a tungsten wire basket. High vacuum (~10^{-4} Torr) is required as under primary vacuum the polymer "burns out". The polymer vapors are deposited under normal incidence on the substrate, distant by ca. 10 cm. The decoration pattern thus

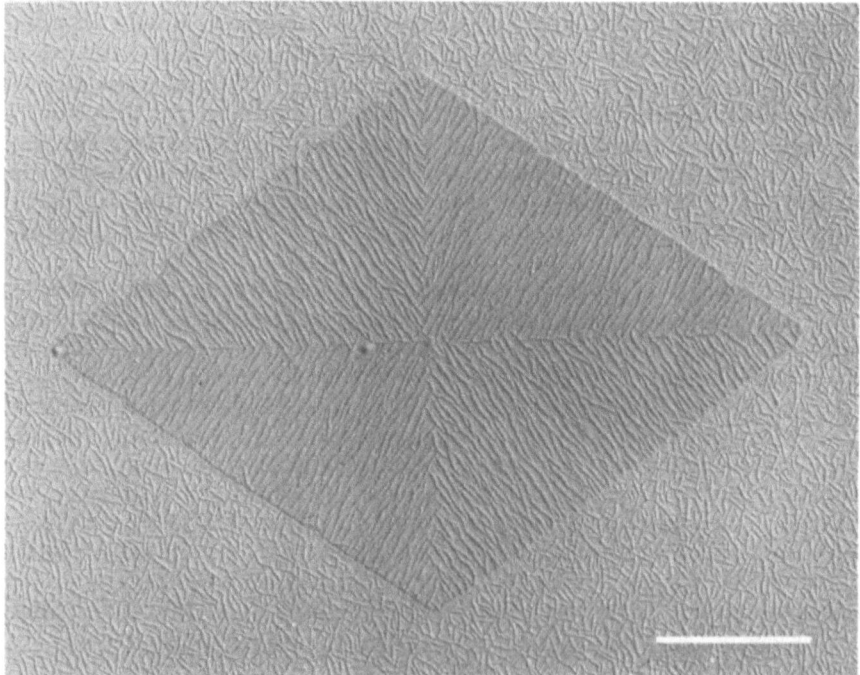

Figure 1. Single crystal of PE (Mw = 6750) grown at 70° from a 0.1% solution in 4/1 p-xylene tetrachloroethylene and decorated with PE vapors ; Pt-C shadowing at tg^{-1} 1/3. Scale bar : 2 μm

produced is usually enhanced by Pt-C shadowing which however introduces an extra orientation component, as it highlights decoration oriented at right angles to the shadowing direction.

The structure and organization of the polymer deposit is best illustrated with a PE single crystal substrate (Fig.1). Whereas rapid crystallization of PE vapors on the cold, amorphous glass substrate yields a random pattern of elongated particles, a regular array of rods is created on the four growth sectors of a PE single crystal. Each rod is actually a thin strip of a polymer lamella seen edge-on, i.e. chains (mostly extended given their low mass (1300) and limited length (100 Å)) are oriented at right angles to the rod direction and parallel to the substrate surface. The crystallographic plane of the decoration in contact with the substrate has been found to vary, depending on the substrate. It is not a major issue in the polymer decoration techique.

Figure 1 highlights the main features of the decoration pattern and the essential differences with gold decoration : (i) the importance of the highly sensitive orientation criterion linked with the anisometry of the polymer marker (compare the orientation of rods in the two pairs of growth sectors) (ii) the polymer decoration extends over the whole surface of each domain, whereas gold decoration only reveals domain edges (e.g. cleavage steps).

Polymer decoration can be performed with a number of polymers. The most convenient one is polyethylene, but other polyolefins (isotactic polypropylene and poly-1-butene) and polyvinylidene fluoride (PVF_2) have been used. Polyamides, polyesters, etc... or e.g. poly (methylene oxide) cannot be used as decorating materials as on heating they give rise to monomeric (gaseous) degradation products. However, these polymers can be decorated by e.g. PE vapors : substrate and decorating polymers need not be identical, i.e. "heterodecoration" is possible (Wittmann and Lotz (1985)).

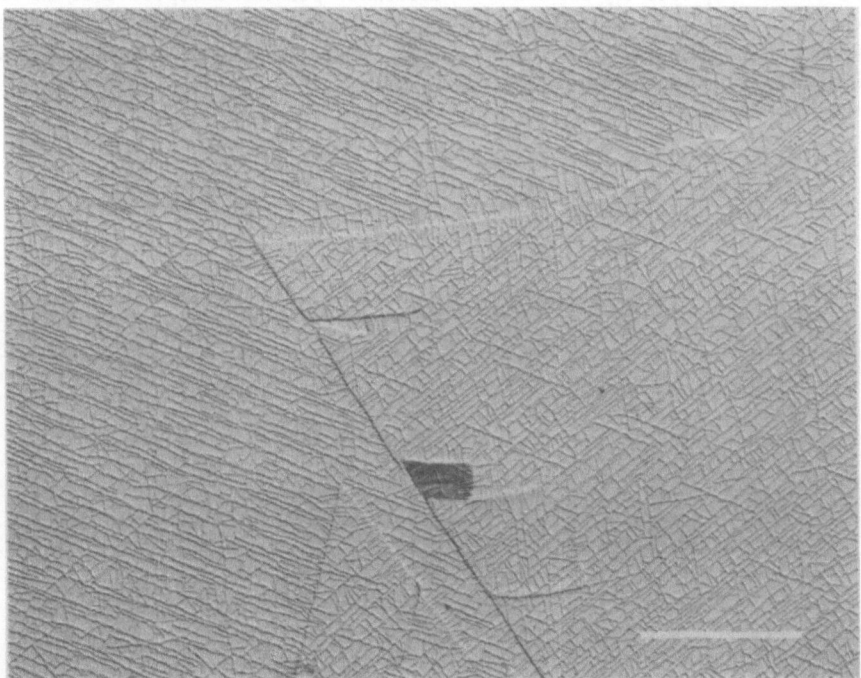

Figure 2. Cleavage step on a cleaved surface of muscovite mica decorated with PE vapors. Pt-C shadowing. Scale bar : 2 μm.

3. Polymer decoration based on epitaxial relationships

Crystallization of PE vapors at high undercooling on a crystallographic substrate is mostly governed by epitaxy. For a substrate with domains, polymer decoration can reveal domains with different structure or with same structure but different orientations. Most illustrative examples relate to the decoration of (1) talc and mica, and (2) ferroelectric domains in a crystal of triglycine sulphate (TGS).

3.1. TALC AND MICA

Talc and mica are phyllosilicates often used as structural or nucleating additives in PE. They have a layered structure made of two hexagonal arrays of silica (SiO_4 tetahedra) that sandwich a layer of brucite. Mica differs from talc mainly by the presence on the outer face of silica, of K^+ ions (Radoslovich (1960)).

Polyethylene decoration of talc flakes (Wittmann and Lotz (1986)) has revealed that, when freshly cleaved, the talc particles induce a single orientation of the PE rods, in spite of the nearly hexagonal arrangement of the surface silica tetrahedra. Cleavage steps are therefore easily revealed by 120° changes in the PE rod orientation.

A similar behaviour is observed for muscovite mica cleavage surfaces (Figure 2). Results are however less reproducible in terms of density and sometimes orientation of rods, etc ... This variability is attributed to the presence of K^+ ions on the cleavage surface. Indeed, washing of the cleavage surface with acids prior to PE decoration results in a considerably higher density of PE rods and a much improved reproducibility of decoration patterns. Irrespective of the last mentioned features, polymer decoration via epitaxy is sensitive enough to discern very small asymmetries of the surface structure : in the present case, slight displacements of the oxygens that built the silica layer of talc and mica (Radoslovich (1960)).

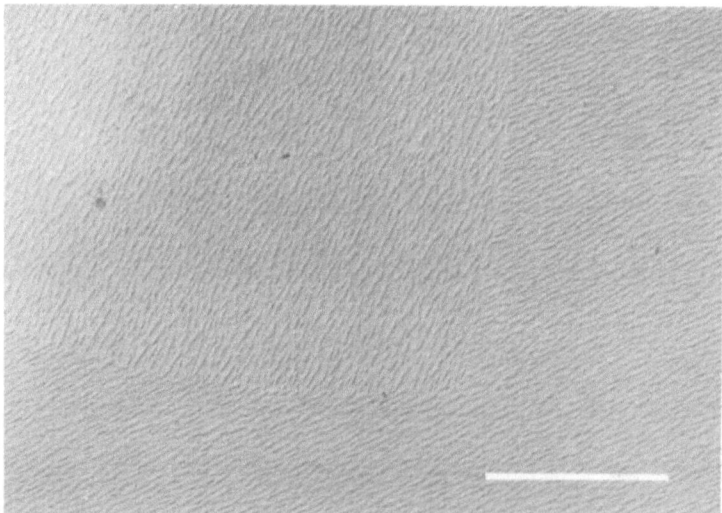

Figure 3. Tip of lenticular domain in the (010) cleavage plane of a TGS crystal. Note the different rod orientation of the polyvinylidene fluoride decorating rods. Scale bar : 2 μm.

3.2. TRIGLYCINE SULPHATE CRYSTALS

The sensitivity just mentioned is also strinkingly demonstrated with TGS. substrates. This ferroelectric compound has a net dipole moment oriented parallel to its b crystallographic axis. Its crystals are spontaneously organized in domains (domain walls parallel to $\underline{b}$) with opposite polarizabilities. (010) cleavage faces are therefore inhomogeneous, being made of domains with net positive or negative charges. On a structural basis however the difference is quite minute, as it rests on a small reorientation of the NH_3^+ ionic part of one only of the three glycine units, while keeping the same unit-cell orientation throughout the cleavage surface.

Decoration of the cleavage surfaces with PVF_2 or PE vapors reveals its domain structure (Wicker et al. (1990)). PVF_2 decoration gives rise to a single orientation of rods in the two domains and does not permit an easy identification of the domain sign (Fig. 3). PE decoration on the contrary gives rise to a single orientation on the positive domain ends but to three (one major, one minor, one very minor) rod orientations on the negative domain ends (the sign of the domains were determined by decorating poled crystals) : PE decoration thus appears as a valuable technique to investigate the domain structure of TGS and, by inference, that of other ferroelectric substrates. Given its very high spatial resolution, it appears of particular value in kinetic investigations, when the domain structure is modified, either spontaneously or when triggered by e.g. applied pressure or electric field, or thermally (on passing from para to ferroelectric phase, i.e. at the Curie transition).

4. Polymer decoration and the structure of chain folded polymer crystals

Polymer decoration has proved to be a very useful technique when investigating the surface structure of polymer lamellae where chain folds (between adjacent stems), loops (between non-adjacent stems) and chain ends are located. As illustrated by Figure 1, these folds, loops or chain ends provide suitable surfaces for the oriented nucleation and growth of polymer vapors. As a rule, the depositing chains become aligned parallel to the underlying folds or loops, i.e. the rods are normal to the local fold direction. Polymer decoration reveals fold orientation with a high spatial resolution (~100 Å) and uses a marker which interacts only with the outermost surface of the layer, mainly through van der Waals forces. However, the decoration patterns needs to be properly interpreted, i.e. "calibrated" with a wide range of model substrates for which the fold structure is well established or can be reasonably infered. We present first some guidelines for this interpretation before turning to specific examples taken from various recent works.

4.1. FOLD SURFACE STRUCTURE AND DECORATION PATTERN

4.1.1 *Sharp folds versus long loops.* A major issue in the structure of crystalline polymers is that of the existence of sharp folds and/or longer loops, and their relative ratio. Whereas sharp folds are known to exist (cf. cyclic paraffins, below), the existence of longer loops is more difficult to assess. Polymer decoration can, under favorable circumstances, help reveal their existence.

The argument is summarized in Figure 4 which represents most chain tilts observed in PE lamellae under different growth conditions, together with the arrays of stem emergence points on the fold plane. It is clear that, whatever the tilt angle, folds between adjacent stems are oriented at $120° + 10°$ to each other (directions defined by the intersection of the densely packed (110), (110) and (100) planes with the fold surface). For any regular growth pattern based on these growth planes, polymer decoration does not and cannot discriminate between different substrate

222

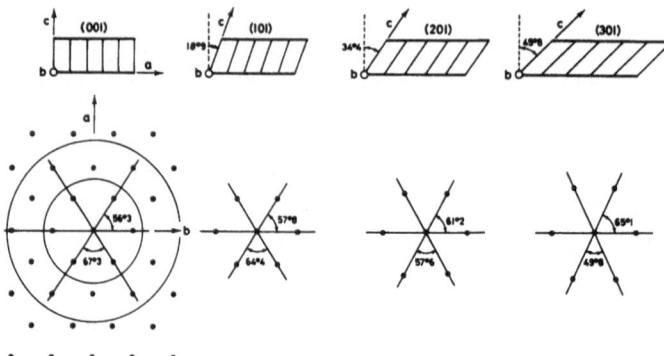

Figure 4. The main stem tilts and corresponding fold surfaces observed in PE crystals (top). Corresponding arrays of stem emergence points on the fold surface, with relative orientation of the three possible tight folds between adjacent stems (bottom).

structures, e.g. made either of several sharp folds or of a long loop, or even their mixture, since they are oriented in the same direction.

When however irregular growth contours are created (as for example in thin film growth in the presence of a paraffinic diluent), the orientation of long loops may deviate from that of sharp folds. Decoration of lamellar crystals with irregular or lenticular contours, and using PEs with a range of different M.w. indicates that:

- for M.w. > 6000, decoration patterns indicative of the presence of long loops are indeed observed (Fig. 5).

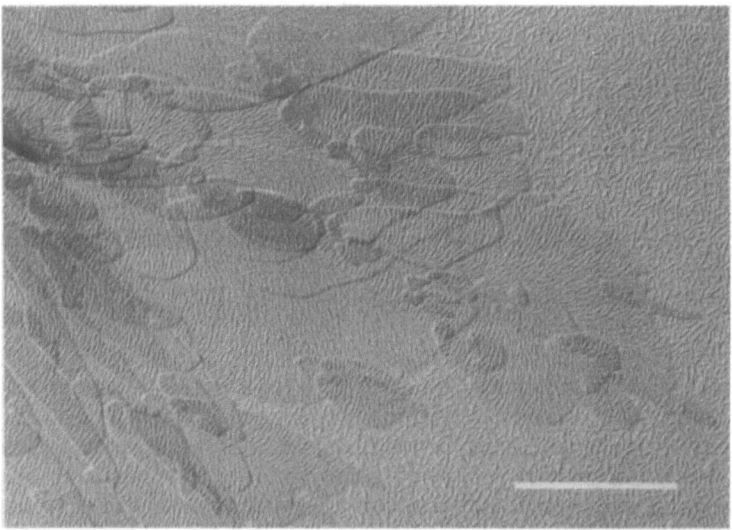

Figure 5. : Polymer decoration of a PE single crystal with irregular contours. Note the arcing of decoration in some growth sectors. Scale bar : 2 μm.

- for M.w. < 6000, a decoration pattern with a mixture of the three orientations defined above is often observed. While stressing the increased importance of sharp folds for low M.w. PEs, this decoration pattern requires further analysis, as examined now.

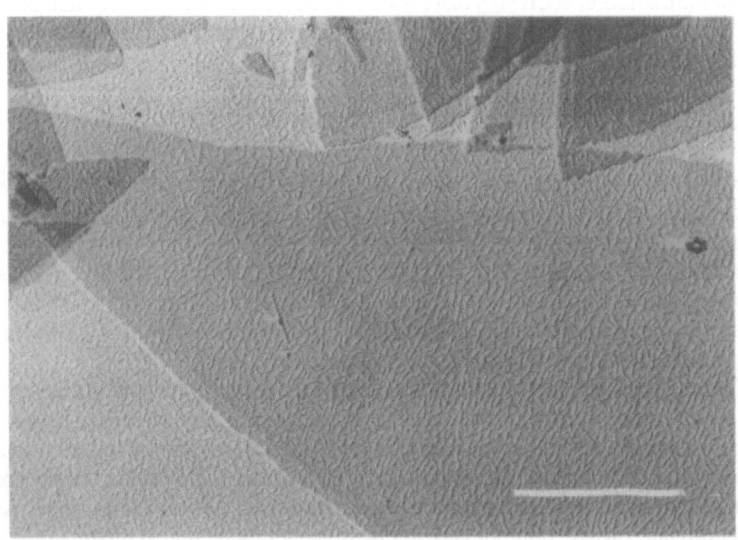

Figure 6. Tip of a large PE single crystal grown in thin film and displaying a PE decoration pattern based on three rod orientations. Scale bar:: 2 µm.

4.1.2 *Decoration patterns based on the mixture of three rod orientations.*. Such patterns are observed, in addition to the above samples, for paraffins and their mixtures. Several surface or structural features may induce these orientations. So far, epitaxy, growth microsectors and surface roughness of various origins have been identified:

- epitaxy on the crystalline array of paraffin chain ends, as observed for $nC_{36}H_{74}$ single crystals, has been analyzed in detail. In this epitaxy, rod orientation normal to (100) is virtually absent

- existence of growth microsectors on macroscopic growth faces clearly disturbs and complicates the decoration pattern, as best illustrated in reentrant growth sectors of twins (Wittmann and Lotz (1985)).

- surface roughness creates additional nucleation sites. For example, in blends of paraffins with different chain length, segregation in clusters of the longer chains is likely to produce cliffs of the chain extremities over the shorter constituant end surfaces, which are likely nucleation sites for the decorating vapors.

These decoration patterns need therefore be interpreted by taking into account the polymer molecular characteristics and growth conditions. Conversely, any alteration or departure from the regular pattern (enhancement of one rod orientation, etc ...) may prove particularly informative.

4.2. DECORATION OF CYCLIC PARAFFINS

The crystal structure of the cycloparaffin $c(CH_2)_{36}$ is based on two slightly twisted trans zig-zag stems linked by two CH_2 groups on both ends. Both stems and folds are located in the (202) plane of the crystal, which results in a "direction of folding" of the cyclic paraffin parellel to the $\underline{b}$ axis. Polymer decoration of solution-grown crystals of $c(CH_2)_{36}$ displays a unique decoration oriented normal to the $\underline{b}$ axis, as expected from the crystal structure (Ihn et al. (1989)). The decoration pattern thus confirms the orientational influence of an array of crystallographically sharp folds. Furthermore, $c(CH_2)_{60}$ displays the same decoration pattern, thus confirming the inference (not previously verified by X-rays) that the two crystal structures are identical, except for the difference in stem length.

4.3. DECORATION OF NORMAL ALKANES AND LOW M.W. PE

Uniform n-alkanes, containing up to almost 400 CH_2 units are model systems to investigate the onset of chain folded crystallization, the formation of lamellar crystals based on once, twice, three, four times folded molecules, their isothermal reorganization and annealing, etc ... Polymer decoration has been used extensively to investigate the fold structure in these model systems and in low M.w. PEs (Organ and Keller (1987), Khoury (1987), Ungar et al. (1988)). It demonstrates in particular that once-folded solution grown crystals with on the average only one fold for four stem sites displays a well oriented decoration pattern.

In a different study, (Dorset et al. (1990)), monolamellar crystals of linear paraffins up to $C_{82}H_{166}$ were found by diffraction contrast to be sectored although the chains do not undergo folding, an observation that ultimately raises the question of the origin of sectorization in polymer crystals. Polymer decoration yields a pattern with the three above mentioned preferred rod orientations. The third orientation, not normally found for shorter n-alkanes, could be associated with "topographic" surface roughness. Indeed, suggestions of regular bands containing these rods are found : they are due to surface discontinuities associated with the collapse of pyramidal crystals on a flat surface. In this case, polymer decoration helps reveal topographic features in a manner very reminiscent of gold decoration. Analysis of the rod orientation was aided by optical Fourier transform of the decoration pattern, and localization of the selected rod orientation was performed with the aid of Fourier peak filtration (Dorset et al. (1990)).

4.4. DECORATION OF BULK-CRYSTALLIZED PE LAMELLAE

Most PE crystals considered so far are highly symmetrical. PE crystals grown from thin films are highly elongated. They possess inherent asymmetries due to a uniform chain tilt in different growth sectors : in thin film growth, lateral growth is slower on faces which overhang the substrate and thus form a reentrant angle with it. Polymer decoration helps clearly locate the crystal center of these asymmetrical crystals and underlines differences in fold surface structure by different decoration densities and orientations (Keith et al. (1989)) (Fig. 7)..

Differences in fold structure associated with chain tilt have been invoked to explain the regular twisting observed in banded polymer spherulites (Keith and Padden (1984)). Specifically, differences in regularity of folding and associated differences in compressive stresses within the fold surfaces must exist, depending upon whether the folds are formed at acute - or obtuse - angled edges of titled growth faces. As a result, a bending moment is built-in in every growth sector with tilted chains ; for lenticular growth habits in PE bulk crystallization, this bending moment affects each elongated crystal half. Further, as symmetrical situations prevail in the opposite sides (halves) of the elongated lamellae, opposing longitudinal components of the bending moment in the crystal halves cause axial twisting of the whole lamella.

Under the crystallization conditions used, a number of edge-on lamellae are generated on one lateral face of the flat-on lamellae (Keith et al. (1989)). The chains in these edge-on lamellae are parallel to the substrate surface, i.e. the lamellae are actually crystal halves considered above.

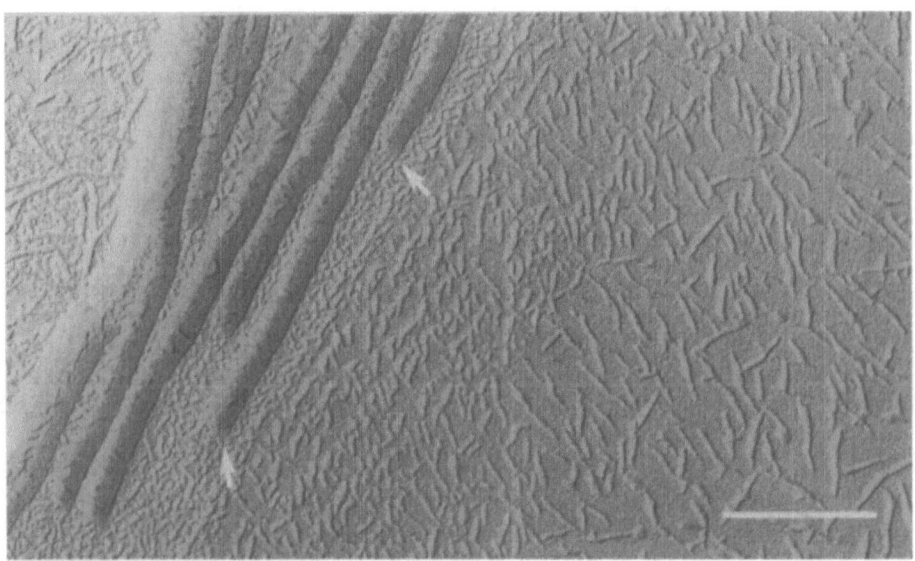

Figure 7. A part of a large PE crystal (T_C = 128°C) with emergence of arced edge-on lamellae (arrowed) and displaying widely different decoration patterns depending on particular stem tilt in the growth sectors. Scale bar : 0.5 μm.

They do indeed exhibit a pronouced regular curvature, of radius ~5 μm. To sum up,the postulated origin of lamellar twisting in polymer spherulites (differences in fold have been experimentally established with the help of the polymer decoration technique and thanks to a favourable growth habit. The combined results yield a considerably improved insight into the micromechanics of chain folded lamellae and spherulitic morphology in melt crystallized polymers.

5. Conclusions and further developments

The vaporization, condensation and oriented crystallization of suitable crystallizable polymers is the basis of an original decoration technique in electron microscopy. The technique has several advantages for surface structure investigations, notably of polymer fold surfaces : it is a non-destructive, local technique based on short range van der Waals interactions, and has a spatial resolution in the 10 nm range. Most importantly, it is based on the use of an anisometric marker and therefore rests on specific orientations of the decoration pattern.

As all decoration techniques however, it remains an indirect technique, i.e. it implies a non-trivial interpretive step of the resulting pattern. Interpretation is fairly straightforward when the decoration rests on specific epitaxies, as illustrated with phyllosilicates or TGS substrates. For decoration of polymer fold surfaces on the opposite, a wide variety of decoration patterns have been observed which underline differences in fold structure. Detailed analysis of that structure from the decoration pattern requires however a "calibration" of the latter, with the help of well characterized model materials (e.g. crystals of cyclic paraffins). Further help may also result from a more quantitative analysis of the pattern, via the Fourier transform of the decoration surface, and from a better control of experimental variables in the vaporization process. Finally, decoration patterns based on oriented crystallization of vapors may be extended to other vaporizable materials. These include other polymers as well as lower Mw materials with anisotropic crystal morphologies (e.g. phthalocyanines) which may help extend the range of substrate temperatures and investigate temperature-dependent structural transitions.

References

Bassett, G.A. (1958) 'A new technique for decoration of cleavage and slip steps on ionic crystal surfaces', Phil. Mag. 3, 1042

Bassett, G.A., Blundell, D.J. and Keller, A. (1967) 'Surface structure of polyethylene crystals as revealed by surface decoration. I. Preliminary survey', J. Macromol. Sci. B1, 161-184

Blundell, D.J. and Keller, A. (1968) 'Controlled crystal growing procedures involving self seeding: some novel twinning habits' J. Macromol. Sci. B2, 337-359

Dorset, D.L., Hanlon, J., Mc Connell, C.H., Fryer, J.R., Lotz, B., Wittmann, J.C., Beckmann, E. and Zemlin, F. (1990) 'Why do polyethylene crystals have sectors?, Proc. Nat. Acad. Sci. USA, in press

Ihn, K.J., Tsuji, M., Isoda, S., Kawaguchi, A. and Katayama, K.I. (1989) 'Crystallization of polyethylene from its vapor phase on cycloparaffin single crystals', Makromol. Chem. Rapid Commun. 10, 185-188

Keith, H.D. and Padden, F.J. (1984) 'Twisting orientation and the role of transient states in polymer crystallization', Polymer 25, 28-42

Keith, H.D., Padden, F.J., Lotz, B. and Wittmann, J.C. (1989) 'Asymmetries of habit in polyethylene crystals grown from the melt', Macromolecules 22, 2230-2238

Khoury, F.,(1987) private communication

Organ, S.J. and Keller, A. (1987) 'The onset of chain folding in ultra long n-alkanes. An electron microscopic study of solution grown crystals', J. Polym. Sci. B. Polym. Phys. 25, 2409-2430

Radoslovich, E. W. (1960) 'The structure of muscovite, K Al$_2$ (Si$_3$ Al) O$_{10}$ (OH)$_2$ ', Acta Cryst. 13, 919-932

Ungar, G., Organ, S.J. and Keller, A. (1988) 'X-ray evidence for sharp chain folds in crystalline linear alkanes', J. Polym. Sci. C Polym. Lett. 26, 259-263

Wicker, A., Lotz, B., Wittmann, J.C. and Legrand, J. F.(1989) 'Electron microscope studies of ferroelectric domain structures of triglycine sulphate crystals using polymer decoration', Polymer Commun. 30, 251-253

Wittmann, J.C. and Lotz, B. (1985) 'Polymer decoration: the orientation of polymer folds as revealed by the crystallization of polymer vapors', J. Polym. Sci., Polym. Phys. Ed. 23, 205-226

Wittmann, J.C. and Lotz, B. (1986) 'Polymer decoration of layer silicates: crystallographic interactions at the polyethylene-talc interface', J. Mat. Sci. 21, 659-668

EPITAXIAL GROWTH OF OCTACYANOMETALPHTHALOCYANINE-METAL COMPLEX CRYSTALS

Michio Ashida
Faculty of Engineering, Kobe University
Rokko, Nada-ku, Kobe 657
Japan

ABSTRACT.   Octacyanometalphthalocyanine-metal [MPc(CN)$_8$-M] complex
crystals are synthesized by vapor-solid reaction of tetracyanobenzene
with metal films and alkali halide crystals. Their crystal structures
and orientations are directly analyzed by high resolution electron
microscopy and electron diffraction. The MPc(CN)$_8$-M complex crystals
are composed of planar MPc(CN)$_8$ molecules packing parallel each other
and intercalated metals. The M$_2$Pc(CN)$_8$-M (M:Na, K, Rb, Cs) complex
crystals have the same interplanar distance of 0.34 nm and their
intercolumn distances increase in the order of Na-< K-< Rb-< Cs-
complex, depending on the radii of intercalated alkali metals. The
MPc(CN)$_8$-M (M:Cu, Ni, Co, Fe) complex crystals are an isomorph of
the K$_2$Pc(CN)$_8$-K complex crystal and their intercolumn distances
become shorter in the order of K-> Fe-> Co-> Ni-> Cu-complex. Thin
films of K$_2$Pc(CN)$_8$-K complex are composed of many crystallites with
double directional orientation intersecting each other by an angle of
about 23°. The crystals take three types of mutual orientations on a
KCl (001) surface and the relative orientation of the deposited
crystal to the KCl substrate is interpreted in terms of multiple
positioning of the K$_2$Pc(CN)$_8$ molecules on an ionic lattice of the KCl
surface.

Introduction

Phthalocyanine(Pc) compounds are representative conjugated organic
molecules and have been attracting interest for new electronic and
optical devices. In order to make a practical architecture of organic
molecular devices, the molecular arrangement must be
crystallographically controlled in thin films. Vacuum-deposited Pc
films on cleaved surface of muscovite and alkali halides exhibit well-
defined epitaxial orientations[1-3]. Since the molecular image of
chlorinated copper phthalocyanine was observed by Uyeda et al.[4],
high resolution electron microscopy (HREM) has been used, not only for
the direct imaging of organic crystal structures[5-7], but also for
molecular structure at atomic level[8,9]. The high resolution image
of chlorinated copper phthalocyanine has been interpreted completely

227

J. R. Fryer and D. L. Dorset (eds.), Electron Crystallography of Organic Molecules, 227–240.
© 1990 *Kluwer Academic Publishers.*

on the basis of computer simulation of the crystal structure[10]. Although HREM has the ability to analyze the molecular arrangement in thin films, the structural study of organic crystals by HREM is limited to some extent by their sensitivity to radiation damage by the electron beam. In order to reduce the radiation damage HREM of organic crystals has been carried out by the minimum dose method[11], in a cold stage, or by encapsulating the specimen[12]. One of the ultimate objectives for HREM is to determine the structure of an unknown reaction product by the direct observation of the atomic arrangement. In previous works[13,14], thin films of octacyanometalphthalocyanine ($MPc(CN)_8$) were synthesized directly on substrates such as metal films and alkali halides by vapor-solid reaction of tetracyanobenzene (TCNB) with substrates as shown in Fig.1.

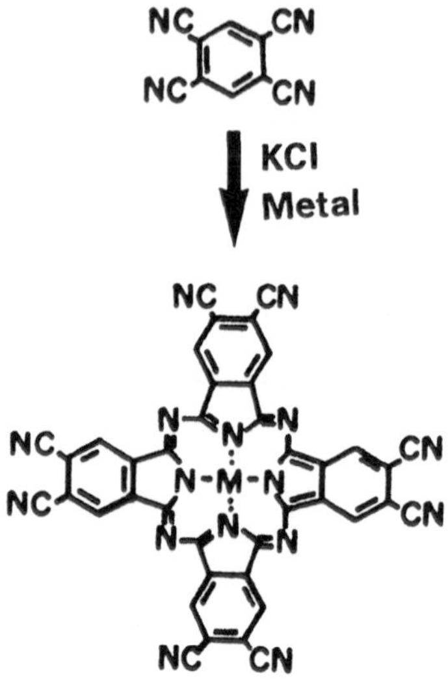

Fig.1. Scheme of synthesis

The present study is concerned with the structural elucidation of films produced from TCNB and metals or alkali halides. The molecular image of the product is observed by atomic level resolution in HREM. The crystal structure and molecular arrangement of the $MPc(CN)_8$-M complexes are discussed on the basis of interaction between the $MPc(CN)_8$ molecules and the intercalated metals. Epitaxial growth of the $MPc(CN)_8$-M complex crystals on substrates is discussed on the basis of the lattice images by HREM and electron diffraction.

Experimental

A sample of TCNB was synthesized by the method of Bailey et al.[15]. The crude TCNB was purified by recrystallization from acetic acid. The purified TCNB powder of about 10 mg and a substrate crystal were sealed in an evacuated pyrex glass tube at about 10 Pa after flushing with argon gas. The substrates used were cleavage faces of alkali halide crystals (NaCl, KCl, RbCl and CsCl) and metallic films. In the case of metallic films, Cu, Ni, Co and Fe were evaporated on a cleavage surface of KCl maintained at a temperature between 350 and 450°C in a vacuum of 0.1 mPa. The tube was heated at a temperature between 300 to 350°C for 10 hr in a hot air oven. The films produced on substrates were separated from the substrates in water and the residual metal films were dissolved away in dilute acid solution. The specimen film was mounted on a microgrid coated with evaporated gold particles. The HREM observations reported here were made in a JEM-

200CX electron microscope equipped with a minimum exposure device. The electron microscope was operated at 200 kV with a microscope parameter of spherical aberration coefficient of either 1.4 or 2.0 mm. The (111) lattice image from gold (d=0.235 nm) provided an accurate magnification calibration. The unit cell dimensions were accurately determined from electron diffraction patterns, using the lattice spacings of gold as reference.

To interpret the experimental images, computer simulation was performed, assuming particular molecular dimensions and arrangement in the crystal. The multislice method was adopted for the calculation of scattering amplitudes using the algorithm developed by Ishizuka and Uyeda[16]. Contrast transfer functions (CTF) were calculated with attenuation by a chromatic defocus spread of 10 nm and a beam convergency of 0.1 mrad.

Results and discussion
1. Crystal structure of $K_2Pc(CN)_8$-K complex

The film formed on the substrate KCl crystal by reaction with TCNB is composed of slender tentacle-like crystallites[17]. The product has been identified as dipotassiumoctacyanophthalocyanin $[K_2Pc(CN)_8]$ by means of infrared and visible spectra and elemental analysis[13,14]. In a very thin film at the first stage, the $K_2Pc(CN)_8$ film is composed of discrete disk-like and lamellar crystallites as shown in Fig.2(a). The electron diffraction pattern from the disk-like crystallite represents the single net pattern with C_{4v}, corresponding to the interplanar spacing of 1.57 nm, as shown in Fig.2(b). The high-resolution electron micrograph from the disk-like crystallite shows a grid pattern with the spacing of 1.57 nm as shown in Fig.3(a). The optical diffraction pattern from the electron micrograph shows C_{4v} symmetry composed of square lattice points of 1.57 nm spacing as shown in Fig.3(b). This pattern is in good agreement with the corresponding electron diffraction pattern. The optical transform confirms that the real image of Fig.3(a) is recorded near the Scherzer focus of 85 nm in consideration of the scattering band from the carbon film superimposed on the crystalline pattern. On the other hand, the selected-area diffraction pattern from the lamellar crystallite represents the fiber pattern which indicates the fiber period of 0.337 nm and the lattice spacing of 1.57 nm. Figure 4 shows a high-resolution micrograph of the lamellar crystal. The structure image indicates that the planar molecules of $K_2Pc(CN)_8$ pile up parallel to each other along the column axis[18]. From the features of growing crystals, it appears that the crystal planes shown in Figs.3 and 4 correspond to the planes normal and parallel to the direction of growth, respectively. From the above relations, the unit-cell dimensions of the crystal can be deduced: $a=b=1.57$ nm and the interplanark distance is 0.337 nm. Figure 5(a) shows a high-resolution micrograph, corresponding to Fig.3(a), which is noise-reduced photographically by a translational multiple-exposure technique. The molecular image indicates many characteristic features: the square network with a distance of 1.57 nm is composed of two nearly parallel lines with a spacing of 0.44 nm and high contrast

230

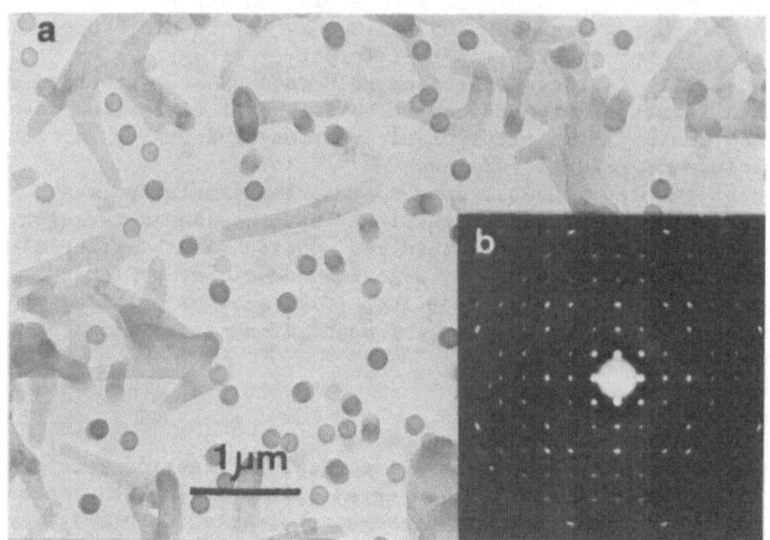

Fig.2. Electron micrograph(a) and electron diffraction pattern(b)
of a thin film of the $K_2Pc(CN)_8$-K complex crystal.

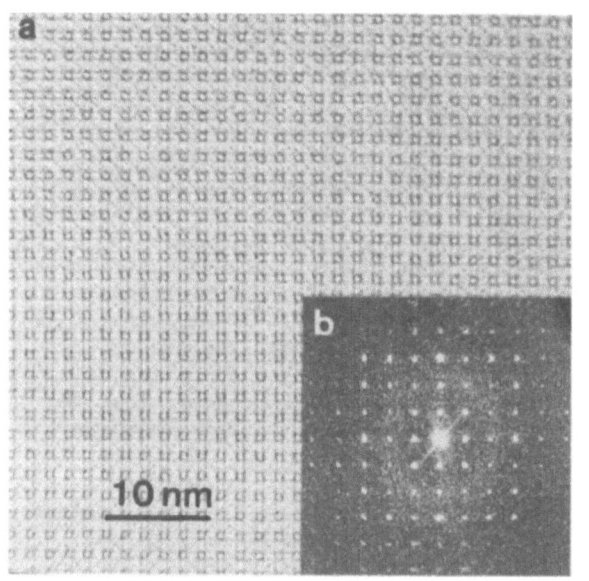

Fig.3. Electron micrograph(a) of a disk-
like crystallite of $K_2Pc(CN)_8$-K complex
and optical diffraction pattern(b).

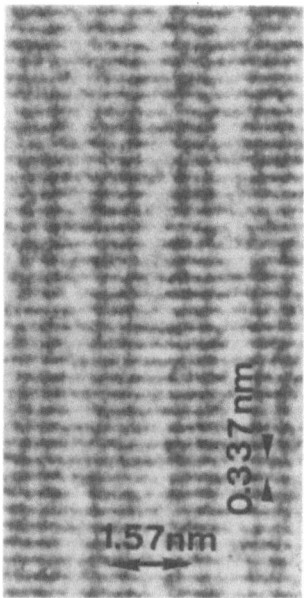

Fig.4. Molecular image of
$H_2Pc(CN)_8$-K complex projected on the (010) plane.

circles of diameter 0.56 nm are situated at each corner of the square.
High-contrast dots are located at the center of each square and four
short rods link the corners to the central dot.   On the basis of
these considerations, the molecular arrangement in the crystal is
illustrated schematically in Fig.6.   The Pc ring image shows a circle
whose inside is blank because $K_2Pc(CN)_8$ molecule is easily demetalized
to metal-free derivative ($H_2Pc(CN)_8$) in water.   The planar molecules
of $H_2Pc(CN)_8$ are in a layer normal to the c axis and are located at
the corners of the square at a distance of 1.57 nm;   the molecules are
rotated alternately from the a axis by $\pm\omega$ in successive layers, as
shown in Figs.6(a) and (b).   When these molecular layers are packed
along the c axis and a K atom is surrounded at the center of the
square by eight CN groups of two successive layers, both molecules and
atoms projected along the c axis are superimposed as shown in
Fig.6(c).   The rotation angle ω was determined to be 27.4° directly
from high-resolution molecular images with the aid of image
simulations [19].   Figure 5(b) shows the computer-simulated image at
70 nm underfocus for a crystal of thickness 4 nm and corresponding
real image: both images coincide well.   Thus, the high-resolution image
of the $H_2Pc(CN)_8$-K complex crystal is interpreted completely by the
proposed crystal structure.

2. Crystal Structures of MPc(CN)$_8$-M complexes

The square lattice images have been observed in films produced
on other alkali halide substrates such as NaCl, RbCl and CsCl[17].
Their high resolution electron micrographs are shown in Fig.7, in which
(a) to (d) and (a') to (d') show the projections of the MPc(CN)$_8$-M
crystals on the (001) and (010) planes, respectively.   An analogy among
those images suggests that octacyanodialkalimetalphthalocyanine-alkali
metal ($M_2Pc(CN)_8$-M) complexes (M:Na, K, Rb and Cs) are produced from
various alkali halides and TCNB by vapor-solid reaction and that they
are an isomorph of the $H_2Pc(CN)_8$-K complex crystal.   Square lattice
fringes shown in Figs.7(a) to 7(d) reveal the molecular columns in
each complex crystal.   These lattice fringes indicate that MPc(CN)$_8$
molecules pile up, taking their molecular planes perpendicular to the
column axis.   The intensity of the dots ascribed to the intercalated
metal increase in the following order corresponding to the atomic
number: Na< K< Rb< Cs.   The open circles of Pc rings suggest that the
$M_2Pc(CN)_8$ molecules are also demetalized to $H_2Pc(CN)_8$ when the films
are dipped in water.   The parallel molecules of all complex crystals
are the same distance of 0.34 nm.   On the contrary, the intercolumn
distances increase in the following order: Na-< K-< Rb-< Cs-complex,
depending on the radii of intercalated alkali metals.   On the other
hand, the high resolution micrograph of films produced on a LiCl
substrate shows two different molecular images.   The one is the same
square lattice image as that of $K_2Pc(CN)_8$-K complex crystal.   The other
is a X-shaped image and a striped image tilted to the column axis, as
shown in Figs.9(a) and (b), respectively.   Such molecular images had
been observed in a ZnPc film[21] and a chlorinated CuPc film[22].
These images indicate that the planar $Li_2Pc(CN)_8$ molecules tilt to the

232

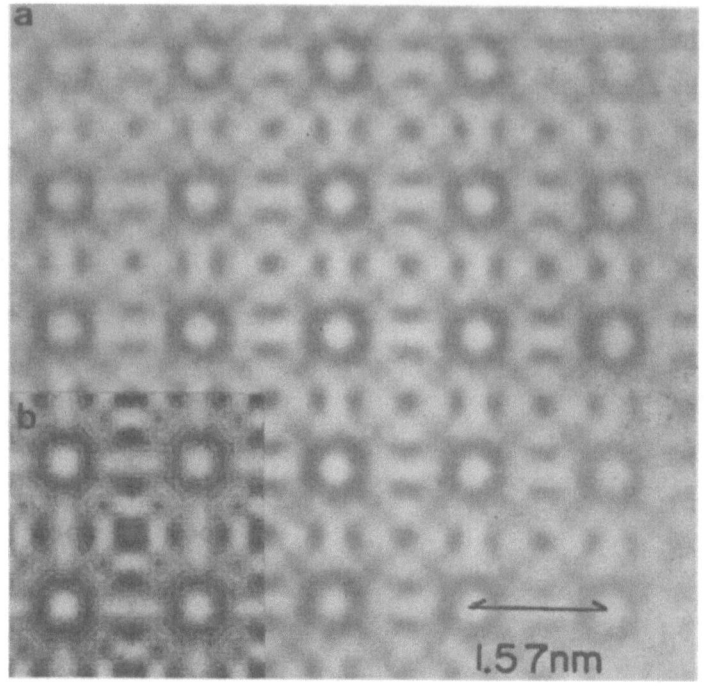

Fig.5. Molecular image(a) and computer simulated image(b) of the $H_2Pc(CN)_8$-K complex projected on the (001) plane.

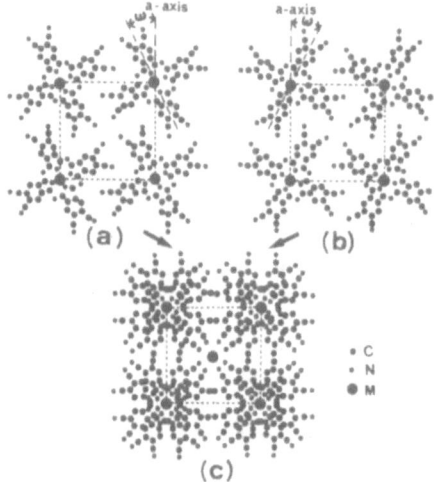

Fig.6. Molecular arrangements in the $K_2Pc(CN)_8$-K complex crystal: (a),(b) molecules in two successive layers;(c)schematic diagram of the molecular overlap projected on the (001) plane.

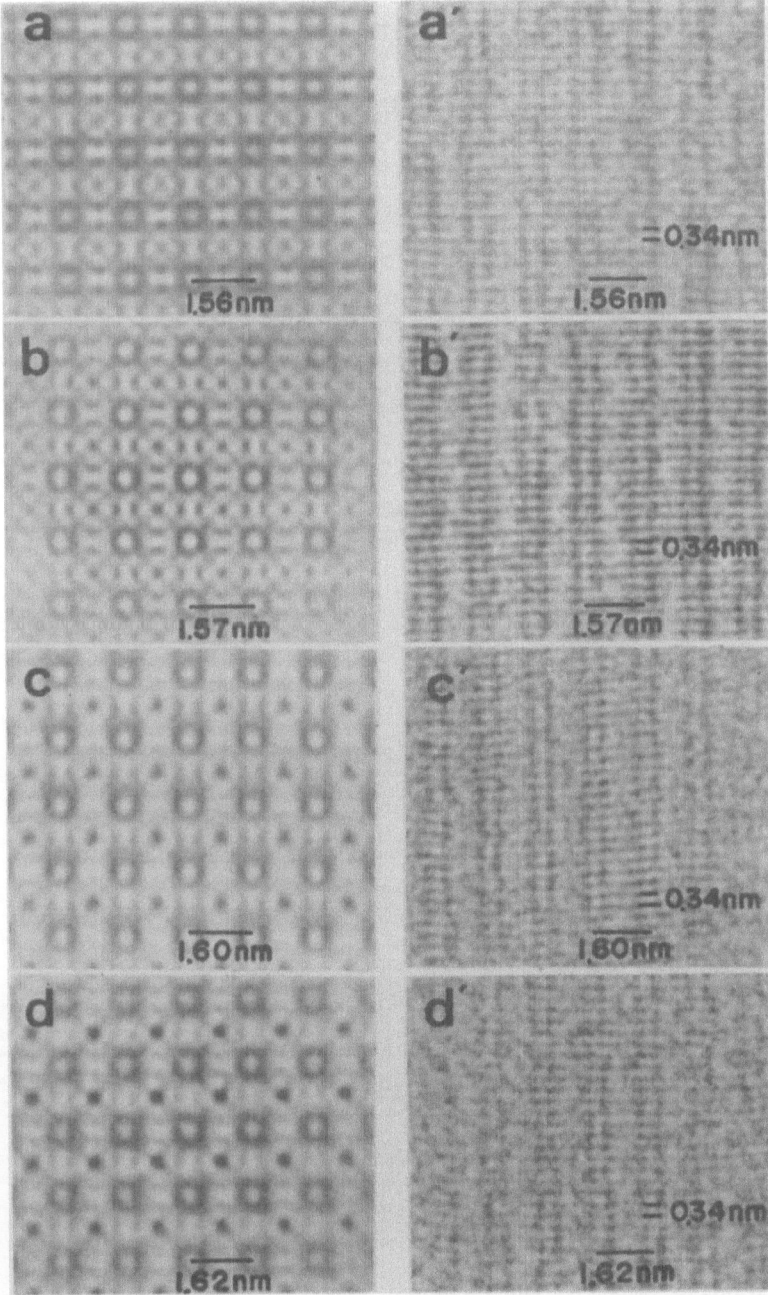

Fig.7. Molecular images of $H_2Pc(CN)_8$-Na(a,a'), $H_2Pc(CN)_8$-K(b,b'), $H_2Pc(CN)_8$-Rb(c,c'), and $H_2Pc(CN)_8$-Cs(d,d') complex.

234

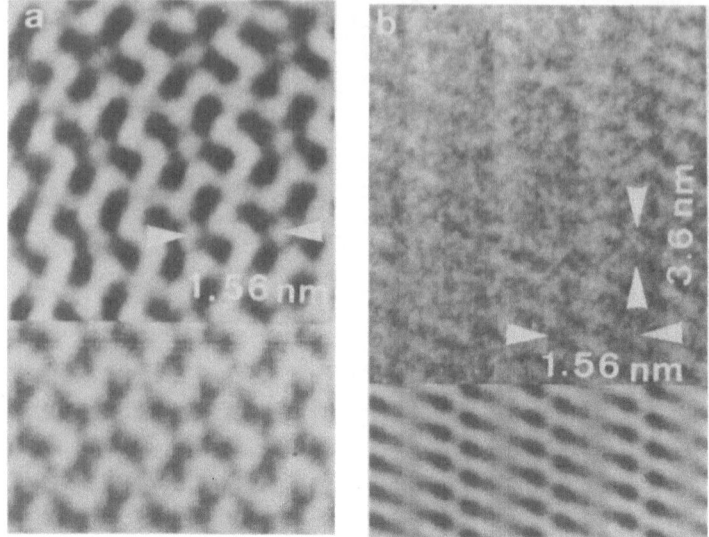

Fig.8. Molecular images of the $Li_2Pc(CN)_8$-projected on the
(001) plane(a) and the (010) plane(b).

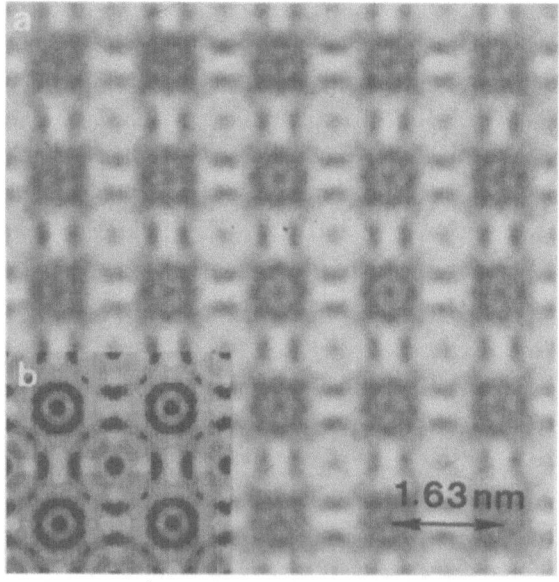

Fig.9. Molecular image(a) and computer-simulated image(b) of
the $CuPc(CN)_8$-Cu complex crystal projected on the (001) plane.

substrate surface and pile up parallel to each other along the
column axis. It is noted that the dots ascribed to the intercalated
atoms are not observed among the $Li_2Pc(CN)_8$ molecules. Therefore, it
seems that some $Li_2Pc(CN)_8$ molecules do not form the complex with Li
metal but pile up tilting to the column axis.

The $CuPc(CN)_8$-Cu complex crystal formed on a copper film shows
the similar square lattice image to that of $K_2Pc(CN)_8$, as shown in
Fig.9. By comparison with a molecular image of $H_2Pc(CN)_8$-K, the rings
of about 0.6 nm in diameter on the corners of the square in Fig.9(a)
are attributed to $CuPc(CN)_8$ molecules stacked parallel on a substrate.
It is noteworthy that a high-contrast dot is observed in the Pc ring of
$CuPc(CN)_8$ molecules. Because MPc's such as CuPc, FePc, CoPc and NiPc
are not demetalized even in acid solutions, the high-contrast dot in
Pc ring is attributed to the metal atom.

3. Epitaxial Growth of $K_2Pc(CN)_8$-K complex crystal

Many crystallites with double-directional orientation are
observed in a $K_2Pc(CN)_8$-K complex film, as shown in Fig.10(a). Each
dark dot aligned in the square lattice corresponds to a molecular
column projected along the c axis. The molecular image in the wedge-
shaped area cannot be clearly observed because molecular stacking may
be disordered at the grain boundary. The a axes of the two crystallites make an angle of 23° to each other. These double-directional
crystals are clearly observed by a selected-area electron diffraction
pattern of the film as shown in Fig.10(b). This pattern is composed of
superposition of two single net patterns with C_{4v} symmetry,
corresponding to a lattice spacing of 1.57 nm. The reflection spots
are schematically represented in Fig.11 which shows two hk0 diagrams
rotated by 23° with respect to each other. It seems that the double-
directional orientation is caused by interaction between the CN
groups of $K_2Pc(CN)_8$ molecules and the KCl crystal, and the intersecting
angle of 23° is related to the crystal structure of the $K_2Pc(CN)_8$-K
complex. Figure 12 shows a schematic diagram of the molecular
arrangement in the $K_2Pc(CN)_8$-K complex crystal projected along the c
axis[18]. Each molecule M_1 in the first layer, indicated by a solid
line, is rotated clockwise by an angle ω and each molecule M_2 of the
second layer, indicated by a dashed line, is rotated counter-clockwise
by the same angle. The superimposed CN groups coordinated to the K
atom are arranged to make an intersecting angle φ as shown with lines
of L_1 and L_2. This angle is calculated to be 22.9° when the rotation
angle ω is 27.4°, where the peripheral CN groups bonded to the benzene
ring are assumed to make an angle of 120° with the C-C bond
corresponding to molecular structure of the TCNB crystal. This value
of φ coincides well to the angle between the double-directional
crystals, 23°. The peripheral CN groups are more electronegative
than other parts of the π-electron

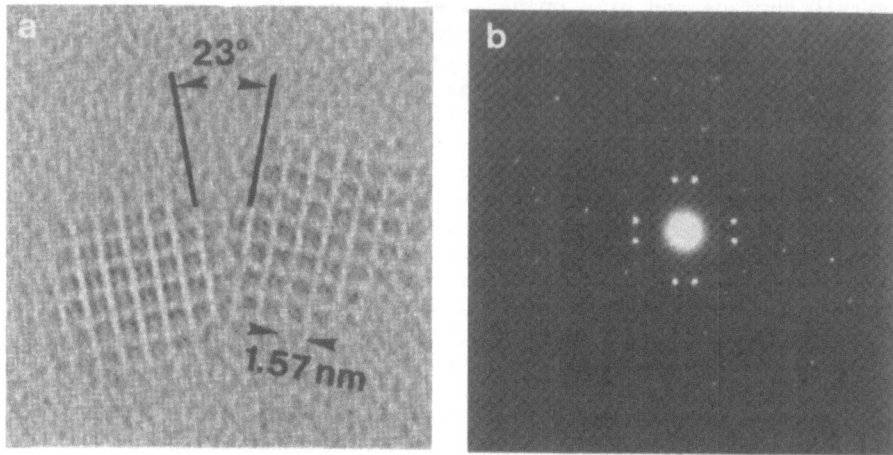

Fig.10. Electron micrograph(a) and electron diffraction pattern(b) of the $K_2Pc(CN)_8$-K crystallite with double-directional orientation.

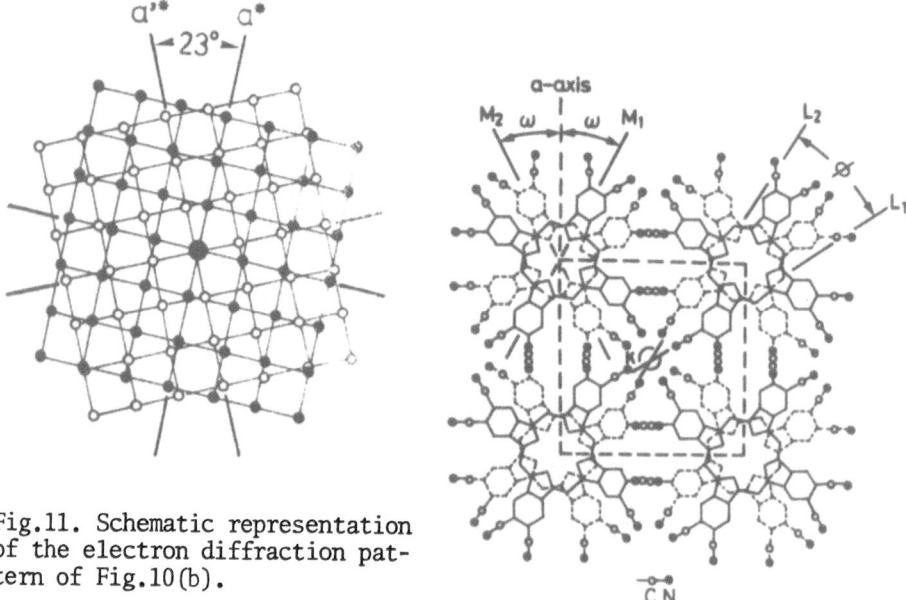

Fig.11. Schematic representation of the electron diffraction pattern of Fig.10(b).

Fig.12. Schematic diagram of the molecular arrangement in the $K_2Pc(CN)_8$-K crystal projected on the (001) plane.

orbitals in a $K_2Pc(CN)_8$ molecule and attract a certain amount of electron charge from a central K atom. A nearly straight row of ionic charges, $C=N^{\delta-}$ $-K^+-N^{\delta-}=C$, is formed between two molecules located at diagonal positions of the square lattice in the $K_2Pc(CN)_8$-K complex. An interaction probably exists between this ionic charge and the KCl substrate. At an initial stage of nucleation, this ionic row is considered to have a very stable fitting with ionic arrangement of the KCl (001) surface. It is assumed that two different orientations of the crystals relative to the KCl lattice arise from two possible contacts of the ionic rows, L_1 and L_2, with the KCl surface. Consequently, the crystals take double-directional orientations so that their a axes make an angle of 23° with respect to each other.

Three different orientations of the square lattices can be observed in a film, as shown in Figs.13(a), (b) and (c), respectively. The crystallites in Fig.13(a) show the double directional orientation as mentioned above. The bisector of the two lattices makes an angle of 45° to the $[100]_{KCl}$ direction, that is, its direction coincides with the $[110]_{KCl}$ direction. The a axis of one square lattice makes an angle of 33.5^O to the $[100]_{KCl}$ direction and the a axes of the crystallites in the Fig.13(b) and 13(c) make angles of 11.5^O and 18° to the $[100]_{KCl}$ direction, respectively.

From these mutual relations between the deposited and substrate crystal lattices, three types of nucleations of the $K_2Pc(CN)_8$-K complex crystal on the KCl(001) surface are assumed as shown schematically in Figs. 13(a'), (b') and (c'), corresponding to the orientations in Figs.13(a), (b) and (c), respectively. In the case of Fig.13(a'), four CN groups (n_1, n_2, n_3, n_4) of four $K_2Pc(CN)_8$ molecules in the first layer come into close contact with four K^+ ions on the KCl (001) plane. When the lines L_1 of CN groups are parallel to the $[100]_{KCl}$ and $[010]_{KCl}$ directions, the a axis of $K_2Pc(CN)_8$-K complex crystal makes an angle of 33.5^O to the $[100]_{KCl}$ direction. A similar interaction between CN groups and K ions has been reported in epitaxial growth of tetracyanoquinodimethane (TCNQ) evaporated on a KCl (001) surface[26]. Another positioning of CN groups is assumed in the case of Fig.13(b'), Four CN groups (n_1, n_2, n_3, n_4) coordinate to a central K^+ ion to form a complex crystal. The CN groups settle in grooves with minimum potential between Cl^- ions because electronegative CN groups tend to minimize repulsion force by Cl^- ions. Consequently, the lines of CN groups are parallel to the $[110]_{KCl}$ and $[1\bar{1}0]_{KCl}$ directions, so that the a axis of the $K_2Pc(CN)_8$-K complex crystal makes an angle of 11.5^O to the $[100]_{KCl}$ direction. The orientation in the case of Fig.13(c) cannot be explained by interaction between CN groups and a KCl crystal. As is well known electrons in a phthalocyanine ring are distributed on the *meso*-bridging N atoms (N_1, N_2, N_3, N_4) with the highest density. These N atoms tend to be located close to K^+ ions as reported for CuPc molecules deposited on a KCl (001) surface[2]. Consequently, the molecular axis M_1 is parallel to the $[110]_{KCl}$ direction and the angle between the a axis of the $K_2Pc(CN)_8$-K complex crystal and the $[100]_{KCl}$ direction is 17.7^O which coincides approximately with the observed

238

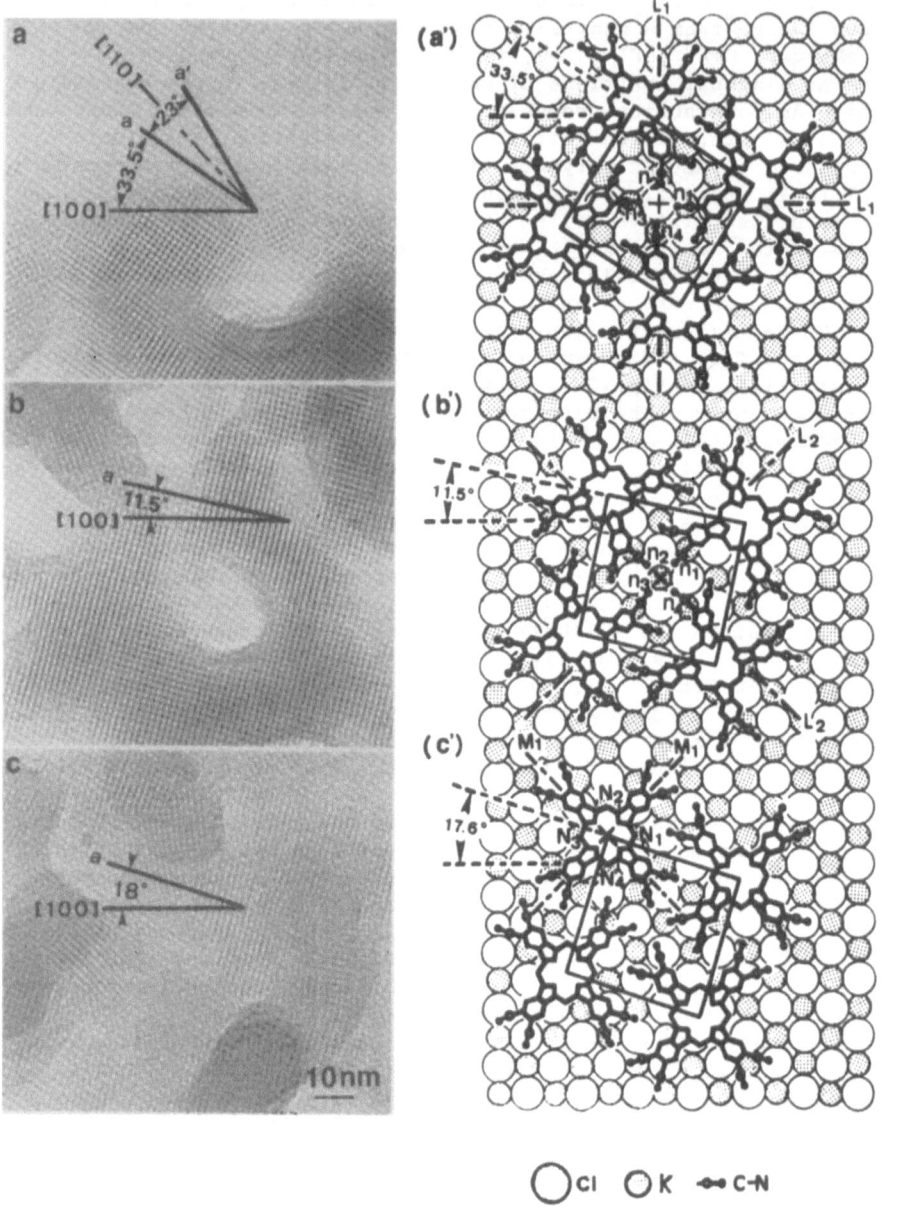

Fig.13. Electron micrographs of oriented crystals(a),(b) and(c) and the corresponding schematic diagrams of mutual relations between the $K_2Pc(CN)_8$-K complex and the KCl (001) surface(a'),(b') and (c').

value of 18°. As this orientation is observed in only a few
crystallites in the film, it seems to be unstable as compared with the
other orientations.

A film of the $CuPc(CN)_8$-Cu complex crystal is synthesized on a Cu
film vacuum-evaporated on a KCl(001) surface. The film shows various
lattice fringes of the $CuPc(CN)_8$-Cu complex crystal which is
identified as an isomorph of the $K_2Pc(CN)_8$-K crystal[24]. The film is
composed of the small crystallites with square lattice oriented in
random directions and the raft-like crystals with various lattice
fringes along the column axis. The finding suggests that the
crystallites of the $CuPc(CN)_8$-Cu complex take random orientations on
the Cu film constructed with single-crystal and there is no specific
relation between the deposited and substrate crystals.

References

1. Uyeda,N., Ashida,M. & Suito,E. (1965) 'Orientation overgrowth of
   condensed polycyclic aromatic compounds vacuum-evaporated
   onto cleaved face of mica', J.Appl.Phys. 36, 1453-1460.
2. Ashida,M.(1966a)'The orientation overgrowth of metal-phthalocyanines
   on the surface of single crystal I. Vacuum-condensed films on
   muscovite', Bull.Chem.Soc.Jpn. 39,2625-2631.
3. Ashida,M.(1966b)'The orientation overgrowth of metal-phthalocyanines
   on the surface of single crystals II. Vacuum condensed films of
   copper-phthalocyanine on alkali halides', Bull.Chem.Soc.Jpn. 39
   2632-2638.
4. Uyeda,N., Kobayashi,T., Suito,E., Harada,Y. & Watanabe,M. (1972)
   'Molecular image resolution in electron microscopy', J.Appl.Phys.43,
   5181-5189.
5. Fryer,J.R.(1978) 'Molecular images of the hydrocarbon $C_{22}H_{12}$-
   anthanthrene', Acta Cryst. A34, 603-307.
6. Fryer,J.R.(1979) 'Molecular images of thin-film polymorphs and phase
   transformations in metal-free phthalocyanine', Acta Cryst. A35,
   327-332.
7. Kobayashi,T., Fujiyoshi,Y. & Uyeda,N.(1983)'High-resolution electron
   microscopy of structural defects in organic crystals',J.Cryst.Growth
   65,511-517.
8. Uyeda,N. & Ishizuka,K.(1974) 'Effect of spherical aberration and
   accelerating voltage on atomic resolution in molecular images',
   J.Electron Microsc. 23, 79-88.
9. Uyeda,N., Kobayashi,T., Ishizuka,K. & Fujiyoshi,Y.(1980). 'Crystal
   sructure of Ag·TCNQ', Nature 285,95-96.
10. Uyeda,N., Kobayashi,T., Ishizuka,K. & Fujiyoshi.Y.(1978-79) 'High
    voltage electron microscopy for image descrimination of constituent
    atoms in crystals and molecules', Chemica Scripta. 14, 47-61.
11. Fujiyoshi,Y., Kobayashi,T., Ishizuka,K., Uyeda,N., Ishida,Y.&
    Harada,Y.(1980) 'A new method for optimal-resolution electron
    microscopy of radiation-sensitive specimens', Ultramicroscopy 5,
    459-468
12. Fryer,J.R. & Holland,F.(1983) 'The reduction of radiation damage in

the electron microscope', Ultramicroscopy **11**, 67-70.
13. Yanagi, H., Ueda, Y. & Ashida, M. (1988) 'Characterization of monomeric and polymeric(octacyanophthalocyaninato)metals in thin films', Bull. Chem. Soc. Jpn. **61**, 2313-2320.
14. Ashida, M., Ueda, Y., Yanagi, H. & Sayo, K. (1989) 'Preparation and characterization of thin films of monomeric and polymeric octacyanophthalocyanines', J. Polym. Sci. Polym. Chem. Ed. **27**, 3883-3893.
15. Bailey, A. S., Henn, R. B. & Langdon, J. M., (1963) 'Molecular complexes of aromatic nitriles', Tetrahedron **19**, 161-167.
16. Ishizuka, K. & Uyeda, N. (1977) 'A New theoretical and practical approach to the multislice method', Acta Cryst. **A33**, 740-749.
17. Yanagi, H., Maeda, S., Hayashi, S. & Ashida, M. (1988). 'Crystal growth of octacyanometalphthalocyanine-metal complexes in thin films' J. Cryst. Growth **92**, 498-506.
18. Ashida, M., Ueda, Y. Yanagi, H., Fujiyoshi, Y., Uyeda, N. & Fryer, J. R. (1988). 'Formation and structure of a 2,3,9,10,16,17,23,24-octacyanophthalocyanine-potassium complex in thin film', Acta Cryst. **B44**, 146-151.
19. Yanagi, H., Maeda, S., Ueda, Y & Ashida. M. (1988). 'Direct structure determination of the octacyanophthalocyanine-potassium complex crystal by HREM image simulation', Ultramicroscopy **25**, 1-12.
20. Ueda, Y., Yanagi, H., Hayashi, H. & Ashida, M. (1989). 'Structure of octacyanodialkalimetalphthalocyanine-alkali metal complex crystals prepared by chemical vapor deposition', J. Electron Microsc. **38**, 101-110.
21. Kobayashi, T., Fujiyoshi, Y., Iwatsu, F. & Uyeda, N. (1981) 'High resolution TEM images of zinc phthalocyanine polymorphs in thin films', Acta Cryst. **A37**, 692-697.
22. Fryer, J. R. & Smith, D. J. (1986) 'High resolution electron microscopy of interfaces in chlorinated phthalocyanine molecular crystals', J. Microsc. **141**, 3-9.
23. Ueda, Y. & Ashida, M. (1989). 'Structure of octacyanocopper-phthalocyanine film prepared by chemical vapor deposition', Bull. Inst. Chem. Res. Kyoto Univ. **66**, 562-571.
24. Yanagi, H., Maeda, S., Ueda, Y. & Ashida, M. (1988). 'High resolution electron microscopy of octacyanometalphthalocyanine-metal complexes in thin film', J. Electron Microsc. **37**, 177-188.
25. Yanagi, H., Hayashi, S., Fujita, N. & Ashida, M. (1989)'Epitaxic Growth of [2,3,9,10,16,17,23,24-octacyano-29H,31H-phthalocyaninato (2-)]metal complex crystals', Acta Cryst. **C45**, 1888-1894.
26. Uyeda, N. Murata, Y., Kobayashi, T. & Suito, E. (1974) 'Epitaxial growth of an organic semiconductor from the vapor phase-TCNQ on potassium chloride', J. Cryst. Growth **26**, 267-276.

EPITAXIAL CRYSTAL GROWTH ON ORGANIC AND POLYMERIC SUBSTRATES.

J. C. WITTMANN and B.LOTZ
Institut Charles Sadron (CNRS-ULP)
6, rue Boussingault
67083 Strasbourg
France

ABSTRACT. In view of their well-adapted chemical and physical structure, organic crystals such as aromatics and oriented polymer films may serve as very efficient substrates for the epitaxial crystallization of long chain molecules. Using various original growth procedures, it is demontrated with two case examples, polyethylene and poly-1-butene, that by a judicious choice of the substrate it is possible to induce the growth of different polymorphs in various projections normal to the chain axis, and to follow by electron diffraction the related phase transformations.These examples are only a few among many of the applications opened by epitaxial growth on organic and polymeric substrates in the field of electron microscopy or synthesis of structurally controlled organic layers.

1. Introduction

The successful crystallization and orientation of organic materials in the form of thin layers (e< 100nm) represents the primary step in their study by electron microscopy and electron crystallography. In the fields of synthetic polymers, biopolymers or shorter chain compounds, epitaxy has been used advantageously as an alternative crystal growth technique to overcome the often unfavourable crystal forms and/or orientations obtained by the more conventional growth procedures. For example, when lamellar crystals grown from solution are deposited on E.M. grids, the molecular axes which are nearly perpendicular to the lamellar surface, parallel the incident electron beam. In epitaxially grown thin films on the contrary, the lamellae stand edge-on i.e. the molecular axes lie in the film plane and become therefore oriented perpendicular to the incident beam. This orientation offers a unit-cell projection yielding the most crystallographic information (Dorset (1985b)).

While epitaxy in general and epitaxy of polymers on inorganic as well as polymeric substrates in particular have been investigated early on, the use of organic substrates has developed only recently. Studies have rapidly concentrated on aromatic compounds with their almost endless possible variations in terms of chemical, physical and crystallographic properties.

After a brief description of the structure and properties of these various classes of substrates, a short survey will be given of the main results obtained with : a) polyethylene (or paraffins) which crystallizes in two different modifications of comparable energy and with the same chain conformation and b) isotactic polybutene-1 which, at ambient pressure, displays three crystal modifications, each with a different chain conformation.

These examples will help to illustrate and discuss the various possibilities offered by epitaxial crystal growth in different domains.

J. R. Fryer and D. L. Dorset (eds.), Electron Crystallography of Organic Molecules, 241–254.
© 1990 *Kluwer Academic Publishers.*

2. The structure and properties of various classes of substrates

2.1. ALKALI HALIDES

Epitaxial growth of long chain molecules on inorganic substrates and on alkali halides in particular is well documented and has been the subject of several reviews (Mauritz et al.(1978), Swei et al.(1986), Wunderlich(1973)). It helped namely to set the rules of polymer epitaxy and to study the influence of different epitaxial parameters (crystalline structure and conformation of the polymer, symmetry and periodicities of the substrate, temperature, concentration, etc...) on the growth processes and the resulting deposit orientations.

Alkali halides offer a series of interesting properties:
-avaibility as macroscopic single crystals of cubic symmetry and with well defined [001] cleavage surfaces,
-possibility to alter the chemical constitution and thus to vary more or less continuously the unit cell parameters,
-owing to their high solubility in water, ease of use for the E.M. preparations...

The epitaxial growth can occur from solution, melt or vapor phase and, more specific to polymers, by epitaxial polymerization i.e. epitaxial crystallization of a monomer followed by solid state polymerization (Rickert et al.(1979), Swei et al.(1986)).

Microscopic studies of the overgrown layers have established that, with few exceptions, the long chain molecules are oriented parallel to the substrate surface i.e. the lamellar crystals stand edge-on in a manner similar to the so-called "bookshelf geometry" of oriented smectic layers (Bradshaw et al.(1987)). However, due to the high symmetry of inorganic substrates, a single orientation of the deposit cannot be reached. The crystals are equally distributed along crystallographically equivalent directions of the substrate and, in view of the high nucleation density, are generally very small.

2.2. ORGANIC SUBSTRATES

Aromatics, on which past studies with organic substrates have rapidly concentrated, built up one of the largest and most varied class of organic compounds ranging from simple aromatic hydrocarbons like naphthalene to multifunctional and polynuclear molecules such as pigments. Among the aromatics, condensed aromatic hydrocarbons, linear polyphenyls or aromatic acids and their various salts form homologous series for which crystal structure data are readily available and can be used for a detailed analysis of the substrate/deposit epitaxial relationships.

2.2.1. *Aromatic hydrocarbons*. In both series of condensed aromatic hydrocarbons (naphthalene, anthracene...) and of linear polyphenyls (biphenyl, p-terphenyl...), the packing of the molecules is such that :
-the long molecular axis in the monoclinic unit cell is approximately parallel to the c axis which thus increases regularly with the number of phenyl rings present,
-both the a and b parameters, which reflect the lateral packing in layers of these planar molecules, remain fairly constant.

Given the weak end to end interactions between layers of molecules, the isostructural (001) plane is therefore the predominant growth face and the exposed surface of the platelet shaped crystals grown from solution or from the melt.

Aromatic hydrocarbons cover also a wide range of melting temperatures since T_m increases systematically with the number of phenyl rings. Given their chemical nature, they are expected to act , above T_m, as good solvents for a variety of long chain molecules and to offer, below T_m, crystal surfaces with adequate compatibility and wettability.

Despite the low symmetry of these organic crystals, a double orientation of the deposit is frequently observed (Fig. 1a) as for alkali halides. The nucleation density is however much smaller, so that fairly large mono-oriented domains can nevertheless be grown and observed individually (Wittmann and Lotz (1981))

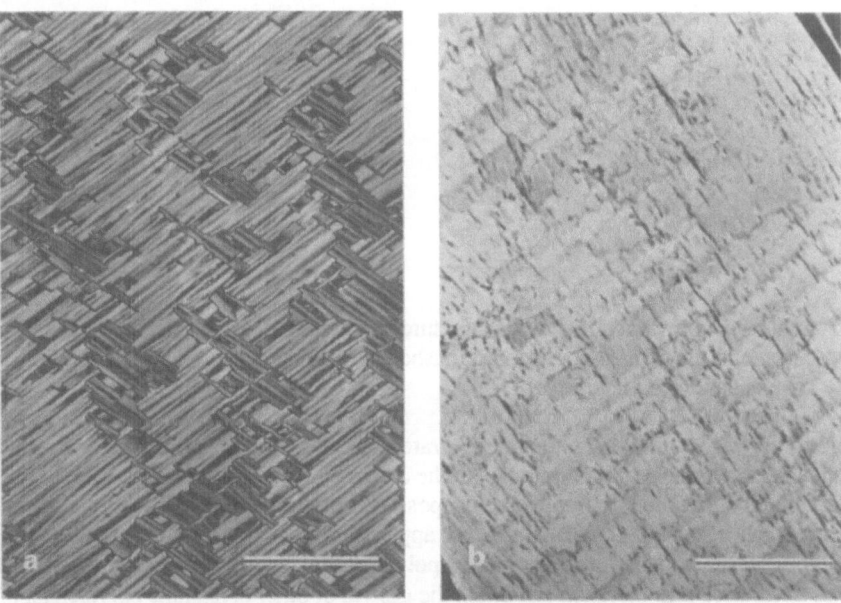

Figure 1. Thin film of hexatriacontane grown epitaxially on (a) naphthalene and (b) benzoic acid by solidification of homogeneous dilute solutions. Scale bars: 40 μm.

Slight alterations of the substrate properties without major changes in the substrate structure can be introduced by using hydrocarbons bearing a substituent (CH_3, Cl, F...) on the phenyl ring. Bi- or polysubstitution affects the crystal structure more drastically, as does the introduction of functionalized molecules as in aromatic acids.

2.2.2. *Aromatic carboxylic acids, hemiacids and salts.* The crystal structure of these series, the simplest of which are benzoic acid or sodium benzoate, is of a "sandwich layered" type. It bears resemblance to the layered structure of aromatic hydrocarbons, with the carboxylic groups or the metal ions forming an inner layer shielded by two apolar layers of aromatic rings. These apolar surfaces, interacting only weakly ,correspond again to the exposed contact surface of the substrate crystals (Fig. 2). As a consequence, large flat crystals or platelets are grown from solution or melt, so that all the diffraction patterns of the deposits are taken as previously normal to the deposit/substrate interface (Wittmann et al.(1983)).

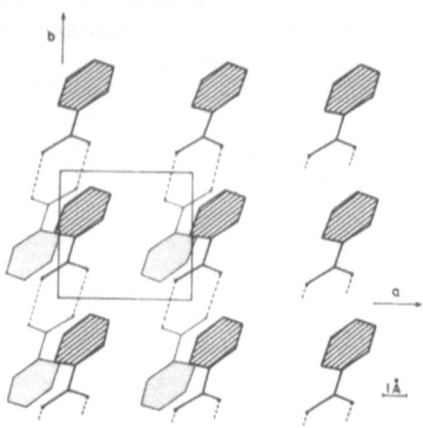

Figure 2. Schematic drawing of the crystal structure of benzoic acid projected on the (001) plane. The molecules form planar hydrogen-bonded (dashed lines) dimers and the rows of benzene rings (striped) are parallel to the a and b axes.

As previously, and unlike inorganic substrates, such surfaces can again be tailored by substitution in para, ortho or meta position(s) to the carboxylic groups, to better match the chemical as well as structural requirements of a given deposit and thus meet the needs (single orientation, given polymorph, etc..(Figs. 1b, 3b)) of specific applications.

2.2.3. *Polymeric substrates*. Although oriented polymer films have been used for a long time as substrates to enhance the nucleation and orient the growth of other crystalline polymers, it is only recently that this series of substrates has been investigated in more detail

The attainable level of deposit orientation depends of course primarily on the perfection and degree of orientation of the substrate itself.Table 1 summarizes polymer/polymer epitaxies and the various possible orientations of the substrate and expected levels of orientation of the deposit. Most frequently,the substrate consists of uniaxially drawn fibres or films which only exceptionally possess a single crystal-like character. Indeed,only the chain axis direction is fixed; as a consequence the contact plane(s) and thus the epitaxy can never be precisely defined.

Table 1. Polymer/polymer epitaxies classified according to the substrate and resulting deposit orientations.

Systems	Observations and possible analysis	Substrate	Deposit
Fiber / Matrix Film / Matrix	Relative orientation of the c axes at the interface		
Fiber / Fiber (oriented blends)	Relative orientation of the c axes Relative orientation of the lamellae		
Oriented film / Film	Relative orientation of the c axes. Contact plane of the deposited film Lamellar orientation		
Single Crystal / Single Crystal	Detailed structural relationship		

More suitable polymeric systems are the high modulus fibers or films drawn from the gel state (Smith et al.(1979), from the virgin state (Smith et al.(1985)), or else single crystalline layers formed by solid state polymerization or even epitaxially grown polymer films with a single unit cell orientation (Lotz and Witmann (1984)).

3. Growth procedures devised for E.M preparations.

Electron microscopy and electron diffraction appear as most appropriate for morphological and structural studies of epitaxially grown layers or crystals. The fact that very thin samples have to be used offers, despite the severe constraints imposed on sample preparation and handling, great advantages since it helps to avoid the often complex growth patterns of more bulky samples (lamellar twisting ...) leading generally to a rapid loss of orientation. As illustrated for the case of polyethylene (PE) below, this thickness limitation makes it also possible to reveal the existence of transient crystal phases induced by the substrate and to fully characterize the related phase transitions (Wittmann and Lotz (1989)).

To study the epitaxial relationship per se, a substrate/deposit bi-crystal should ideally be used whenever the substrate withstands the E.M. vacuum. Only with a composite diffraction pattern such as the one shown in Fig.3a, can the relative unit cell orientations and the respective contact planes be univoquely determined. When the substrate must be removed (because of its high vapor

246

pressure, excessive thickness...) the epitaxial relationship can only be deduced indirectly by correlating the deposit orientation with the substrate growth habit.

Figure 3. (a) Composite diffraction pattern of a thin PE film oriented on the potassium salt of benzylpenicillin and (b) Diffraction pattern of PE film vapor-deposited on the potassium salt of p-bromobenzoic acid after selective dissolution of the substrate. The monoclinic 010 and 111 reflections are arrowed on patterns (a) and (b) respectively.

The various substrates can be broadly divided into two different classes depending on whether or not the substrate can be used in the molten state as solvent for a given deposit. The first class includes almost exclusively organic substrates like the lower members of aromatic hydrocarbons (naphthalene, p-terphenyl...) or aromatic acids (benzoic acid and p-substituted homologues) and the second class encompasses the higher members and the salts of the previous series as well as the polymeric and inorganic substrates.

As already mentioned for alkali halides, epitaxial crystallization on the latter class of substrates, which stay in the solid state during the entire growth process, can be achieved in classical ways from solution, melt or vapor phase. Noteworthy, vapor phase deposition can also be applied to some polymers and namely polyolefins (cf. paper by Lotz and Wittmann on polymer decoration).

More original growth procedures have been devised with organic substrates having melting temperatures T_{ms} higher than or comparable to the deposit melting temperature T_{md}. Advantage is taken of the fact that, on cooling deposit/substrate homogeneous solutions or hypoeutectic mixtures (when $T_{ms} \approx T_{md}$), uncontaminated substrate crystals can be formed in-situ prior to the crystallization of the host compound concentrated in the vicinity of the growth front.

In practice, dilute deposit/substrate solutions are formed between two glass slides by co-melting crystals of the two species.at $T>T_{ms}$ The thin molten layer is then solidified for example by moving it towards the cooler end of a Köffler hot stage in order to favor the growth of large, mono-oriented substrate crystals and the subsequent epitaxial growth of the deposit on top of them.

After complete solidification, the glass slides are separated mechanically and the substrate crystals removed when necessary by sublimation in vacuum or selective dissolution. The oriented films of the deposit left on both slides have the proper thickness for E.M. investigations, provided that the concentration of the starting solution is kept low enough.

In alternative methods elaborated by Fryer (1981) or Dorset (1986), carbon coated grids are either dipped directly in the molten mixtures or placed between glass slides or mica sheets prior to the co-melting and subsequent solidification processes.

These various techniques have been successfully used for polyolefins and paraffins, linear and aromatic polyesters, polyamides, phospholipids... in association with aromatic hydrocarbons, aromatic acids, quinones etc...

4. Case examples of polymer epitaxy.

4.1. POLYETHYLENE.

The most important studies on epitaxy of long chain molecules have focused on PE (and paraffins) for a number of reasons, mainly experimental and structural as e.g.:
- comparatively low melting temperature; good chemical and thermal stability with nevertheless the possibility to realize vapor phase deposition in vacuum; solubility in various organic solvents etc...
- a planar zigzag all-trans chain conformation and a simple linear molecular shape close to a cylindrical rod,
- two isoconformational crystal polymorphs: a stable orthorhombic and an unstable monoclinic modification offering as shown in Fig. 4 a wide range (0.404 to 0.918 nm) of interchain distances, a parameter which plays a major role in polymer epitaxy.

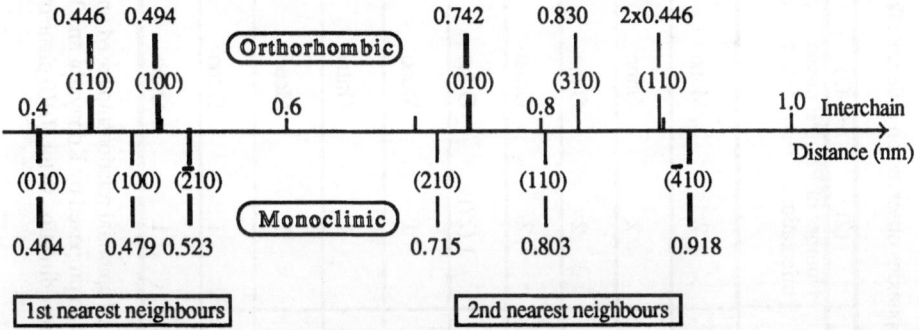

Figure 4. Linear scale of PE interchain distances in the orthorhombic and monoclinic phases. Heavy bars correspond to all the observed contact planes.

Some of the most representative examples of PE epitaxies on ionic, organic and polymeric substrates known today are reported in Table 2.

A detailed analysis of these PE epitaxies (Wittmann and Lotz (1989)) has led to important conclusions summarized below and concerning the impact of the growth procedure used, the substrate structure and periodicities determinant for the polymer orientation(s), the issue of lattice matching, the polymorphic transitions etc...

Table 2 : PE epitaxies observed on different alkali halides, organic and polymeric substrates.

(1) Substrate	(2) Number of PE orientations	(3) PE phase in contact	(4) Substrate directions	(5) PE / substrate contact planes	(6) d(PE)/d(subs.), Disregistry Δ (in %)		(7) Monoclinic/ orthorhombic transition
NaCl	2	Mono	<110>	$(010)_m/(100)$	0.404/0.398	+1.53	$(010)_m \rightarrow (110)_o$
KBr	2	Ortho	<110>	$(110)_o/(100)$	0.446/0.465	-4.15	–
Anthracene	2	Ortho	<110>	$(100)_o/(001)$	0.495/0.493	+0.32	–
p-Terphenyl	2	Ortho	<110>	$(110)_o/(001)$	0.446/0.460	-3.09	–
Benzoic acid	1 (2?)	Ortho	[100]	$(100)_o/(001)$	0.495/0.514	-3.77	–
p-Br Benzoic acid	1	Mono	[001]	$(210)_m/(100)$	0.523/0.560	-6.53	$(210)_m \rightarrow (100)_o$
p-Phenyl Benzoic acid	1	Ortho	[001]*	$(010)_o/(010)$ *	0.747/0.744 *	+0.40	–
K Benzyl Penicillin	1	Mono	[010]	$(\bar{4}10)_m/(001)$	0.918/0.360	-1.90	$(\bar{4}10)_m \rightarrow (310)_o$
Polyoxy-methylene**	1	Ortho	[001]	$(110)_o/(10\bar{1}0)$	0.443/0.446	-0.01	–
Isotactic Polypropylene	1	Ortho	[101]***	$(110)_o/(010)$	0.494/0.505	-0.02	–

(*) Hypothetical epitaxial relationship based on the substrate crystal structure.
(**) Relationship proposed by Kobayashi and Takahashi(1964).
(***) Methyl row direction in the (010) plane of iPP.

248

i) All the epitaxial relationships analyzed so far (Table 2) rest on a close dimensional matching between the polymer interchain distance d_{PE} and the substrate periodicity d_{subst} in their respective contact planes (cf.Fig.5), with a maximum of ~6.5% disregistry Δ expressed as usual by:

$$\Delta = 100(d_{PE}-d_{subst}) / d_{subst}$$

As clearly apparent, this matching controls not only the PE crystal orientation but also the crystal modification. Dimensional matching in a direction normal to the previous one, i.e. along the chain axis, may also exist but appears less determining for the deposit orientation.

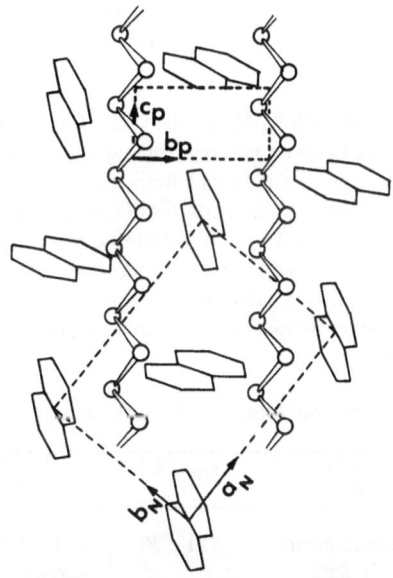

Figure 5. Epitaxial relationship between PE and aromatic hydrocarbons. The PE chain is oriented along properly spaced "furrows" formed by the aromatic rings on the (001) contact surface.

ii) Double or single orientations are obtained depending on the substrate structure and related surface symmetry which, for the very efficient aromatic substrates, is dictated by the packing of the phenyl rings in the exposed apolar layers (Fig. 2).

Furthermore, except for the very particular case of iPP (Lotz and Wittmann (1986)) a unique deposit orientation is generally obtained on oriented polymeric substrates as a result of chain axes parallelism.

iii) The observed PE orientations -three different ones for each polymorph- involve mostly densely packed planes with first nearest neighbour PE chains (cf. Table 2 column 5 and Fig. 4).

Epitaxies which involve the much larger interchain distances corresponding to second nearest neighbours (Fig.4) are of course energetically less favourable. Their occurence on substrates like the rather exotic potassium salt of benzylpenicillin (periodicity of ~0.9 nm) is probably promoted or better "forced" by the unusual conditions (very high supercooling) used for the crystallization of PE deposited from the vapor phase on crystals held at room temperature.

iiii) In agreement with the earlier findings of Wellinghoff et al. (1974), the epitaxially induced monoclinic phase transforms systematically, within a short distance from the interface, into the more stable orthorhombic form. As can be deduced from the relative orientations of the two phases, the mono / ortho transition (column 7 in Table 2) seems to follow a single pathway identical to Mode I described by Seto et al. (1968) for a similar, mechanically induced phase transition.

All the above findings provide strong support in favour of at least a one dimensional lattice match between deposit and substrate. Accomodation of chain molecules to different substrates may require a reorientation of the deposit lattice i.e. different contact planes, the formation when possible of isoconformational polymorphs or even of crystal modifications with different chain conformations as is the case for Poly-1-butene (PB-1) discussed next.

4.2. POLY-1-BUTENE.

Isotactic poly-1-butene is a most interesting polymeric system for structural studies and represents also a real challenge in the field of polymer epitaxy. Indeed, it crystallizes at ambient pressure in three different modifications each with a different helix conformation (Table 3). Interestingly enough, the unstable tetragonal form II transforms spontaneously into the stable rhombohedric form I on storage at room temperature. The orthorhombic form III is metastable and cannot therefore be oriented as usual by stretching for X-ray structural investigations. Until now, the structure of this form has only been investigated by X-ray powder diffraction techniques (Cojazzi et al.(1976)) or partially by electron diffraction on solution grown single crystals (Holland and Miller (1964))

Table 3. Polybutene-1 crystal structures.

Crystal modification	Symmetry and conformation	T_m (°C) *	Unit cell parameters (nm)	Occurence
Stable form I Natta et al.(1960)	Rhombohedric 3_1 helix	121-136	a=1.77,b=1.77 c=0.65, γ=120°	Solution
Unstable form II Turner Jones(1963)	Tetragonal 11_3 helix	100-120	a=1.49, c=2.09	Solution, melt
Metastable form III Cojazzi et al.(1976)	Orthorhombic 4_1 helix	~96	a=1.24, b=0.89 c=0.76	Solution, solvent evaporation

* From Luciani et al.(1988)

Current investigations show that by a judicious choice of aromatic substrates, oriented thin films of all three modifications can be successfully grown by epitaxy. Although details of the different epitaxial relationships have still to be worked out the following informations are available:
- The unstable, kinetically favoured form II has been oriented on benzoic acid (Lotz and Wittmann (1984)). However, in contrast to polymethylene chains on the same substrate, a puzzling geometry with two PB-1 chain axis orientations 94° apart is observed. The tetragonal form gradually transforms to form I with time or irradiation under the electron beam.
- Thin oriented films of the stable form I, which is seldom obtained directly, and of the metastable form III can be obtained on 4-chloro benzoic acid and 2-hydroxyquinoxaline crystals

respectively, both with a unique orientation of the helix axis.

These original samples offer not only the possibility to better characterize the various crystal structures (especially that of the metastable form), but also to follow by electron microscopy and electron diffraction the various spontaneous or thermally induced phase transformations of PB-1.

5. Uses and potential.

Alkali halides are substrates still often used, in view of their ease of use for E.M. preparations, for exploratory experiments such as the epitaxial polymerization first proposed for $(SN)_x$ (Rickert et al.(1979)) and subsequently applied to polydiacetylenes (Rickertet al.(1983)) or poly paraxylylene (Isoda (1984)).

However, organic substrates offer significant improvements over inorganic ones and lead to an ever increasing number of applications regarding the types of long chain molecules oriented on them, the use of such oriented samples for morphological and structural analyses and, last but not least, the elaboration of highly structured thin films with appropriate physical or physico-chemical properties.

The examples described above underline the ability and high versatility of aromatics to orient polymers of different chemical constitution, conformation or crystal structure (Wittmann and Lotz (1985,1990)). As pioneered by Dorset (1985,1987) and by Dorset and coworkers (1983,1988,1989) a straightforward extension of the above can evidently be made to paraffins, but also to an impressive catalog of shorter chain molecules including cycloalkanes, diglycerides, phospholipids ...The catalog of structural studies carried out on single crystal-like films in various projections normal to the chain axes is equally impressive and covers crystal structure analysis(Dorset (1985b,1987), Dorset et al.(1988)) direct imaging of lamellar structures (Fryer(1981), Fryer et al. (1987)), crystal-crystal transitions as exemplified above, formation of solid solutions and eutectics (Dorset(1985a,1986)) etc...

Epitaxial growth on organic substrates offers also a means of tailoring the polymer film structures in the search of new specific properties e.g. optical or electronical properties; the first example is provided by the epitaxial synthesis of cis-polyacetylene on aromatic crystals containing dissolved catalyst (Woermer et al. (1982)).

Finally, oriented polymeric substrates have since long been used as substrates not only for the study of polymer-polymer interactions (Willems(1973),Takahashi et al. (1970,1976), Broza et al. (1985), Lotz and Wittmann (1986),Petermann et al. (1987)) but also for the alignment of nematic or smectic liquid crystal materials (e.g. Geary et al. (1987)) and recently also for the oriented growth of low molecular weight organics (Parikh and Philips (1985) or even metals (Schultz and Peneva (1986), Petermann and Broza (1987)).Unfortunately, the quasi-general use of uniaxially drawn films or, even worse, films mechanically rubbed or buffed, has severely hindered a detailed analysis of the polymer / deposit relationships or of the alignment mechanisms. However, rapid progress is very likely to be made in this field since processes like drawing from the gel state, solid state polymerization or epitaxy itself offer new methods for growing large, single crystalline polymeric substrates.

6. Conclusion.

The success achieved in the field of epitaxial growth of long chain molecules on organic or polymeric substrates is undoubtely due to the major improvements brought about by such substrates: adjustable chemical compatibility and to a certain extent crystalline structure, wide range

252

of lattice periodicities allowing (via lattice matching) a great number of different orientations or even different polymorphs for a given polymer, low symmetry often resulting in a single and well defined orientation of the deposit, etc...

The original growth procedures developed, especially those taking advantage of the deposit / substrate reciprocal solubility, have thus led to highly structured long chain molecule layers which have already found a number of valuable applications in morphological and structural investigations by electron microscopy, in polymer decoration or in the processing of organic materials for specific applications requiring a good structural control.

A better understanding of the alignment mechanisms on polymeric substrates will certainly help to widen the spectrum of the foregoing applications.

References

Bradshaw, M.J., Brimmell, V. and Raynes, E.P. (1987) 'A novel alignment technique for ferroelectric smectics', Liquid Crystals 2, 107-110

Broza, G., Rieck, U., Kawaguchi, A. and Petermann, J. (1985) 'Epitaxial Crystallization of Polyethylene and Paraffin on Oriented Polypropylene', J. Polym. Sci., Polym. Phys. Ed. 23, 2623-2627

Cojazzi, G., Malta, V., Celotti, G. and Zanetti, R. (1976) 'Crystal structure of form III of isotactic poly-1-butene', Makromol. Chem 177,915-926

Dorset, D.L., Pangborn, W.A. and Hancock, A.J. (1983) 'Epitaxial crystallization of alkane chain lipids for electron diffraction analysis', J. Biochem. Biophys. Meth. 8, 29-40

Dorset, D.L. (1985a) 'Crystal structure of n-paraffin solid solutions, an electron diffraction study', Macromolecules 18, 2158-2163

Dorset, D.L. (1985b) 'Electron crystal structure analysis of small organic molecules', J. Electron Microsc. Techn. 2, 89-128

Dorset, D.L. (1986) 'Crystal structure of lamellar paraffin eutectics', Macromolecules 19, 2965-2973

Dorset, D.L. (1987) 'Electron diffraction structure analysis of phospholipids', J. Electron Microsc. Techn. 7, 35-46

Dorset, D.L., Wittmann, J.C., Lotz, B. and Fryer, J.R. (1988) 'Epitaxy and the crystal structure analysis of linear molecules in the electron microscope', Inst. Phys. Conf. Ser. 93, 9-14

Dorset, D.L. and Hsu, S-L. (1989) 'Polymethylene chain packing in epitaxially crystallized cycloalkanes: an electron diffraction study', Polymer 30, 1596-1602

Fryer, J. R. (1981) 'Crystallization and lattice resolution of a straight paraffin nC36H74 and its adducts', Inst. Phys. Conf. Ser. 61, 19-22

Fryer, J. R. and Dorset,D. L. (1987) 'Direct imaging of paracrystalline phospholipid structure in the electron microscope', J. Microscopy 1, 45-61

Geary, J.M., Goodby, J.W., Kmetz, A.R. and Patel, J.S. (1987) 'The mechanism of polymer alignment of liquid-crystal materials' J. Appl. Phys. 62, 4100-4108

Holland, V.F. and Miller, R.L. (1964) 'Isotactic polybutene-1 single crystals: morphology', J. Appl. Phys. 35, 3241-3248

Isoda, S. (1984) 'Epitaxial synthesis of poly(p-xylylene)', Polymer 25, 615-624

Kobayashi, K. and Takahashi, T. (1964) 'Epitaxial growth of crystalline polymers' in 'The theory of crimp of textile fibers' Memoirs Faculty Eng. Kyoto Univ. 26, 247-256

Lotz, B. and Wittmann, J.C. (1984) 'Epitaxy of helical polyolefins: polymer blends and

253

polymer -nucleating agent systems', Makromol. Chem 185, 2043-2052

Lotz, B. and Wittmann, J.C. (1986) 'Structural relationships in blends of isotactic polypropylene and polymers with aliphatic sequences', J. Polym. Sci. Polym. Phys. Ed 24, 1559-1575

Luciani, L., Seppälä, J. and Löfgren, B. (1988) 'Poly-1-butene: its preparation, properties and challenges', Prog. Polym. Sci. 13, 37-62

Mauritz, K.A., Baer, E. and Hopfinger, A.J. (1978) 'The epitaxial crystallization of macromolecules', J. Polymer Sci., Macromol. Rev. 13, 1-61

Natta, G., Corradini, P. and Bassi, I.W. (1960) 'Crystal structure of isotactic poly-α-butene', Nuovo Cimento, Suppl. 15, 52-53

Parikh, D. and Philips, P.J. (1985) 'The mechanism of orientation of acridine in oriented polyethylene', J. Chem. Phys. 83, 1948-1951

Petermann, J. and Broza, G. (1987) 'Epitaxial deposition of metals on uniaxial oriented semi-crystalline polymers', J. Mat. Sci. 22, 1108-1112

Petermann, J., Broza, G., Rieck, U. and Kawaguchi, A. (1987) 'Epitaxial interfaces in semi-crystalline polymers and their applications', J. Mat. Sci. 22, 1477-1481

Rickert, S.E., Lando, J.B., Hopfinger, A.J. and Baer, E. (1979) 'Epitaxial polymerization of $(SN)_x$. 1. Struture and morphology of single crystals on alkali halide substrates', Macromolecules 12, 1053-1057

Rickert, S.E., Lando, J.B and Ching, S. (1983) 'The formation of single crystal films of polydiacetylenes' Mol. Cryst. Liq. Cryst. 93, 307-314

Schultz, J.M. and Peneva, S.K. (1986) 'RHEED investigation of thin tin films on polypropylene', Bull. Am. Phys. Soc. 31, 511

Seto, T., Hara, T. and Tanaka, K. (1968) 'Phase transformation and deformation processes in oriented polyethylene', Jpn. J. Appl. Phys. 7, 31

Smith, P., Lemstra, P.J., Kalb, B. and Pennings, A.J. (1979) Ultrahigh-strenght polyethylene filaments by solution spinning and hot drawing' Polymer Bull. 1, 733-736

Smith, P., Chanzy, H.D. and Rotzinger, B.P. (1985) 'Drawing of virgin ultrahigh molecular weight polyethylene: an alternative route to high strenght fibres', Polymer 26, 258-260

Swei, G. S., Lando, J. B.,Rickert , S. E. and Mauritz, K.A (1986) 'Epitaxial processes', Encyclopedia of Polymer Science and Engineering vol. 6, p 209

Takahashi, T., Inamura, M. and Tsujimoto, I. (1970) 'Epitaxial growth of polymer crystals on uniaxially drawn polymers', J. Polymer Sci.; Polym. Lett. Ed. 8, 651-657

Takahashi, T., Teraoka, F. and Tsujimoto, I. (1976) 'Epitaxial crystallization of crystalline polymers on the surface of drawn polytetrafluoroethylene', J. Macromol. Sci.-Phys. B12, 303-315

Turner-Jones, A. (1963) 'Polybutene-1 - Type II crystalline form', J. Polym. Sci. part B 1, 455-456

Wellinghoff, S., Rybnikar, F. and Baer, E. (1974) 'Epitaxial crystallization of polyethylene', J. Macromol. Sci.-Phys. B10, 1-41

Willems, J. (1973) 'Orientierte Aufwachsung von Polyathylen und Paraffin auf Polyamiden', Kolloid-Z. u. Z. Polym. 251, 496-497

Wittmann, J.C. and Lotz , B. (1981) 'Epitaxial crystallization of polyethylene on organic substrates : a reappraisal of the mode of action of selected nucleating agents', J. Polym. Sci., Polym. Phys. Ed. 19, 1837-1851

Wittmann, J.C., Hodge, A.M. and Lotz, B. (1983) 'Epitaxial crystallization of polymers onto benzoic acid: polyethylene and paraffins, aliphatic polyesters, polyamides', J. Polym. Sci.

Polym. Phys. Ed. 21, 2495-2509

Wittmann, J.C. and Lotz, B. (1985) 'Epitaxial crystallization of long chain molecules: morphological and structural investigations and their applications', Inst. Phys. Conf. Ser. 78, 417-422

Wittmann, J.C. and Lotz, B. (1989) 'Epitaxial crystallization of monoclinic and orthorhombic polyethylene phases', Polymer 30, 27-34

Wittmann, J.C. and Lotz, B. (1990) 'Epitaxial crystallization of polymers on organic and polymeric substrates', submitted to Prog. Polym. Sci.

Woerner, T., MacDiarmid, A.G. and Heeger, A.J. (1982) 'Direct synthesis of oriented polyacetylene, $(CH)_x$, from acetylene gas' J. Polym. Sci. Polym. Lett. Ed. 20, 305-309

Wunderlich, B. (1973) Macromolecular Physics, vol. 1, Academic Press, New York and London,

SOLVING DIFFICULT STRUCTURES INCLUDING THOSE FROM POWDER AND ELECTRON DIFFRACTION DATA

C.J. Gilmore, Department of Chemistry,
University of Glasgow, Glasgow G12 8QQ, Scotland.

1.0 Introduction

This chapter outlines some of the options available to you if routine direct methods have failed to solve a structure. I am going to include data from powder and electron diffraction, since the *ab-initio* solution of crystal structures from these experiments is becoming a topic of increasing importance. The possibilities outlined in the following sections are not quoted in any order of preference, except perhaps for normalisation where the most scope exists for easy ways to alter the phasing path taken by the direct methods program you are using.

The first necessity is to have available *several* direct methods packages and not just one. All programs have their strengths and weaknesses, and one structure may yield readily to one program whilst being quite difficult for another. It is also important to *read the output* generated by these programs. This may sound facetious, but such is the level of automation in direct methods, there is a temptation to try as many options as quickly as possible without scanning the output carefully.

Why are some structures difficult? There are some obvious reasons and these are usually known *before* data collection begins:

(i) The structure is large and/or there are several molecules in the asymmetric unit. The latter are often particularly difficult; the presence of multiple molecules in the asymmetric unit gives rise to phase relationships between certain phase angles which can distort the phasing process.
(ii) The crystal diffracts weakly and there is a paucity of high angle reflections - in particular those which violate Sheldrick's Rule:

"If less than 50% of the theoretically observable reflections in the resolution range 1.1-1.2 A are observed (F>4.0σ(F)) then the structure will be difficult to solve by conventional direct methods."

J. R. Fryer and D. L. Dorset (eds.), Electron Crystallography of Organic Molecules, 255–263.
© 1990 Kluwer Academic Publishers.

256

(iii) The data are sparse - this applies to powder and electron diffraction data. In the former case peak overlaps effectively reduce resolution, and in the latter the use of tilted slices and the methods by which the sample is prepared limits the accessibility of reciprocal space.
(iv) Pseudosymmetry may be present giving rise to subsets of reflections which are systematically weak.
(v) Structures with a high degree of internal regularity.
(vi) The sheer perversity of nature.

Many structures give problems in most of these categories - macromolecules for example. In general, however, these problems are known *before* data collection and it is important to use this information from the outset; in other words data collection becomes an important aspect of solving the structure.

2.0 Data Collection.

The rules are simple to write, but often difficult to practice:

(i) Spend as much time as possible to get good intensity measurements of the *weak* reflections. They intervene in phasing in three ways: as data for normalisation, in the ψ_0 and NQEST figures of merit, and in the estimation of negative quartets for active use of four-phase invariants. If you have pseudosymmetry present collect the systematically weak reflections separately from the strong and scale the two data sets. Never use options that pre-scan peaks, and to use this information as a filter by which the weak reflections are not measured. The Bayesian approach pioneered by French and Wilson (French & Wilson, 1978) provides an optimal approach to data processing.
(ii) Use low temperature if possible. It lowers the background and hence improves signal to noise ratios and so can extend the data resolution.
(iii) Collect equivalent reflections.
(iv) In powder diffraction the highest possible resolution is of paramount importance; obviously synchrotron data is optimal.

3.0 Normalisation.

(i) Check the data. If there are many duplicates or systematic absences list them all. Then investigate this list with suspicion - are you sure about the space group? Have you collected at least the unique data? How good is the resolution of the data? It is important not to put duplicate reflections into a direct methods program which does not test for them.

(ii) Read the normalisation output very carefully. Are the statistics sensible? Do the large E-magnitudes form a readily identifiable subset where some parity groups are missing? If some of the E-magnitudes are very large (say >3.5-4.0), it is often useful to reduce or remove them. The variation of $E^2 - 1.0$ as a function of $\sin \partial/\lambda$ is a good guide to the applicability of the calculated temperature factor. If there is a fall-off in this average as $\sin \partial/\lambda$ increases, use the relevant command to input a larger value of B than that calculated by the normalisation module. If some or all the stereochemistry is known, then input this as a randomly oriented, randomly positioned fragment if your program permits. This is especially important with molecules containing planar or nearly planar fused rings.

(iii) If you have generated the best set of E-magnitudes that you can, and the structure still will not solve, then a systematic distortion of the E's can be successful. Changes in the E-magnitudes cause changes in relative weights of the invariants, and these in turn give rise to drastic modifications of the convergence map, and the subsequent phasing path. There are several ways of doing this:

(a) Modify the unit cell contents-doubling the contents is the best starting point.
(b) Use artificially raised or lowered temperature factors. Often only a small change in B gives rise to drastic changes in the E's. This is especially recommended for situations where the resolution is less than the Cu sphere.
(c) Insert a molecular fragment that does not correspond to any group expected in the molecule.

(iv) Downweight or remove the high angle E-magnitudes. These are often the E's with the largest associated standard deviations. Triplets involving three such E's are especially prone to error.

(v) Try using another normalising program that has different facilities. For example the NORMAL module in the X-ray system permits the use of an overall anisotropic temperature factor. Nixon (1978), has described an alternative approach to normalisation via the Patterson function.

The problems of normalising sparse or restricted data are very acute since there is an implicit assumption in the process that all non-input data is zero. Under these circumstances it is usually better to define a temperature factor and let the program scale although this still introduces errors. A Bayesian approach to normalisation can be much more powerful here especially when prior knowledge is used. This is an option in the new version of MITHRIL (Gilmore and Brown, 1988)

4.0 Invariants and Seminvariants

There are two problems which can arise here. One concerns a paucity of suitable relationships which can be common in situations of

low symmetry.   The other concerns the accuracy of the invariants
themselves.

   (i) If there is a paucity of triplets:

(a) Use quartets as well actively in the phasing process.   Try just the
negative ones first, then add the positives if this is unsuccessful but
remember the correlations which exist between triplets and positive
quartets for which allowance must be made.
(b) Increase the number of reflections for which triplets are generated.
(c) Reduce the minimum κ-value ($\kappa = |E_h E_k E_l|/\sqrt{N}$) for which triplets are
used.   This will, however, introduce a number of very unreliable
relationships.

   (ii) If there is a paucity of quartets:

(a) Increase the number of reflections for which quartets are generated.
This is usually the best way.
(b) Invoke the third neighbourhood although this increases the required
computer time.
(c) Ask for positive quartets as well.   This can also be useful when
triplets are scarce.
(d) Allow more missing second (and third) neighbours; in this case the
missing E-magnitudes are given values of unity.

   (iii) If there is still a paucity of good negative invariants,
then it can be worth considering quintets but it is very time
consuming and of dubious value.

   (iv) Even if the space group is not symmorphic, quartets often have
a very beneficial effect on a direct methods analysis, and can be
recommended as an option to try early in the list of weapons in the
armoury.   Only the negative quartets should be tried first, since they
are independent of the triplets.   Even a few four-phase invariants
can drastically alter the phasing path.

   (v) The MDKS or similar formulae which attempt to estimate the
triplet cosine, coupled with triplet weighting also has a drastic
effect.   However, the MDKS estimates themselves are very time
consuming to calculate, and are very inaccurate.   The P_{10} formula
(Cascanaro, Giacovazzo, Camalli, Burla, Nunzi & Polidori 1984) is more
reliable.

   In the case of powder or electron diffraction, quartets are
essential and it may prove necessary to attempt the phasing of those
reflections having $|E_h| < 1.0$.

5.0 The Starting Set and Convergence Mapping.

Convergence mapping lies at the very heart of the multisolution approach to direct methods. With the rise of random phasing procedures, it is less significant that hitherto, but it is still a very informative procedure, since it shows how individual reflections are linked *via* the invariants. It is important to examine the convergence map carefully in cases of difficult, even if random phasing is being used.

(i) Make sure that the starting set is a good one with all the starting set reflections used early in the phase determination. This can be checked by examining the bottom of the convergence map. If a starting set reflection is not used at all early on, then a better starting point can often be obtained by including at least one other reflection whose phase depends on that of the late starter. Introducing quartets or quintets may also have a similar effect.

(ii) If there are gaps near the bottom of the convergence map (i.e. reflections with a zero estimated α and no invariants contributing), or the map is very 'thin' with many phases determined by only one or two relationships, then the phasing often fails. This can be remedied by increasing the size of the starting set or introducing higher invariants, particularly quartets.

(iii) Be wary of the Σ_1 determined phases. If they play a major role in the early stages of phasing, it is often worthwhile excluding them. They can be given permuted phases if really necessary. If the MDKS or a similar checking equation is available, perform a Σ_1 triplet analysis. The special triplets should have estimated cosiness close to unity.

(iv) If it still proves impossible to obtain a suitable convergence map without a massive amount of computer time, then several options are possible:

(a) Run MAGEX or symbolic addition. Apart from origin and enantio-morph definition, they can largely ignore the convergence map.
(b) Run YZARC. Try both least-squares and steepest descents - they give different results.
(c) Run a random tangent procedure instead of regular tangent refinement.

(v) Check to see if all the reflections at the bottom of the map have something in common e.g. they all have h even or k+1 divisible by 3. If so, then make sure that the average value of E^2 - 1.0 is unity for such reflections. Renormalisation may be necessary. It may be possible to use editing facilities at normalisation time to juggle these magnitudes. Try introducing new reflections into the starting set which do not belong to these groups.

(vi) Altering the origin and enantiomorph is often unsuccessful, particularly if only small changes are made.   The same relationships are still used in the early stages of phasing, but in a different form.   For example, the triplet $\Psi_1 - \Psi_2 + \Psi_3$ may appear in one map generating Ψ_1 from Ψ_2 and Ψ_3.   If the origin is partially re-defined by the user, this triplet may well appear again in a critical place but this time generating Ψ_2 from Ψ_1 and Ψ_3.   If this triplet is erroneous it will be erroneous however it is used.   This said, juggling with the starting set can be successful on some occasions, and is worth a try.

(vii) Symbolic addition, even in a limited form, can give rise to possible relationships between phases and these can be introduced into the convergence map if your program permits.   The relationships linking two phases (the pair relationships) are the most valuable. The inclusion of only one or two with high associated κ-value will drastically alter a convergence map.   There is the added bonus that symbolic addition can give valuable insights into the causes of phasing difficulties. (Karle and Karle 1966).   Do not use the convergence map for symbolic addition; get a list of triplets and work with this.   The convergence procedure has too many weak relationships early in the phasing path.   Symbolic addition can be very useful for heavy atom structures where it can prevent the overconsistency problem of the tangent formula.

6.0 Phase Expansion and Refinement

The same formula is used for both phase extension and refinement. This is usually the tangent fomula, but Woolfson and co-workers have pioneered the use of Sayre's equation in the SAYTAN package, and Sheldrick has developed a method based on simulated annealing which is a powerful optimisation technique now finding increasing use in a wide variety of such problems.   There is also the maximum entropy method which is discussed in section 8.

The only way to monitor phase expansion and refinement in traditional packages is by inspecting the final figures of merit, so it is important to examine these closely in difficult cases.   In particular do not just inspect the final combined figure or merit (CFOM), but look also at its individual contributors:

(a) ABSFOM is the least reliable.   If the ABSFOM values all tend to be large then the refined phases are over-consistent.   Using Hull-Irwin weights will often give better results.
(b) NQEST and ψ_0 are the most reliable figures of merit provided that the weak reflections have been accurately measured.   Do not expect NQEST to be very negative, particularly if quartets are to be used actively in phase refinement.   In these circumstances values around −0.1 are often satisfactory.
(c) Heavy atom cases often give extreme figures of merit.   The

correct solution may well be present even if the figures of merit seem unrealistic.
(d) If all the phase sets have similar figures of merit too few invariants may be present. See section 2 for a remedy.

In the case of pseudo-symmetry, the presence of heavy atoms or substantial planar moieties in the structure, use the Hull-Irwin scheme. Do not forget weighting schemes. If your system has more than one, try them all - if one does not work, the other may, even using the same starting set and convergence map.

Be careful of early stop options in which the program stops tangent refinement finds a solution with an obviously optimal set of figures of merit. It is often found that the set selected by the tangent refinement modules during refinement as being an obvious solution is not the correct one. Similarly, if you are using early figures of merit, and most sets are getting rejected, it is possible that the correct set is also being discarded. Turn off early figures of merit for all difficult structures.

Do not confine your attentions to the one or two phase sets with the highest CFOM's. It may be necessary to examine maps with quite low associated CFOM's. The CFOM's themselves are dependent on the relative weights of the individual figures of merit. Adjusting the weights to reflect your own intuition concerning the contributing figures of merit will often result in a drastic re-ranking of the solutions.

Sparse data sets will give figures of merit that are often worthless; it is necessary to examine all the solutions produced. An alternative in all these cases is to use a maximum entropy-likelihood method.

7.0 Maps

Look at the resulting E-maps carefully. Remember that the interpretation assumes that you have well-resolved peaks, and this may not be the case. The routines which perform the chemical interpretation may be quite sophisticated, but they are never as good as a trained crystallographer. Do not, therefore, accept the given interpretations as the only possibilities. Some other points to note are:

(a) If the map contains one or two large peaks and no heavy atoms are expected, the phases are probably incorrect - but not always. Some-times something can be salvaged. If heavy atoms are present direct methods will probably only produce these atoms.
(b) If the E-maps show pseudosymmetry switch to Hull-Irwin weights in tangent refinement or use simulated annealing or the Sayre approach.
(c) One or two missing peaks coupled with one or more spurious ones,

can quickly make a map uninterpretable. Increasing the number of peaks can make parts of the map more readily interpretable at the expense of producing more noise peaks.

In the case of powder and electron diffraction data the contoured maps themselves must be examined. It is worthwhile taking a leaf from the protein crystallographer's book and transferring maps to a graphics device capable of running FRODO/TOM to examine them and to carry out model fitting exercises. This is of obvious applicability to small single crystal structures as well.

In very difficult cases be tenacious. If just a small portion of the expected structure is found, stick with it through all the possible recycling schemes. It may be correctly oriented but misplaced, so the use of the relevant group type (3 in MULTAN) useful, but this procedure is not infallible. It sometimes does not work even with correct information. In addition, the use of a rotation-translation function package will prove indispensible. The use of vector methods in SHELX has been discussed by Eggert (Eggert 1983). It is an approach that could be very relevant when dealing with sparse data sets. C.C. Wilson has discussed the limits of data quality in vector methods (Wilson 1989).

As an object lesson in tenacity and persistence see Karle, Karle, Mastropaolo, Camerman and Camerman 1983.

8.0 Maximum Entropy

Bricogne and Gilmore have shown in two recent papers (Bricogne & Gilmore 1990; Gilmore, Bricogne & Bannister, 1990) how Bayesian methods can be used to solve crystal structures by using a maximum entropy prior combined with likelihood which serves, in part, as a figure of merit in a multisolution environment of great power. One feature of the method is stability at all ranges of data resolution and data sampling. The method has been used to solve $KAlP_2O_7$ from its powder diffraction pattern using ca. 60 non-overlapped reflections in routine calculations. This stability extends to other data sets, indeed the method is stable with just meridional reflections from fibre diffraction. It seems likely that the method offers the best approach to sparse data sets.

References

Bricogne, G and Gilmore, C.J. (1990) Acta Cryst., A46, 284-297
Burla, M.C., Cascanaro, G., Giacovazzo, C., Nunzi, A., and Polidori, G. (1987), A43, 370-374.
Cascanaro, G., Giacovazzo, C., Camalli, M., Burla, M.C., Nunzi, A. and Polidori, G. (1984) Acta Cryst. A40, 278-283
Eggert, E. (1983) Acta Cryst. A39, 936-940
French, S. and Wilson, K. (1978) Acta Cryst. A34, 517-525.

Gilmore, C.J., Bricogne, G and Bannister, C. <u>Acta</u> <u>Cryst</u>. (1990) A46, 297-308.

Karle, J. and Karle, I.L. (1966) <u>Acta</u> <u>Cryst</u>. 21, 849-859.   Karle, I.L

Karle, J., Mastropaolo, D., Camerman, A. and Camerman, N. (1983) <u>Acta</u> <u>Cryst</u>. B39, 625-637

Nixon, P.E., (1978) <u>Acta</u> <u>Cryst</u>. A34, 450-453

Wilson, C.C. (1989) <u>Acta</u> <u>Cryst</u>. A45, 833-839.

DIRECT PHASE DETERMINATION IN THE ELECTRON CRYSTALLOGRAPHY OF ORGANIC

COMPOUNDS

Douglas L. Dorset

Medical Foundation of Buffalo, Inc.
73 High Street
Buffalo, New York   14203   U.S.A.

ABSTRACT. Although other, more traditional, methods for determination
of crystallographic phases have been often used with electron
diffraction intensity data from organic crystals, direct phasing based
on the probabilistic estimate of structure invariants has been found
recently to be generally applicable for such structure analyses. New
structure solutions include the lamellar packings of several
phospholipids, a previously undetermined polymorph of an odd-chain
paraffin, the solid solution of two paraffins, and the three-dimensional
crystal structure of polyethylene. In some cases, low-dose high-
resolution electron microscope images provide phase information useful
for the elucidation of the total crystal structure, especially when the
diffraction pattern is dominated by the scattering from a sublattice
structure.

INTRODUCTION

The use of electron diffraction intensity data for quantitative
determination of organic crystal structures has been an interesting
alternative to x-ray crystallographic methods, especially for materials
which are most easily grown as thin microcrystals. This technique has
been used at least since the initial attempt to solve an n-paraffin
crystal structure in the 1930's by Rigamonti[1]. With the derivation of
reasonably accurate atomic scattering factors for electrons[2], the
technique was adapted to other small molecule problems[3-5], in addition
to the n-paraffin structure determination[6]. Most recently it has been
often used for the elucidation of linear polymer crystal structures[7],
since the lamellar crystals are most easily obtained by self-seeding
using a dilute solution in a poor solvent.

     Various techniques have been used to obtain crystallographic phase
information to carry out the structure analysis. Often a model is
suggested by a related x-ray crystal structure. For linear polymers, an
oligomer structure can be concatenated to form a model of the infinite
polymer so that a conformational search around specified linkage bonds
can be tested both against the crystallographic residual and also a
minimized non-bonded potential energy[8]. For simpler structures, the
interpretation of Patterson functions[9] has enabled the construction of
phasing models. In some optimal cases, details of the atomic

J. R. Fryer and D. L. Dorset (eds.), Electron Crystallography of Organic Molecules, 265–271.
© 1990 *Kluwer Academic Publishers.*

266

arrangement are visible in electron microscope "lattice images" of the crystalline array[10].

The major objection to using model structures for phasing electron data is that the number of refineable parameters has the same order of magnitude as the number of observed data. As is well known, the crystallographic residual in this case is a very imprecise figure of merit so that models with slightly different atomic arrays may not be distinguished at the confidence level often assumed in x-ray crystal structure determinations[11]. High-resolution electron microscope images would overcome this problem but the resolution limits imposed by the transfer function of the objective lens, and also by radiation damage, can be a serious problem.

DIRECT PHASING WITH STRUCTURE INVARIANTS

Polymethylene Chains in Projection Along the Long Axis

The first actual use of traditional direct phasing techniques based on probabilistic estimates of three- and four-phase structure invariants was reported by Dorset and Hauptman[12] in 1976. In their analysis of an n-paraffin crystal structure in a projection down the long chain axis, the phases of 24 of 42 measured reflections were correctly determined from Σ_2-triples and quartets. The resulting electrostatic potential map clearly showed the positions of the carbon atoms (Fig. 1) and these were found to be close to the positions found earlier from the interpretation of the Patterson function.

n-hexatriacontane

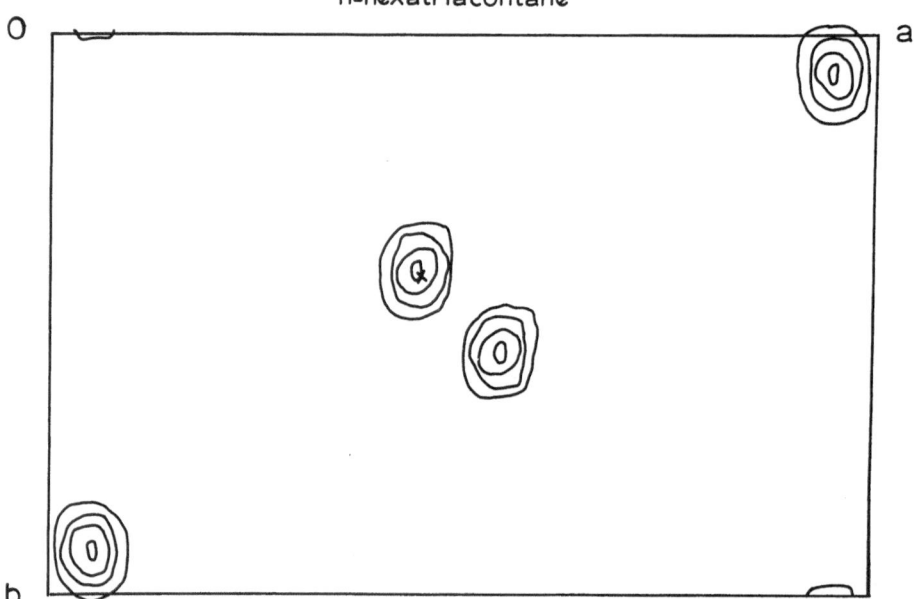

Figure 1. Crystal structure of a n-hexatriacontane monolamellar layer determined by direct phasing.

A second methylene chain packing was analyzed using electron diffraction data from a phospholipid . In this case 29 of 42 phases were determined unambiguously from the structure invariants and three additional phases were inter-related by an unresolved ambiguity. Calculation of two maps allowed the correct structure to be discerned.

Model Studies on Aromatics

In order to test the effect of dynamical scattering on the success of structure analysis by direct phasing, multislice n-beam calculations of structure factors were carried out for two aromatic structures[13] at various simulated crystal thicknesses and at two electron accelerating voltages. For the smaller crystal structure of cytosine, a 76Å thick crystal scattering 100kV electrons was found to be correctly analyzed by direct phasing to produce an interpretable structure map. At 1000kV, data from a 300Å crystal could be used. A similar thickness limit was found at 100kV for a larger molecule, disodium 4-oxypyrimidine-2-sulfinate hexahydrate, but, in this instance, no improvement was found at 1000kV, in order to be equivalent to the cytosine example. The maximum number of phase errors allowed for the calculation of a structurally reasonable electrostatic potential map was 4 in 69 reflections where $|E| > 1.0$.

A similar analysis was carried out for the cytosine crystal structure simulating the perturbation of the diffraction data from bend-deformed crystals[14]. For a projected unit cell axis dimension c=3.82Å, corresponding, perhaps, to an epitaxially crystallized aromatic, a crystal bend deformation of 7.5° could be tolerated and still allow a correct structure to be determined. The situation would be less favorable for structures with larger projected unit cell axial dimensions.

More recent simulations have been carried out with calculated image and electron diffraction data from epitaxially-oriented phthalocyanines[15]. Hence, the likelihood of phase extension beyond the resolution limit defined by the electron microscope objective lens transfer function was evaluated by using the crystallographic phases determined from the image as a basis set to phase the higher resolution electron diffraction data. The phase refinement used constraints on real and reciprocal space defined by Gassmann and Zechmeister[16].

Lamellar Structures

The effectiveness of phase extension realized by the use of structure invariants was demonstrated with measured lamellar data from epitaxially oriented phospholipids[17]. In one case, 1,2-dihexadecyl-*sn*-glycerophosphoethanolamine (DHPE), direct low dose high resolution lattice images photographed using an electron microscope with a helium-cooled superconducting objective lens were found to diffract to 6Å resolution on an optical bench[18]. Image analysis demonstrated that the phases of the first seven lamellar orders correspond exactly to the values found in an earlier crystal structure analysis based on a molecular packing model. Calculation of three- and four-phase structure invariants permitted the phases of the remaining 9 reflections to be

268

determined, resulting in a one-dimensional electrostatic potential map
at 3.4Å resolution (Fig. 2) with features very similar to those seen in
electron density maps determined from the phasing of x-ray data by
various means (e.g. swelling of lamellae in water or model building).

   If the electron microscope image is assumed to be unavailable and
if one specifies the value of one phase to define the unit cell origin,
the analysis of structure invariants correctly phases 13 of 16
reflections for DHPE. In this case the electrostatic potential map
again closely resembles the complete Fourier transform. With this
encouragement, electron diffraction data from four other phospholipids
were directly phased to produce maps from which the structure can be
interpreted. These analyses include two structures for which no x-ray
determination has been reported. It is also possible to analyze similar
lamellar x-ray data from phospholipids with equally positive results.

Epitaxially-Oriented n-Paraffins

When n-paraffins are epitaxially-oriented on substrates such as benzoic
acid, it is possible to collect three-dimensional electron diffraction
data suitable for *ab initio* crystal structure analysis[19]. The three-
dimensional data from n-hexatriacontane was analyzed by combining phase
information from high resolution electron microscope images[18] and the
use of structure invariants. The electron microscope images obtained on
the cryomicroscope mentioned above diffract on an optical bench to 5Å
resolution, enabling the phases of lamellar reflections to

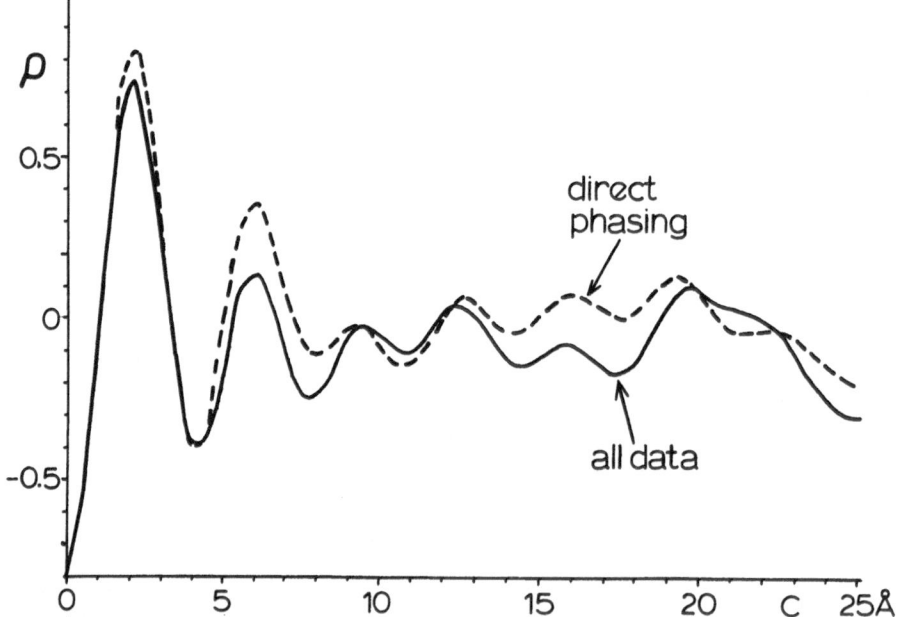

Figure 2. Lamellar structure of epitaxially-crystallized 1,2-
dihexadecyl-*sn*-glycerophosphoethanolamine. ("All data" combines both
electron microscopy and direct phasing.)

be determined by image analysis. Structure invariant relationships phase the so-called "polyethylene" reflections, permitting the major reflections of the unit cell transform to be determined and thus the electrostatic potential map can be compared favorably to the x-ray crystal structure determined by Teare[20]. A similar analysis carried out for the B-form of the odd-chain paraffin, n-tritriacontane, determines the phase values of seventeen key $0k\ell$ reflections to produce the structure map in Figure 3. The positions of all 33 carbon atoms in a chain can be discerned. Along with the earlier use of a packing model[21] this represents the first quantitative structure analysis of this polymorphic crystal form.

More interesting perhaps is the determination of the crystal structure of a 1:1 solid solution of $nC_{32}H_{66}/nC_{36}H_{74}$ with structure invariant relationships. An earlier use of a superlattice model[22] was found to be somewhat incorrect, since the crystal structure analysis distinctly requires an average lamellar packing like that of $nC_{34}H_{70}$, except that the outer chain atoms have non-unitary occupancies. Refinement of this occupancy factor lowers the R-value to 0.23.

Polyethylene

Recently the crystal structure analysis[23] of polyethylene based on three-dimensional data collected from epitaxially-oriented and solution-crystallized samples was reported. Of the 50 measured reflections with intensity significantly above background, use of structure invariants enables one to determine 40 unique phase values. Calculation of the electrostatic potential map supports the notion that the chain setting angle is nearly the same as the value found for even-chain n-paraffins.

CONCLUSIONS

Although electron diffraction data have been demonstrated to be prone to perturbations resulting from multiple scattering and crystal bend deformation, conditions can be manipulated to permit *ab initio* analyses yielding new crystal structures which are chemically reasonable. Among the arsenal of phasing techniques, use of direct methods is now found to be as useful potentially as in their application to x-ray diffraction data. As demonstrated here, a variety of different structure types can be analyzed including polymethylene compounds, the structures of which are often difficult to determine due to the dominance of the diffraction pattern by the strong sublattice scattering. In this case, electron microscope images provide the complementary phase information needed to permit the elucidation of the total crystal structure.

Figure 3. Crystal structure of epitaxially-oriented n-tritriacontane in a projection down the unit cell a=7.5Å axis.

270

REFERENCES

1. R. Rigamonti, 1936, La struttura della catena parafinnica studiata mediante i raggi di elettroni, Gazz. Chim. Ital., 66:174-182.
2. B.K. Vainshtein, 1964, "Structure Analysis by Electron Diffraction", Pergamon Press, Oxford.
3. B.K. Vainshtein, 1955, Elektronograficheskoe issledovanie diketopiperazina, Zh. Fis. Khim., 29:327-344.
4. A.N. Lobachev and B.K. Vainshtein, 1961, An electron diffraction study of urea, Sov. Phys.-Crystallogr., 6:313-317.
5. I.A. D'yakon, L.N. Kairyzk, A.V. Ablov, and L.F. Chapurina, 1977, Atomic structure of the copper salt of two different amino acids. Cu(DL-Ala) (DL-Ser), Dokl. Akad. Nauk. SSSR, 236:103-105 (translation journal p. 481-483).
6. B.K. Vainshtein, A.N. Lobachev, and M.M. Stasova, 1958, Electron diffraction determination of the C-H distance in some paraffins, Sov. Phys. Crystallogr., 3:452-459.
7. D.L. Dorset, 1989, Electron diffraction from crystalline polymers, in "Comprehensive Polymer Science, Vol. 1," Sir. G. Allen, ed., Pergamon Press, Oxford, Chapter 29, pp. 651-668.
8. S. Perez, M. Roux, J.F. Revol, and R.H. Marchessault, 1979, Dehydration of nigeran crystals: Crystal structure and morphological aspects, J. Mol. Biol., 129:113-133.
9. D.L. Dorset, 1976, Aliphatic chain packing in three crystalline polymorphs of a saturated racemic phosphatidylethanolamine. A quantitative electron diffraction study, Biochim. Biophys. Acta, 424:396-403.
10. N. Uyeda, T. Kobayashi, K. Ishizuka, and Y. Fujiyoshi, 1979, High voltage electron microscopy for image discrimination constituent atoms in crystals and molecules, Chem. Scripta, 44:47-61.
11. W.C. Hamilton, 1964, "Statistics in Physical Science," Ronald Publ. Co., New York, p. 157-162.
12. D.L. Dorset and H.A. Hauptman, 1976, Direct phase determination for quasi-kinematical electron diffraction intensity data from organic microcrystals, Ultramicroscopy, 1:195-201.
13. D.L. Dorset, B.K. Jap, M.-H., Ho, and R.M. Glaeser, 1979, Phasing of electron diffraction data from organic crystals. The effect of n-beam dynamical scattering, Acta Cryst., A35:1001-1009.
14. B. Moss and D.L. Dorset, 1982, Effect of crystal bending on direct phasing of electron diffraction data from cytosine, Acta Cryst., A38:207-211.
15. K. Ishizuka, M. Miyazaki, and Uyeda, N., 1982, Improvement of electron microscope images by the direct phasing method, Acta Cryst., A38:408-413.
16. J. Gassmann and K. Zechmeister, 1972, Limits of phase expansion in direct methods, Acta Cryst., A28:270-280.
17. D.L. Dorset, 1987, Electron diffraction structure analysis of phospholipids, J. Electron Microsc. Techn.,7:35-46.
18. F. Zemlin, E. Beckmann, and D.L. Dorset, in preparation.
19. B. Moss, D.L. Dorset, J.C. Wittmann, and B. Lotz, 1984, Electron crystallography of epitaxially-grown paraffin, J. Polym. Sci.-Polym. Phys. Ed., 22:1919-1929.
20. P.W. Teare, 1959, The crystal structure of orthorhombic hexatriacontane $C_{36}H_{74}$, Acta Cryst., 12:294-300.

21. D.L. Dorset, 1986, Electron diffraction structure analysis of epitaxially crystallized n-paraffins, J. Polym. Sci.-Polym. Phys. Ed., 24:79-87.
22. D.L. Dorset, 1985, Crystal structure of n-paraffin solid solutions: an electron diffraction study, Macromolecules, 18:2158.
23. H. Hu and D.L. Dorset, 1989, Three-dimensional electron diffraction structure analysis of polyethylene, Acta Cryst., B45:283-290.

STRUCTURE DETERMINATION AT ATOMIC RESOLUTION BY ELECTRON CRYSTALLOGRAPHY

Sven Hovmöller
Structural Chemistry
University of Stockholm
S-106 91 Stockholm
Sweden

ABSTRACT. The reolution needed for structure determination to atomic resolution is about 2.5 Å. In order to reach this goal, good crystals are needed, but also well-determined amplitudes and phases for all diffraction peaks up to this resolution. Electron microscopy has two major advantages over X-ray diffraction in this respect; phases are obtained experimentally and only extremely small crystals are needed for the analysis. On the other hand radiation damage, multiple diffraction and electron optical distortions of phases and amplitudes are serious problems that have to be overcome. Crystallographic Image Processing (CIP) by Fourier transformation of digitized images and subsequent imposing crystallographic constraints on amplitudes and phases can often restore them to nearly ideal, error-free values. In such cases unknown structures can be solved and atomic positions may be determined with an accuracy of 0.1 Å or better.

Crystallographic image processing may be further strengtened by the use of amplitudes from electron diffraction patterns, and phases for strong reflections may be phased by so-called crystallographic direct methods.

1. Electron versus X-ray crystallography

Structure determination by electron microscopy is emerging as a serious competitor to the classical method of X-ray diffraction. The main advantage of EM is that the phase information is restored in the image, whereas with X-ray diffraction it is lost.

Electrons interact over 1000 times stronger with matter than do X-rays, making it possible to study very small crystals and even non-crystalline material by EM. On the other hand, the strong interactions cause severe problems of multiple scattering if the objects are not extremely thin. Furthermore when real atomic resolution microscopy is aimed at, additional problems arise from optical distortions of the image. This paper will describe the conditions and procedures needed for structure determination to atomic resolution of a crystalline or non-crystalline object, be it an inorganic or an organic compound.

J. R. Fryer and D. L. Dorset (eds.), Electron Crystallography of Organic Molecules, 273–281.

2. Atomic resolution at 2.5 Å

Distances between covalently bound (non-hydrogen) atoms are about 1.5 Å, while non-bonded atoms are about 3.5 Å apart. The volume occupied by each non-hydrogen atom is close to 18 Å^3. The resolution needed for resolving individual atoms is the cube root of 18, i.e. about 2.5 Å. In order to see individual atoms it is necessary to obtain correct amplitudes and phases to at least this resolution. There are several difficulties connected with this.

2.1. FACTORS LIMITING THE RESOLUTION

2.1.1 *Radiation damage.* For organic samples the main limitation is radiation damage. Proteins do not withstand electron doses higher than about 1 electron per Å^2. For this reason well-ordered crystals of more than 100 by 100 unit cells are needed, in order to reduce the radiation dose given to each unit cell. Aromatic compounds tend to be more stable in the electron beam than aliphatic compounds.

2.1.2.*Multiple scattering.* Most inorganic compounds are quite stable in the electron beam; metal oxides for example can survive an electron dose a million times higher than that which will destroy a protein completely. The structure determination of an inorganic crystal by electron microscopy would be trivial if the micrograph was just a magnified image of the electron density in the crystal. Unfortunately it is not. A limited number of non-linear effects distort the image. These effects can be divided into two parts; those arising already in the crystal and those that are due to optical distortions.

　　For a crystal one unit cell thick the wave front at the exit surface of the crystal is a linear function of the electron density within the unit cell. Images taken along crystal axes are usually the most easily interpreted, but also images taken along diagonals or any other directions will contain useful information. As the crystals become thicker more and more of the diffracted electrons will have been mutliply diffracted, making the image no longer a linear function of the electron density. However, for crystals thinner than half the mean free path for electrons, most of the scattered electrons are still scattered only once, making the image interpretable.

2.1.3.*Electron optical distortions.* The optical distortions are caused by the contrast transfer function, astigmatism and beam tilt. Even a perfectly aligned electron microscope, i.e. without astigmatism and beam tilt, will not give an undistorted image of the wave plane at the exit surface of the crystal. However, images taken under optimal defocus conditions, the so-called Scherzer focus condition, will have all phases correct and most of the amplitudes only slightly attenuated inside the Scherzer focus limit, i.e. inside the first cross-over point of the contrast transfer function. For a sufficiently thin crystal an electron micrograph taken at Scherzer focus in a well aligned electron microscope is directly interpretable as a magnified picture of the electron density in the crystal.

In practice almost all micrographs suffer from more or less severe distortions. In order to solve the crystal structure it is therefore necessary to reconstruct a distortion-free image. This involves restoring the distorted amplitudes and phases to their true undistorted values. The amplitudes and phases of the image can be measured experimentally in the Fourier transform of the digitized image. The different distortions affect amplitudes and phases differently, and it is possible to reconstruct the correct image, provided the distortions are not too large. Electron beam tilt will cause phase errors proportional to the square of the resolution, but will not affect the amplitudes. On the other hand crystal tilt will mainly attenuate the amplitudes, but not change the phases for thin crystals and small tilt angles.

3. Reconstructing a distortion-free image

3.1. WITHOUT IMAGE PROCESSING

3.1.1. *Analytical compensation for distortions.* In principle the reconstruction of correct amplitudes and phases can be done by applying corrections for the quantitatively known distortions. In practice, however, this is not possible since we do not know exactly the defocus, astigmatism etc.

3.1.2. *Image simulation.* The most commonly used technique for structure determination of inorganic compounds by high resolution electron microscopy (HREM) is image simulation (also called image calculation). A simulated image is calculated by computer, using a proposed structural model and estimated optical conditions (defocus etc.) and crystal orientation and thickness as input.

There are several problems connected with the approach of image simulation. Firstly if the aim is to determine an unknown structure it is a very severe limitation that a tentative model must be used in the input. While this is possible in some cases of variation of well established structural themes, like for example with niobium oxides, it can often be impossible to guess the correct structure, and then image simulation is not possible to do. Secondly the optical conditions, as well as crystal tilt and crystal thickness are not known, and therefore a very large number of images may have to be calculated until one with the correct conditions is found. Finally it is rather unsatisfactory that the last step in this procedure is not a quantitative comparison of the experimental and simulated images, but just a, perhaps subjective, judgement by eye.

3.2. CRYSTALLOGRAPHIC IMAGE PROCESSING (CIP)

When the crystal structure is unknown it is usually necessary to have a virtually distortion-free image to 2.5 Å resolution in order to solve the structure. Crystallographic image processing (CIP) is a powerful tool both for assessing the quality of different areas of crystal images, and for improving the quality of the experimental data.

3.2.1.*Optical diffraction*. The first step of CIP is to analyse several electron micrographs in the optical diffractometer. This is a fast way to find the best area among several pictures. The best areas are those with the highest number of diffraction points, and those with the best preserved symmetry of the diffraction pattern.

3.2.2.*Film scanning*. The best area or areas of about $1cm^2$ are scanned in a microdensitometer. Until very recently this had to be done using very expensive ($100,000 or more) equipment, but recently CCD cameras at prices around $1000 have become available. The quality of the best CCD cameras is just sufficient for image processing. It is important that the resolution is high enough (10 to 25 μm), the magnification along x- and y-directions of the scan must be equal, and the optical density should be followed linearly.

3.2.3.*Fourier transformation*. The scanned image is transferred to a computer, where the Fourier transform is calculated. The amplitude part of the transform is shown on a computer graphics display. Here the same diffraction spots that were seen in the optical diffractometer should again appear. The diffraction pattern is indexed, and amplitudes and phases are extracted automatically from the Fourier transform, given the coordinates of the unit vectors describing the lattice.

3.2.4.*Phase extraction*. It must be emphasized that the Fourier transform is a complex function, i.e. all the data is of the form $(A+iB)$, where i is the square root of -1. Such numbers can also be described as vectors with an amplitude and a phase. **This is exactly the same amplitude and phase as those X-ray crystallographers refer to as the structure factor F.** Amplitudes are best estimated by interpolating over a few (3 by 3) points nearest to the expected position of each lattice point. The phase of a reflection is taken as the phase value of the point in the Fourier transform closest to the expected center of the peak.

Phases describe positions, in this case the position of the maximum of a cosine-shaped wave of electron density. The phase value of each such wave, or Bragg-plane, depends on the reference point in the crystal, i.e. the origin. If the origin is shifted by a distance (x,y) all phases will change by $360^{\circ}(hx + ky)$, where (x,y) is a shift in real space and (h,k) are the indices of a reflection. The first phase values should be relative to the center of the processed area, but later we will try to find an optimal place in the unit cell for the phase origin. The optimal choice of origin depends on the space group, and can be seen from the International Tables for Crystallography. Usually the origin is chosen to coincide with a center of symmetry or a rotation axis.

3.2.5.*Origin refinement*. Most images of crystals are centrosymmetric projections. This is true also for most chiral molecules, like for example proteins, since projections down 2-, 4- and 6- fold screw- or rotation axes are centrosymmetric. In these cases it is optimal to position the origin on a center of symmetry, since there we can be sure that all phases must take on one of the two values 0 or 180 degrees.

A computer program reads the indices and phases of all reflections, as they come out of the Fourier transform as mentioned above. The program then shifts around the phase origin throughout one full unit cell. At each position it calculates the phase residual, defined as the average deviation from 0 or 180 degrees, for all reflections. The position with the lowest phase residual is considered as the correct position of the origin. The highest phase residual for any reflection is 90 degrees, and thus the average phase residual at a random place in the unit cell is expected to be 45 degrees. Very good images may reach phase residuals under 10 degrees, but values up to 20 degrees are common and will often prove sufficiently good.

Another computer program then shifts the origin to the one found.

4. Space group determination

In order to make full use of the information it is essential to correctly determine the space group of the crystal. This is a science in itself, taught as part of any course in X-ray crystallography. However it should be pointed out that although the symmetry of the crystal is the same whether looked at by X-rays or electrons, and the principles of space group determination are also the same, the practical procedures differ. This is so because the electron microscopy data includes the phase values which are not available by X-ray diffraction, while on the other hand the quality of the amplitudes obtained by X-ray diffraction is superior to that obtained by EM.

4.1. CRYSTALLOGRAPHIC CONSTRAINTS ON AMPLITUDES AND PHASES

Once the space group has been determined and the origin specified there are important constraints on both amplitude and phase values. The constraints are different for every space group, and the constrained phase values are also dependent on the choice of origin. A detailed description of these questions can be found in Hovmöller (1981).

4.1.1. *Amplitudes of symmetry-related reflections.* Amplitudes of symmetry-related reflections must always be identical. While this is quite evident in X-ray diffraction (deviations of 10 % are considered very high), it is not uncommon to find pairs of symmetry-related reflections differing by a factor of 5 or more in electron microscopy. The reasons for this are crystal tilt and astigmatism. The most simple way to overcome this problem is to assign the average amplitude value of all the symmetry-related reflections to each reflection. In orthorhombic structures, for example, pairs of reflections (h,k) and (-h,-k) are symmetry-related and should be given the same amplitude.

A further improvement in amplitudes may be achieved by correcting for the attenuation of the contrast transfer function. It is also possible to replace the amplitudes calculated from the image by amplitudes obtained by electron diffraction. These amplitudes have the advantage of not being distorted by the objective lens in the EM, but on the other hand they may suffer from other problems, for example multiple diffraction. Inorganic crystals are usually wedge-shaped, and only thin very close to the edge. It is possible to apply CIP on a small area near the edge, where the crystal is sufficiently thin, but the electron diffraction normally comes from a much larger part of the crystal, and so has contributions both from the thin and the thick regions. For this reason the electron diffraction data is often corrupted by a strong contribution of multiply scattered electrons.

4.1.2.*Phase restrictions.* Most crystals have a centrosymmetric projection. In such projections only phases of 0 or 180 degrees are allowed. If the distortions are not too severe it is then possible to deduce the true, undistorted phase of all reflections out to about 2.5 Å. If the symmetry is even higher, for example 4- or 6-fold, then there will be a very useful redundancy in the data, since pairs (or more) of reflections will have related phase values. This gives a possibility to check if the deduced phase values are consistent. For crystals of lower symmetry it may be necessary to digitize several images in order to confirm that the deduced phases are indeed correct.

Which reflections that have restricted phases, which possible phase values they may have, and how phases of symmetry-related reflections are related can be deduced algebraically from the equivalent positions in any space group. This is all described in detail in Hovmöller (1981) for the 230 space groups, and in Hovmöller (1986) for the 17 two-sided plane groups that are possible for chiral molecules, such as proteins.

4.1.3. *Assignment of phases.* For images with low phase residuals, say below 15 degrees, it is trivial to assign the correct phase values to all reflections (except perhaps a few very weak ones, which anyway do not contribute significantly to the electron density). For centrosymmetric projections the phases are simply set to 0 if they are between −90 and +90 degrees, or else they are set to 180 degrees. As the phase residual increases it becomes more and more difficult to assign phases, and one has to consider the possiblity that mistakes are introduced. For crystals of high symmetry it is a very useful help that two or more reflections have phases with known relations between them. Reflections related by rotation axes or mirror planes will always have equal phases, while some reflections related by screw axes or glide planes may have different phases, for example if one is 0 then the other one must be 180 degrees. The high symmetry will allow two or more independent measurements for each such group of symmetry-related reflections. An example of the use of high symmetry can be found in Wang et al. 1988, where a complicated niobium oxide with the space group P4bm was solved in this way.

If the unit cell is larger than about 20Å in both directions, then the reciprocal lattice points are sufficiently close to make it possible to follow the gradual increase of distortions with resolution. For smaller unit cells it may be very difficult to reconstruct the correct amplitudes and phases.

5. Reconstructing the crystal structure

When a set of amplitudes and phases have been obtained in the above way, with the crystallographic constraints applied, they are used as input to a computer program which calculates the inverse Fourier transform. The result is an electron density map. If the phases have been correctly assigned and the resolution is sufficient, then this density map can be directly interpreted in terms of atomic structure (Hovmöller et al. 1984). For metal oxides typical metal-oxygen distances are around 1.8 Å, while metal-metal distances are about 3.5 Å. At a resolution of 2.5 Å we shall expect to see well-resolved symmetrical peaks representing the MeO_6 octahedra, MeO_4 tetrahedra etc., but the resolution is just not sufficient for resolving the oxygen atoms. Several metal oxides have now been determined in this way, and in all cases where it was later possible to compare the results with the same or an isomorphous structure, solved by X-ray crystallography, the positions of the metal atoms in the unit cells were correct to within 0.1 Å (Wang et al. 1988).

A summary of the different steps in the procedure of crystallographic image processing, from the electron micrograph to the electron density map, is shown in Figure 1.

6. 3D reconstructions of organic structures

In some favorable cases, like niobium oxides and phtalocyanides, the structures are very flat, and it is possible to solve the structure from a single projection. In general this is not the case, and it becomes necessary to make a full 3D reconstruction of the electron density in the unit cell in order to solve the structure. In priniciple the procedure will then be the same as above, but a number of projections must be processed and merged into a single 3D map. The number of projections needed rises with the desired resolution. For 3-dimensional crystals the procedure closely resembles that of X-ray crystallography, but for 2D crystals, as for example membrane proteins (Hovmöller 1986) or bacterial surface proteins (Baumeister and Engelhardt 1987, Hovmöller, Sjögren and Wang 1988), the Fourier transform is a set of continuous spikes and special procedures are adopted, as developed by Henderson and Unwin 1975.

7. Non-crystalline material

Unfortunately the electron optical distortions are equally severe in the case of non-crystalline materials. However in such cases we are much worse off since it is not possible to say anything beforehand about neither amplitudes nor phases. This makes it almost impossible to quantitatively determine the values of the distortions, and correct for these. One possibility may be to prepare samples where crystalline and amorphous material are found adjacently in the material, and take images where both types of structures are present in the same image. The distortions could then be estimated from the crystalline areas and applied to the entire picture since the optical parameters are nearly identical over the whole image (if it is untilted). A further problem for amorphous samples is that they give rise to a continuous Fourier transform, and so it is necessary to apply corrections to every point in the Fourier transform and not only correct the values at the few diffraction points, as is the case with crystals.

8. Acknowledgements

This work has been carried out with financial support from The Swedish Science Research Council (NFR).

9. References

Baumeister, W. and Engelhardt, H. (1987) Three-Dimensional Structure of Bacterial Surface Layers. In Electron Microscopy of Proteins, Vol. 6, Ed. Harris, Academic Press, London, 110-154.

Henderson, R. and Unwin, P.N.T. (1975) Three-dimensional Model of Purple Membrane Obtained by Electron Microscopy. Nature 257, 28-32.

Hovmöller, S. (1981) Rotation Matrices and Translation Vectors in Crystallography. No.9 in Pamphlet Series issued by the International Union of Crystallography Commission on Crystallographic Teaching. University College Cardiff Press.

Hovmöller, S. (1986) 3-Dimensional Structure of Membrane Proteins. In: Techniques for the Analysis of Membrane Proteins, Eds. Cherry and Ragan, Chapman & Hall, 315-344.

Hovmöller, S., Sjögren, A., Farrants, G., Sundberg, M and Marinder, B.-O. (1984) Accurate Atomic Positions from Electron Microscopy. Nature 311, 238-241.

Hovmöller, S., Sjögren, A. and Wang, D.N. (1988) The Structure of Crystalline Bacterial Surface Layers. Prog. Biophys. molec. Biol. 51, 131.163.

International Tables for X-ray Crystallography, Vol. A. (1983) Riedel, Dordrecht.

Wang, D.N., Hovmöller, S., Kihlborg, L. and Sundberg, M. (1988) Structure Determination and Correction for Distortions in HREM by Crystallographic Image Processing. Ultramicroscopy 25, 303-316.

CRYSTALLOGRAPHIC IMAGE PROCESSING

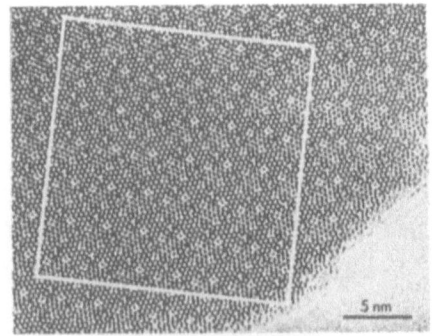

1. Find best area of electron micrograph by optical diffraction.
   Scan the area for example as 256 by 256 pixels of 40 by 40 μm.

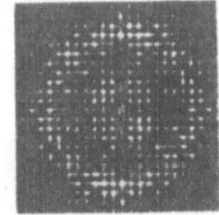

2. Compute the Fourier transform.

Index	Observed phases		Deduced phases	
h k	(h,k)	(h,-k)	(h,k)	(h,-k)
1 1	147	161	180	180
2 1	158	327	180	0
2 2	338	337	0	0
3 1	328	330	0	0
3 2	134	308	180	0
...				

3. Extract amplitudes and phases from the Fourier transform. Correct the phases and amplitudes for distortions and impose the correct symmetry.

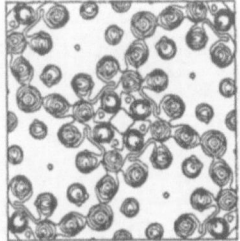

4. Calculate the electron density map.

Figure 1. Schematic summary of the procedures used in crystallographic image processing.

ELECTRON CRYSTALLOGRAPHY OF BACTERIAL SURFACE PROTEINS

W. BAUMEISTER, G. LEMBCKE, R. DÜRR and B. PHIPPS
Max-Planck-Institute for Biochemistry
Department of Structural Biology
D-8033 Martinsried
Federal Republic of Germany

ABSTRACT. Bacterial surface (glyco)proteins forming two-dimensional crystalline arrays are a common feature of the cell envelopes of eubacteria and archaebacteria. Electron crystallography is the primary method which has been used for determination of their structures. We describe some general structural features and principles of spatial organization which have emerged from these studies. A distinct advantage of real space averaging techniques is the ability to detect and analyse deviations from perfect crystallinity which may have biological significance. The extremely well-ordered protein arrays of *Thermoproteus tenax* and *Pyrobaculum islandicum* exhibit a defined lattice orientation and particular pattern of lattice defects which suggest a model for the growth of the array that is consistent with its putative role in cell shape determination. The use of real space averaging techniques to analyse the surface layer of *Sulfolobus acidocaldarius*, previously characterized as an array with p6 symmetry, enabled us to determine that the layer is actually composed of domains of two distinct kinds of p3 structure separated by twin boundaries. We conclude by suggesting ways in which analysis of lattice distortions could lead to higher resolution structures and a better understanding of the mechanical properties of proteins in crystalline arrays.

1. Introduction

Proteins or glycoproteins which constitute more or less regularly arrayed surface networks ('S-layers') are commonly found in eubacteria and they are ubiquitous in the archaebacterial kingdom where, tightly associated with the plasma membrane, they represent the major macromolecular component of the cell envelope. Surface proteins are thought to be involved in phenomena as diverse as molecular sieving, adhesion, cell-cell interactions, and shape maintenance or determination (Baumeister and Engelhardt 1987; Hovmöller et al. 1988; Sleytr and Messner 1983; Smit 1986). Knowledge of their three-dimensional architecture is indispensible if a satisfactory understanding of their function is to be achieved. Electron microscopy is by far the most powerful physical technique available today for elucidating the spatial organization of complex supramolecular systems, i.e. macromolecules in their native topological situation. Electron crystallography is at the brink of establishing itself as a technique capable of analysing macromolecular structure with a precision that had hitherto been considered exclusive to the realm of x-ray crystallography (Baldwin et al. 1988; Jap 1989; Sass et al.

J. R. Fryer and D. L. Dorset (eds.), Electron Crystallography of Organic Molecules, 283–296.
© 1990 *Kluwer Academic Publishers.*

1989). Further progress can be expected to come from the improvement of existing techniques (this pertains to specimen preparation, image recording, and image analysis) as well as from the introduction of emerging technologies. Fast on-line image processing and microscope control may serve as an example of the latter; this technology will stimulate the development of sophisticated (3-D) data collection schemes (involving imaging modes such as 'spot scan' or 'dynamic (auto)focussing') beyond the practicability of manual EM operation.

Recent advances in and the prospects of crystallographic electron microscopy should not cause us to overlook the fact that electron micrographs provide a means of *directly* visualizing structures at molecular dimensions. It is this facet of electron microscopy which enables us to detect and characterize in real space deviations from perfect crystallinity such as dislocations, twin boundaries, and local stress and strain fields. Such defects must not only be taken into account in the course of image processing in order to retrieve a maximum of information or to avoid false results; they may in fact have biological relevance *per se*, especially in the case of two-dimensional arrays enclosing a cell which undergoes continuous growth and gross morphological changes during the cell cycle.

2. Bacterial Surface Proteins: Some General Features and Principles of Organization

Regular surface proteins, or 'S-layers' as they are often referred to, are defined simply by their location at the surface of the prokaryotic cell without implying that they belong to one family of proteins. They form more or less regular surface lattices, often difficult to distinguish from regularly arrayed outer membrane proteins (rOmps) which occur in many Gram-negative eubacteria. An operational criterion for identifying a surface array as an S-layer is a lateral interaction strong enough to maintain the integrity of the lattice without relying on a membranous support. For true rOmps, on the other hand, the lipid matrix is an essential structural component of the supramolecular assembly and its removal, e.g. by detergent extraction, will cause the array to dissociate.

A few well established cases apart, where two distinct layers composed of different polypeptides form a composite (Yamada et al. 1981), those layers investigated in greater detail appear to be made up of a single polypeptide species (Koval and Murray 1984). Primary structures are available for one archaebacterial and a few eubacterial surface proteins (Lechner and Sumper 1987; Peters et al. 1987; Peters et al. 1989; Tsuboi et al. 1986; Tsuboi et al. 1988); they exhibit only a few, weak indications for relatedness. One perhaps significant common feature is the relatively high content of the hydroxyamino acids serine and threonine which often occur clustered in the sequence; such motifs also occur in many eukaryotic adhesion proteins (Dickson et al. 1987; Noegel et al. 1986). Most archaebacterial and many eubacterial surface proteins appear to be glycoproteins, although detailed information about the structure of the glycan chains and their linkage to the protein is rather scarce; notable exceptions in this respect are the S-layers of *Halobacterium salinarium* (Sumper 1987) and *Bacillus stearothermophilus* (Messner and Sleytr 1988). According to spectroscopic data (IR, CD) a ß-sheet content of approximately 30% is characteristic for most S-layers investigated so far, while there is no indication of appreciable amounts of α-helix (Baumeister et al. 1982; Phipps et al. 1983).

Our knowledge of the three-dimensional spatial organization of bacterial surface layers has been obtained almost exclusively through the agency of electron crystallography. In archaebacteria a great variety of architecture is observed, ranging from well-ordered and rigid layers (e.g. *Thermoproteus tenax* and *Pyrobaculum islandicum* [Phipps et al. 1990; Wildhaber and Baumeister 1987]) to rather loose and disordered networks (e.g. *Desulfurococcus mobilis* [Wildhaber et al. 1987]). The structures of several archaebacterial S-layers are shown in Figure 1.

In contrast, some unifying principles of organization have emerged from three-dimensional analyses of eubacterial S-layers. Not surprisingly, not all of the 17 two-dimensional space groups possible with (chiral) protein molecules are found in S-layers, presumably because several of them would result in a layer with no vectorial properties; to date only p2, p3, p4, and p6 lattices have been found. Describing the protein monomers rather simplistically as composed of a heavy domain and light domain(s), the disposition of these two domains relative to the crystallographic axes provides the basis for a classification scheme (Saxton and Baumeister 1986). It is remarkable that, invariably, the heavy domains are grouped around the higher symmetry axis to form a massive core, while the light domains provide connectivity at one of the lower symmetry axes. The result is a relatively conserved, often funnel-shaped core, and surface-directed connective domains whose mass disposition varies between organisms.

In eubacteria the S-layer protein interacts with the underlying peptidoglycan layer or outer membrane by means relatively weak noncovalent interactions. There are only a few notable exceptions, amongst them the HPI layer of *Deinococcus radiodurans*, which is tightly associated with the outer membrane via hydrophobic interactions and probably via a covalently attached fatty acid membrane anchor as well (Peters et al. 1988).

Most archaebacteria have a much simpler cell envelope structure, consisting solely of the plasma membrane and the S-layer intimately associated with it. In micrographs of thin sections or freeze-fractures, the S-layers often appear to be located at a uniform distance above the membrane surface and to be linked to it via long pillar-like protrusions. Unfortunately it has been difficult to analyse the structure of these protrusions as they are nearly always invisible or truncated in three-dimensional reconstructions as a consequence of their flexibility; they occupy variable spatial positions and are therefore averaged out in the course of a reconstruction. Investigations of the structure of the *Thermoproteus tenax* cell envelope provided evidence that the distal ends of these protrusions, which are an integral part of the surface array, penetrate or even traverse the plasma membrane (Baumeister et al. 1989). From the primary structure of the surface glycoprotein of *Halobacterium salinarium* it was postulated that the C-terminus serves as a membrane anchor (Lechner and Sumper 1987). Hence archaebacterial surface proteins should be classified as membrane proteins.

This membrane protein character is particularly well-illustrated in the case of *Staphylothermus marinus*. Here the surface protein, which has extraordinary dimensions, forms a canopy of parabolic elements supported by filiform stalks approximately 65 nm above the plasma membrane. While the surface network is resistant to harsh detergent treatments, it dissociates upon exposure to glycerol and the released (probably tetrameric) protomers spontaneously form 'micelles' in order to shield the hydrophobic membrane domains. Adding detergent or propanol at this stage leads to dissociation of the micelles; the process is reversible upon dialysis (Baumeister et al., manuscript in preparation).

Data of the kind mentioned above have led to the proposal that a common feature of archaebacteria is a type of periplasmic space, delimited by the plasma membrane and the interconnected surface domains of the S-layer protein and maintained by elongated spacer elements. Schematic models of some archaebacterial cell envelopes are presented in Figure 2.

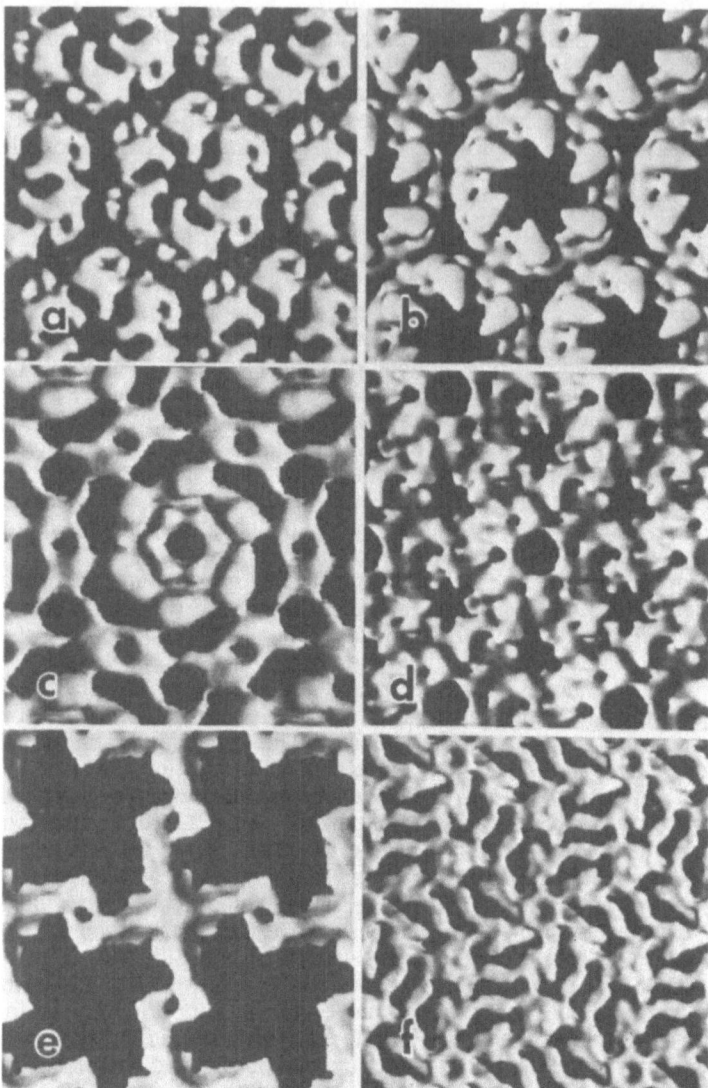

FIG. 1. Views of the outer surfaces of archaebacterial S-layers. The 3D structure of each layer was determined by applying the hybrid Fourier space / real space method of reconstruction to correlation averages of negative stain tilt series micrographs. The number following each strain name is the lattice centre-to-centre spacing in nm. (a) *Archaeoglobus fulgidus* (17.5). (b) *Pyrodictium occultum* (21.8). (c) *Hyperthermus butylicus* (25.8). (d) *Sulfolobus brierleyi* (19.3). (e) *Desulfurococcus mobilis* (18.0). (f) *Pyrobaculum islandicum* (29.9). The outer surface was identified from freeze-etching and shadowing data except in the case of *Pyrodictium occultum*, for which the putative outer surface was assigned by structural analogy with other archaebacterial layers.

287

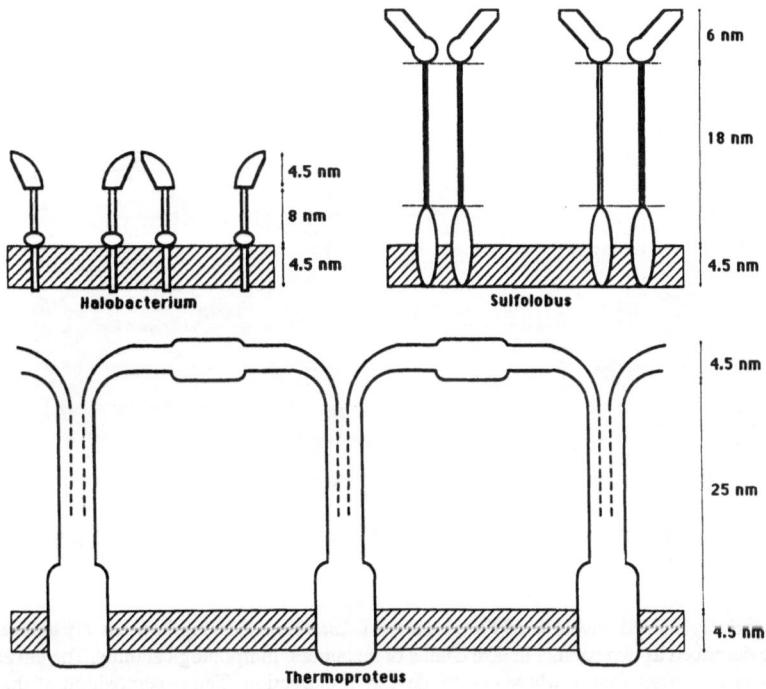

FIG. 2. Schematic diagrams of cell envelope structures in archaebacteria. Hatched areas represent the plasma membrane. Between the membrane and porous outer canopy of the S-layer is an interspace of defined width which is maintained by spacer elements; this may serve as a kind of 'periplasmic space'. The S-layer protein forms all or part of the spacer element and may be anchored in the membrane (*Halobacterium, Thermoproteus*) or interact with a separate membrane-embedded protein, as might be the case for *Sulfolobus*.

3. The highly ordered surface protein arrays of *Thermoproteus* and *Pyrobaculum*

A clear-cut case in which the pattern of lattice defects and the orientation of the lattice observed in real space led to a cogent hypothesis for the growth of a surface protein layer is found in the extremely thermophilic archaebacteria *Thermoproteus tenax* and *Pyrobaculum islandicum*. The S-layers of these organisms are very similar. They are extremely well-ordered p6 lattices which apparently determine the remarkably uniform diameter of these rod-shaped bacteria. The layers are composed of elongated, multidomain protein subunits which are highly interconnected and have so far resisted even the most vigorous attempts to dissociate them (Fig. 1f). A long narrow domain at 6-fold crystallographic axes appears to insert into the underlying plasma membrane, defining a large interspace between the membrane and the porous canopy formed by the main body of the S-layer (Messner et al. 1986; Phipps et al. 1990; Wildhaber and Baumeister 1987).

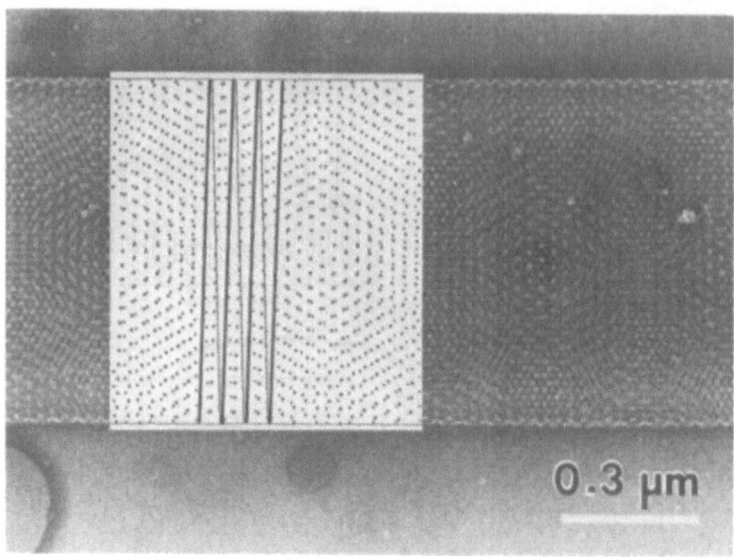

FIG. 3. Negatively-stained envelope of *Pyrobaculum islandicum* showing how the crystalline protein layer can be described as two parallel helical chains of hexameric morphological units. The envelope was isolated essentially intact from a whole cell by detergent extraction. The superposition of the top and bottom p6 lattices creates a hexagonal moiré. The course of a single helix is indicated, with the thick and thin lines corresponding to opposite sides of the envelope.

Analysis of this lattice revealed that one of the lattice base vectors is invariably offset from the perpendicular to the long axis of the cell by a small angle, 3-4° in the case of *T. tenax* and 0.5-4° for *P. islandicum*. Consequently the layer can be described as 1-3 parallel helical strands, consisting of protein hexamers, which wrap around the girth of the cell at a shallow angle (Fig. 3). The order of the array over the cylindrical portion of the cell is remarkable; no lattice defects of any kind are seen. However, pentagonal wedge disclinations have been clearly visualized at the poles of *Thermoproteus* cells by Messner *et al.* (1986), using polycationic ferritin (PCF) to label the S-layer; furthermore, they observed no other defects (e.g. edge dislocations, other types of disclinations) at the poles, and calculated the average number of wedge disclinations per polar cap to be six, the minimum number required to fit a p6 lattice on a hemisphere (Caspar and Klug 1962).

Based on these data, Wildhaber and Baumeister (1987) proposed the 'helical template model' for growth of the protein array in which intussusception of new morphological units (monomers, trimers, hexamers?) occurs at the end of the helical strands where these meet the polar caps, generating edge dislocations which are absorbed by migration of the wedge disclinations (Harris and Scriven 1970). In principle, the insertion of new units could occur anywhere in the polar region, but the inevitable dislocation must be rapidly deleted by movement of a disclination. Thus new subunits are added only at the ends of the cells and the cells grow by elongation whilst maintaining a strictly defined diameter, as observed.

Examination of unit cell positions in images of PCF-labelled (Messner et al. 1986) and negatively-stained envelopes suggest that some degree of stretching distributed over many unit cells can occur in the polar regions. Since these envelopes are cylinders with rounded ends which have become flattened on an artificial support, this probably represents at least in part an artefactual situation, although such stretching may be present *in vivo* as well. The important point is that this array appears to be *capable* of undergoing smoothly-distributed long range deformation, regardless of whether it normally does so *in vivo*. If borne out by further analysis, this means that the (glyco)protein subunits can 'flex' to a certain degree without abrogating intersubunit contacts which are essential for the integrity of the array. It should be possible to test this idea by applying strain analysis to the kinds of images mentioned above, and to images of freeze-etched cells (where the envelope ought to maintain its native curvature); deformed regions would be identified by unit cells with a high level of strain.

The concept of flexibility and deformability within two-dimensional protein crystals may well be an important one. Such crystals can be expected to experience potentially disruptive forces as a result of growth, attachment to surfaces, the shearing action of a turbulent fluid environment, etc.. The filiform, multidomain nature of archaebacterial surface layer proteins may allow these molecules to undergo elbow-type bending, such as occurs in immunoglobulins and other extended proteins (Huber 1988). Another way to provide this property would be to introduce covalently-attached carbohydrate chains to mediate intersubunit interactions. The inherent flexibility of carbohydrate chains (Homans et al. 1987; Imberty et al. 1990) could permit a certain amount of unit cell distortion. As mentioned above, many bacterial surface proteins are believed to be glycoproteins. The fact that high concentrations of glycerol disrupt the interactions between the 'arms' of the surface protein of *Staphylothermus marinus* suggests that the contacts are mediated by carbohydrate. Treatment of *Halobacterium salinarium* with bacitracin inhibits the glycosylation of the surface array protein. Although the protein is still present on the cell surface, the order of the array is lost and the normally rod-shaped cells round up, indicating that the carbohydrate moieties are necessary for correct lattice-maintaining interactions (Mescher and Strominger 1976).

4. Analysing the surface protein of *Sulfolobus acidocaldarius*

Sulfolobus acidocaldarius is another example of an extremely thermophilic archaebacterium; the *Sulfolobales* are actually relatively closely related to the *Thermoproteales* which include *Thermoproteus*, *Pyrobaculum*, and *Staphylothermus* (Klenk et al. 1986). All of these organisms have the same basic cell envelope architecture, consisting of an S-layer tightly associated with the plasma membrane via filiform linker elements spanning a 'quasi-periplasmic' interspace which, in the case of *Sulfolobus*, is 18 nm in width (Fig. 2; Baumeister et al. 1989; Baumeister et al. 1988). Unlike *Thermoproteus* or *Pyrobaculum*, however, the cell shape of *Sulfolobus* is not well-defined; cells appear as rather irregular cocci with edges and lobes. As early as 1982 the *Sulfolobus acidocaldarius* S-layer was investigated by electron crystallographic techniques and a three-dimensional reconstruction was performed, revealing a spongy structure with a network of channels and caves created by multidomain protomers apparently arranged on a p6 lattice (Taylor et al. 1982). We recently began to reinvestigate this structure, on one hand as part of a broader attempt to compare the surface proteins of several species of *Sulfolobus* in order to identify common and variable features in their design, and on the other hand with the goal in mind of attaining significantly higher resolution.

Surface layers, like many other two-dimensional crystals, natural as well as synthetic, suffer notoriously from lattice imperfections and distortions. This led us to introduce and to apply correlation averaging techniques for the analysis of electron micrographs of such crystals in preference to the then commonly used Fourier-filtration methods, which rely on ideal crystallinity. In correlation averaging, unit cell displacements from an ideal lattice are determined by cross-correlation and the averaging of precisely aligned motifs prevents the lateral displacements from having a detrimental effect on resolution (Saxton and Baumeister 1982). Knowledge of the unit cell displacements can, of course, be used to reinterpolate images prior to Fourier-filtration, to correct for distortions; under the name of 'lattice unbending' this approach has become a fashionable alternative to correlation averaging (Henderson et al. 1986). Both routes in their standard forms compensate lateral (x,y) displacements but ignore unit cell rotations, although in principle both can be adapted to include compensation for rotation as well. It has been shown, however, that lateral displacements are by far the most important factor limiting resolution. Moreover, high levels of local rotation often correlate with internal distortions of the unit cell (Hegerl and Baumeister 1988). While in some respects lattice unbending may be the more convenient approach, local real space averaging clearly offers more options for selective averaging; this will be of utmost importance for analysing and dealing with microheterogeneities in crystals.

When analysing micrographs of several S-layer fragments of *Sulfolobus acidocaldarius*, the results were alarmingly inconsistent with regard to symmetry. Independent of the method used (Fourier filtration with or without unbending, correlation averaging) some averages showed almost perfect p6 symmetry while others were clearly p3 and many were somewhere between these two extremes. Careful examination of cross-correlation functions of large arrays gave a first hint; they showed local variations with characteristic interpenetrating domains of intrinsically uniform peak height and shape. Deliberately extracting reference patches from neighbouring domains converted domains with high cross-correlation peaks to low peak levels and *vice versa*. Congruent domain patterns were obtained by an entirely independent approach, namely multivariate statistical analysis of eigenvector-eigenvalue data (van Heel and Frank 1981). The map in Figure 4 displays the results of such an analysis. Selective intra-domain averages yielded clear p3 structures and it turned out that a 60° rotation of the trimeric motif, indicative of the existence of twin boundaries, was causing the heterogeneity observed in the cross-correlation function (Fig. 5). Obviously, global averaging over image areas with interpenetrating domains can create a higher (p6) symmetry, as seen in Figure 5 (Bragg 1962). This emphasizes the power of real space averaging techniques when used in conjunction with pattern recognition and image classification methods. Apart from being essential in this particular case for obtaining a correct structure and being a prerequisite for proceeding to higher resolution (Lembcke et al., manuscript in preparation), the same techniques might help to determine the real structure of the twin boundary itself. This, in turn, could provide useful information with regard to the delineation of monomer boundaries or the mechanisms of insertion of new units into the lattice during growth.

5. Some thoughts about future image processing strategies

As outlined above, the image processing techniques currently being used for combatting the effects of lattice distortions only take lateral displacements of unit cells into account, at

FIG. 4. Classification and mapping of motifs according to correspondence analysis in a negatively-stained fragment of *Sulfolobus acidocaldarius* S-layer. A Fourier-filtered p3 unit cell extracted from the image was used as a reference to identify motifs in the image by cross-correlation. A single unit cell was extracted at each motif position and the set of images classified by correspondence analysis. The 2 classes found are represented by different symbols. Note the clear segregation of the classes into domains separated by a sharp (twin) boundary.

least in the first instance. A quantitative analysis of the distortions in various bacterial surface layers indicates that this is in fact sufficient if the resolution target is approximately 1 nm (R. Dürr, manuscript in preparation). While other distortions such as unit cell rotations or stretching have only marginal effects at this level of resolution, they have to be compensated if we aim for significantly higher resolutions (R. Dürr, manuscript in preparation). Whether or not this can be accomplished depends ultimately on the accuracy in determining the displacement field; hence there is some incentive to refine these procedures.

There is one aspect which has been ignored so far, although it might be relevant, especially when dealing with large multidomain proteins: The geometry of lattice distortions may not be related in a simple manner to distortions within the unit cell itself, i.e. the lattice may not behave like a homogeneous rubber sheet. It is actually more likely

292

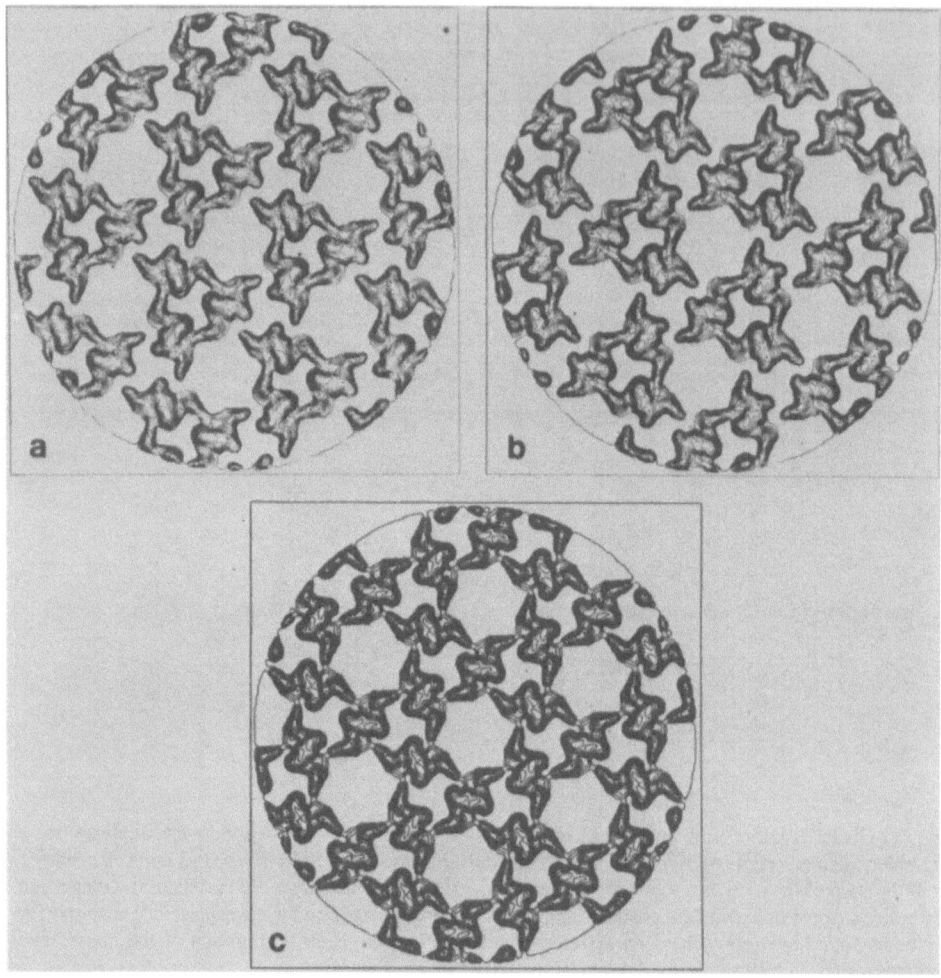

FIG. 5. Real space averages of the *Sulfolobus acidocaldarius* S-layer. (a),(b) Selective averages obtained by combining unit cells belonging to one or the other of the classes identified by correspondence analysis. The p3-symmetric structures are identical except for a 60° rotation of the trimeric motif. (c) Global correlation average of the same frame without regard for heterogeneity in the crystal, with resultant p6 symmetry.

that unit cells of such proteins are composed of rigid domains linked by relatively flexible hinge regions prone to distortions. In this case it is no longer possible to deduce unit cell distortions from displacements at lattice sites. However, knowledge of the displacement field allows us to perform detailed deformation analysis specifying local changes of unit cell area, shape, etc. (R. Dürr, manuscript in preparation). This type of analysis reveals

FIG. 6. Strain map of the fragment of *Sulfolobus acidocaldarius* S-layer represented in Fig. 4. Strain is a measure of the change of shape (deformation) of a local area element. The lines indicate the direction of maximum strain for each unit cell.

domains of uniform distortion in the lattice (Fig. 6). Within the environment of such domains, unit cells can be expected to be distorted in an identical fashion and hence selective averages created from them will allow us to analyse the mechanical properties at the unit cell level. Provided that a sufficient number of uniformly distorted unit cells can be accumulated, an improvement in resolution can also be expected. Another strategy that might be worth exploring is the use of sub-unit-cell-sized references comprising only the more rigid elements of the unit cell in the course of a refinement cycle of correlation averaging, in order to locate and align them more accurately.

6. Acknowledgements

We gratefully acknowledge the financial support of the Deutsche Forschungsgemeinschaft (SFB 266 to W.B.), the Fonds der Chemischen Industrie, and the Medical Research Council of Canada (postdoctoral fellowship to B.M.P.).

294

7. References

Baldwin, J.M., Henderson, R., Beckman, E., and Zemlin, F. (1988) Images of purple membrane at 2.8 Å resolution obtained by cryo-electron microscopy, *J. Mol. Biol.* **202**, 585-591.

Baumeister, W., and Engelhardt, H. (1987) Three-dimensional structure of bacterial surface layers, in J.R. Harris and R.W. Horne (eds.), *Electron microscopy of proteins*, vol. 6, *Membranous structures*, Academic Press, London, p. 109-154.

Baumeister, W., Karrenberg, F., Rachel, R., Engel, A., Ten Heggeler, B., and Saxton, W.O. (1982) The major cell envelope protein of *Micrococcus radiodurans* (R1): Structural and chemical characterization, *Eur. J. Biochem.* **125**, 535-544.

Baumeister, W., Wildhaber, I., and Engelhardt, H. (1988) Bacterial surface proteins: Some structural, functional, and evolutionary aspects, *Biophys. Chem.* **29**, 39-49.

Baumeister, W., Wildhaber, I., and Phipps, B.M. (1989) Principles of organization in eubacterial and archaebacterial surface proteins, *Can. J. Microbiol.* **35**, 215-227.

Bragg, L. (1962) *The crystalline state*, vol. 1, *A general survey*, G. Bell and Sons, London, 352 pp.

Caspar, D.L.D., and Klug, A. (1962) Physical principles in the construction of regular viruses, *CSH Symp. Quant. Biol.* **27**, 1-24.

Dickson, G., Gower, H.J., Barton, C.H., Prentice, H.M., Elsom, V.L., Moore, S.E., Cox, R.D., Quinn, C., Putt, W., and Walsh, F.S. (1987) Human muscle neural cell adhesion molecule (N-CAM): Identification of a muscle-specific sequence in the extracellular domain, *Cell* **50**, 1119-1130.

Harris, W.F., and Scriven, L.E. (1970) Function of dislocations in cell walls and membranes, *Nature* **228**, 827-829.

Hegerl, R., and Baumeister, W. (1988) Correlation averaging of a badly distorted lattice: The surface protein of *Pyrodictium occultum*, *J. Electron Microsc. Technique* **9**, 413-419.

Henderson, R., Baldwin, J.M., Downing, K.H., Lepault, J., and Zemlin, F. (1986) Structure of purple membrane from *Halobacterium halobium*: Recording, measurement and evaluation of electron micrographs at 3.5 Å resolution, *Ultramicroscopy* **19**, 147-178.

Homans, S.W., Dwek, R.A., and Rademacher, T.W. (1987) Solution conformations of N-linked oligosaccharides, *Biochemistry* **26**, 6571-6578.

Hovmöller, S., Sjögren, A., and Wang, D.N. (1988) The structure of crystalline bacterial surface layers, *Prog. Biophys. Molec. Biol.* **51**, 131-163.

Huber, R. (1988) Flexibility and rigidity of proteins and protein-pigment complexes, *Angew. Chem. Int. Ed. Engl.* **27**, 79-88.

Imberty, A., Gerber, S., Tran, V., and Pérez, S. (1990) Data bank of three-dimensional structures of disaccharides, a tool to build 3-D structures of oligosaccharides: Part I. Oligo-mannose type N-glycans, *Glycoconjugate J.* **7**, 27-54.

Jap, B.K. (1989) Molecular design of phoE porin and its functional consequences, *J. Mol. Biol.* **205**, 407-419.

Klenk, H.-P., Haas, B., Schwass, V., and Zillig, W. (1986) Hybridization homology: A new parameter for the analysis of phylogenetic relations, demonstrated with the urkingdom of the archaebacteria, *J. Mol. Evol.* **24**, 167-173.

Koval, S.F., and Murray, R.G.E. (1984) The isolation of surface array proteins from bacteria, *Can. J. Biochem. Cell Biol.* **62**, 1181-1189.

Lechner, J., and Sumper, M. (1987) The primary structure of a procaryotic glycoprotein:

Cloning and sequencing of the cell surface glycoprotein gene of Halobacteria, *J. Biol. Chem.* **262**, 9724-9729.

Mescher, M.F., and Strominger, J.L. (1976) Structural (shape-maintaining) role of the cell surface glycoprotein of *Halobacterium salinarium*, *Proc. Natl. Acad. Sci. USA* **73**, 2687-2691.

Messner, P., and Sleytr, U.B. (1988) Asparaginyl-rhamnose: A novel type of protein-carbohydrate linkage in a eubacterial surface-layer glycoprotein, *FEBS Lett.* **228**, 317-320.

Messner, P., Pum, D., Sára, M., Stetter, K.O., and Sleytr, U.B. (1986) Ultrastructure of the cell envelope of the archaebacteria *Thermoproteus tenax* and *Thermoproteus neutrophilus*, *J. Bacteriol.* **166**, 1046-1054.

Noegel, A., Gerisch, G., Stadler, J., and Westphal, M. (1986) Complete sequence and transcript regulation of a cell adhesion protein from aggregating *Dictyostelium* cells, *EMBO J.* **5**, 1473-1476.

Peters, J., Peters, M., Lottspeich, F., Schäfer, W., and Baumeister, W. (1987) Nucleotide sequence analysis of the gene encoding the *Deinococcus radiodurans* surface protein, derived amino acid sequence, and complementary protein chemical studies, *J. Bacteriol.* **169**, 5216-5223.

Peters, J., Peters, M., Lottspeich, F., Schäfer, W., Cejka, Z., and Baumeister, W. (1988) The primary structure of the HPI-layer polypeptide of *Deinococcus radiodurans*, in U.B. Sleytr, P. Messner, D. Pum, and M. Sára (eds.), *Crystalline bacterial cell surface layers*, Springer-Verlag, Berlin, pp. 140-144.

Peters, J., Peters, M., Lottspeich, F., and Baumeister, W. (1989) S-layer protein gene of *Acetogenium kivui*: Cloning and expression in Escherichia coli and determination of the nucleotide sequence, *J. Bacteriol.* **171**, 6307-6315.

Phipps, B.M., Trust, T.J., Ishiguro, E.E., and Kay, W.W. (1983) Purification and characterization of the cell surface virulent A protein from *Aeromonas salmonicida*, *Biochemistry* **22**, 2934-2939.

Phipps, B.M., Engelhardt, H., Huber, R., and Baumeister, W. (1990) Three-dimensional structure of the crystalline protein envelope layer of the hyperthermophilic archaebacterium *Pyrobaculum islandicum*, *J. Struct. Biol.* (in press).

Sass, H.J., Büldt, G., Beckmann, E., Zemlin, F., van Heel, M., Zeitler, E., Rosenbusch, J.P., Dorset, D.L., and Massalski, A. (1989) Densely packed ß-structure at the protein-lipid interface of porin is revealed by high-resolution cryo-electron microscopy, *J. Mol. Biol.* **209**, 171-175.

Saxton, W.O., and Baumeister, W. (1982) The correlation averaging of a regularly arranged bacterial cell envelope protein, *J. Microsc.* **127**, 127-138.

Saxton, W.O., and Baumeister, W. (1986) Principles of organization in S layers, *J. Mol. Biol.* **187**, 251-253.

Sleytr, U.B., and Messner, P. (1983) Crystalline surface layers on bacteria, *Annu. Rev. Microbiol.* **37**, 311-339.

Smit, J. (1986) Protein surface layers of bacteria, in M. Inouye (ed.), *Bacterial outer membranes as model systems*, Wiley, New York, pp. 343-376.

Sumper, M. (1987) Halobacterial glycoprotein biosynthesis, *Biochim. Biophys. Acta* **906**, 69-79.

Taylor, K.A., Deatherage, J.F., and Amos, L.A. (1982) Structure of the S-layer of *Sulfolobus acidocaldarius*, *Nature* **299**, 840-842.

Tsuboi, A., Uchihi, R., Tabata, R., Takahashi, Y., Hashiba, H., Sasaki, T., Yamagata, H., Tsukagoshi, N., and Udaka, S. (1986) Characterization of the genes coding for the two major cell wall proteins from protein-producing *Bacillus brevis* 47: Complete nucleotide sequence of the outer wall protein gene, *J. Bacteriol.* **168**, 365-373.

Tsuboi, A., Uchihi, R., Adachi, T., Sasaki, T., Hayakawa, S., Yamagata, H., Tsukagoshi, N., and Udaka, S. (1988) Characterization of the genes for the hexagonally arranged surface layer proteins in protein-producing *Bacillus brevis* 47: Complete nucleotide sequence of the middle wall protein gene, *J. Bacteriol.* **170**, 935-945.

van Heel, M., and Frank, J. (1981) Use of multivariate statistics in analysing the images of biological macromolecules, *Ultramicroscopy* **6**, 187-194.

Wildhaber, I., and Baumeister, W. (1987) The cell envelope of *Thermoproteus tenax*: Three-dimensional structure of the surface layer and its role in shape maintenance, *EMBO J.* **6**, 1475-1480.

Wildhaber, I., Santarius, U., and Baumeister, W. (1987) Three-dimensional structure of the surface protein of *Desulfurococcus mobilis*, *J. Bacteriol.* **169**, 5563-5568.

Yamada, H., Tsukagoshi, N., and Udaka, S. (1981) Morphological alterations of cell wall concomitant with protein release in a protein-producing bacterium, *Bacillus brevis* 47, *J. Bacteriol.* **148**, 322-332.

STRUCTURE OF HYDROGENASE FROM *Thiocapsa roseopersicina*.

M.B. SHERMAN, E.V. ORLOVA
Institute of Crystallography USSR Acad. of Sci.
59 Leninsky pr.
117333 Moscow
USSR

and

S. HOVMOLLER
Structural Chemistry
Stockholm University
S-106 91 Stockholm
Sweden

ABSTRACT.   The three-dimensional structure of the nickel-containing
enzyme - hydrogenase from *T. roseopersicina* has been determined by
electron microscopy and image processing on microcrystals of the
enzyme.   The enzyme forms a large, open, ring-shaped complex,
containing 6 molecules each of the large (62 kDa) and small (26 kDa)
subunits.   The molecule is dumb-bell shaped.

Introduction

Hydrogenase are proteins which catalises the reversible oxidoreduction
of molecular hydrogen in many microorganisms. The enzyme from the
phototrophic bacterium *Thiocapsa roseopersicina* contains nickel which
has only recently been recognized as an essential micronutrient for
many microorganisms. Nickel is found in enzymes involved in
ureolysis, hydrogen metabolism, methane biogenesis and acetogenesis.
It is known that different hydrogenases vary in molecular masses,
number of subunits, active center structure and other properties [1].
All hydrogenases contain one or several Fe-S clusters and many of them
contain nickel as well [2].   These redox centers are important for
the action of the enzymes.   The hydrogenase molecule from *Thiocapsa
roseopersicina* is believed to consist of two different subunits, α and
β, having masses of 62 kDa (α) and 26 kDa (β)[3].   It is very stable
and shows a high activity at 70°C over a long time period.
        The only structural information on the above mentioned enzymes
comes from electron microscopy of individual molecules of F-420-
reducing hydrogenase[4].   Recently two hydrogenases (from
Desulfovibrio gigas [5] and *Desulfovibrio vulgaris* [6]) have been

297

J. R. Fryer and D. L. Dorset (eds.), Electron Crystallography of Organic Molecules, 297–304.
© 1990 *Kluwer Academic Publishers.*

crystallized in macrocrystals suitable for X-ray analysis. In the
present paper the medium-resolution 3D structure of a bacterial
hydrogenase, the hydrogenase from *Thiocapsa roseopersicina* is
discussed. The structure was obtained by electron microscopy and a
3-D reconstruction from tilted projections of the crystalline enzyme
layers.

Materials and Methods.

The hydrogenase was purified from the cells of phototrophic bacterium
Thiocapsa roseopersicina, strain BBS, by the procedure described elsewhere [7] with minor modifications.
    The layers were crystallized by the hanging drop vapour
diffusion method. A 2.5% protein solution buffered in 0.1 TAPS buffer,
pH 7.0 was used for crystallization. The concentration of ammonium
sulphate in the protein solution was 1M and 2.2M in the outer
reservoir. After several weeks crystals were formed. The crystal
suspension was applied to the carbon coated grids and negatively
stained with 2% uranyl acetate. Grids were thoroughly washed with
several drops of staining solution to remove excess ammonium sulphate.
    Electron micrographs were taken at a magnification of 47000
times on a Philips EM 420. Crystals obtained were mainly large thin
platelets and their size varied from tens to thousands nm. Many of
them were several unit cells thick but it was possible to locate a
reasonable quantity of mono- or double layered crystals. It was
possible to discriminate between mono- and double-layered crystals

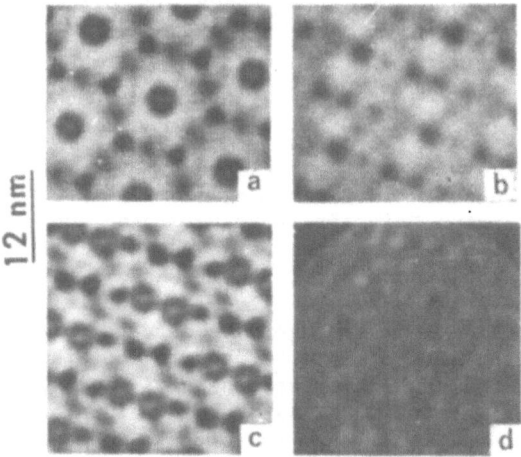

Figure 1. a) -monolayer-type filtered image of *T. roseopersicina* hydrogenase; b)-bilayer-type filtered image; c) -'bilayer' image simulated from a); d)-difference of b) and c).

because their visual appearance was rather different (Fig.1a,b). It
was possible to simulate 'b'-type image from 'a'-type by simply
summing two images of 'a'

type together with a certain directional shift applied (Fig.1c). One can see the rather structureless appearance of Fig.1d representing the difference between the real image (Fig.1b) and the simulated one (Fig.1c). Only those crystals which proved to be monolayers were selected for the 3-D analysis. They corresponded to images similar in appearance to Fig.1a and their contrast was sufficiently low (similar to that of individual hydrogenase particles). For the 3-D analysis tilt series of up to 45° angles of tilt were taken. Electron micrographs were analysed in an optical diffractometer and well focused images of good quality crystals were further processed.

Selected well-ordered areas of the crystals were digitized on a Joyce Loebl Microdensitometer 6 using a raster size of 40 micrometers. Scanned areas of 256 by 256 pixels contained about 200 unit cells (see Fig. 2a). Calculations were done on a VAX 11/750 computer and a Digital VS11 color graphics display was used for visualization of the results. The general idea of 3-D reconstruction was similar to that of Henderson and Unwin [8]. The diffraction patterns (Fig.2b) were indexed and lattice vectors refined, using a least-squares procedure. Amplitudes and phases were extracted from Fourier transforms and the phase origin refined as described in [9].

Results and Discussion

The repeat motif is formed of ring-shaped particles having a diameter of about 110 Å. The layers were hexagonal with $a = b = 122$ Å $\gamma = 120°$. Three tilt series of 22 images altogether were used for 3-D reconstruction, covering -43 ÷ 49° tilt range. The real tilt angle values of individual images were calculated by a least-squares procedure described in [10].

There was no indication of significant deviations from P6 symmetry in the zero tilt projection. So, all the projection data were merged together and symmetrized according to P6 symmetry. And this was the lowest possible symmetry so no artefacts could arise from the process. Yet, (h,k) and (k,h) reflections had very similar amplitudes and phases and it was quite possible that real symmetry could be even higher, namely two-fold axes could occur in the layer plane. But in order to make an unbiassed determination of the 3D structure only P6 symmetry was imposed. And these merged data were used as the reference for further analysis.

As a first stage 3D data were merged together in individual series. After that it became clear that real symmetry was not consistent with 6-fold axis because for some reflections phase curves for different data series were very similar in values but with opposite sign slopes. Analysing the data, P32 symmetry was considered as the real symmetry. After this symmetry correction all the data were merged together in P321 two-sided group. In merging average phase error was about 30°.

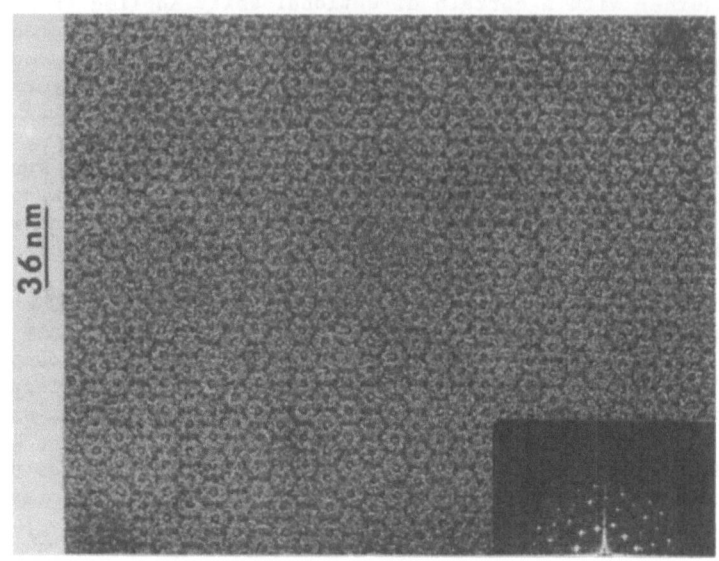

Figure 2. Original image of hydrogenase *T. roseopersicina*; insert - its optical diffraction pattern.

We observed rather a specific amplitude dependence on z^* for several reflections. In Fig. 3 one of these reflections is presented, namely (0,2). One can see a region in z^* in which amplitudes fall to zero level. Such a behaviour could be caused by a multilayered structure diffraction. It is known that for a quasiperiodic function g(x), having N periods of value h:

$$g(x) = f(x) \; \bullet \; \sum_{l=0}^{N-1} \delta(x-lh), \; {}^\dagger \qquad (1)$$

its Fourier spectrum could be written as follows:

$$F(g) = F(f) \; \cdot \; \sum_{m=0}^{N-1} \exp(-2\pi mkh), \qquad (2)$$

where h is the real space period value for function g, F means Fourier transformation and k is the coordinate in reciprocal (spectral) space. For amplitudes of $F(g)$ and $F(f)$ another equation can be written:

---

† $\bullet$ is a convolution sign

$$F(g) \quad = \quad F(f) \quad \cdot H(N,y) \qquad (3)$$

where

$$H(N,y) = \frac{\sin(Ny)}{\sin(y)} \qquad y = 2\pi kh, \qquad (4)$$

or, in other words, $F(f)$ amplitudes should be modulated with the $H(N,y)$ function to obtain the $F(g)$ amplitudes. Graphs of the $H(N,y)$ function for $N = 2$ and 3 are presented in Fig. 4. The presence of the amplitude zeroes of such reflections is confirmed by their phases: in same regions phases abruptly change their values from, say, 180° to 0° and vice versa, i.e. the sign of the scattering amplitude changed. If the whole set of reflections were modulated in the same manner, then it would be a consequence of the multi-layered crystal structure. Moreover, having reasonably good quality data, it would be possible

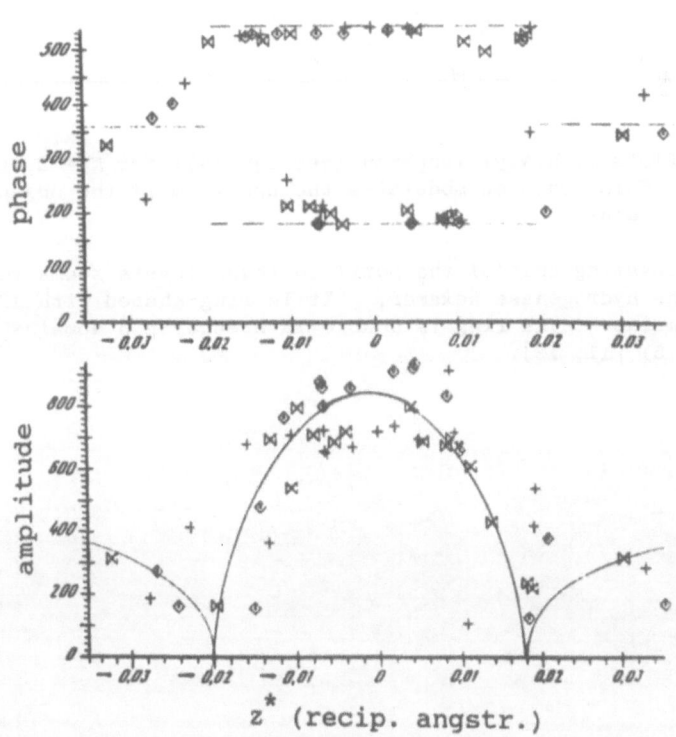

Figure 3. Plots of phases and amplitudes of (0,2) reflection versus z^*. Data from different tilt series are plotted with different symbols.

to determine the number of layers constituting the crystal simply by calculating the number of zeroes in the reflection amplitudes. But in our case only several reflections were modulated, so it could not be the influence of crystal structure but should be attributed to the unit cell content itself. Summarizing all the above it is quite possible that the object itself has a double-layered density distribution.

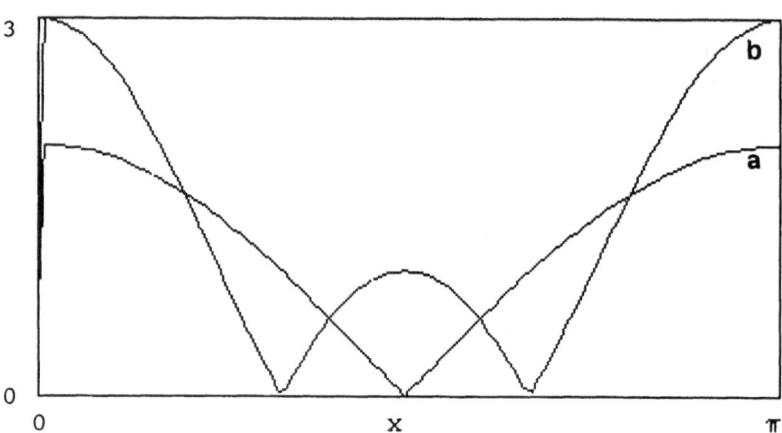

Figure 4. Plots of H(N,y) function (see eqn (4)) for $N = 2$ (a), and $N = 3$ (b). This function modulates the spectrum if the object has N-layered structure.

The repeating unit of the motif in these layers seems to represent the hydrogenase hexamer. It is ring-shaped with 12 distinct domains. The ring is double-layered with 6 domains in each layer (Fig. 5) [11, 12].

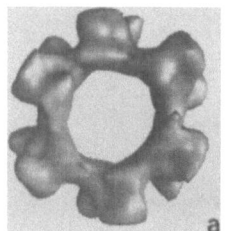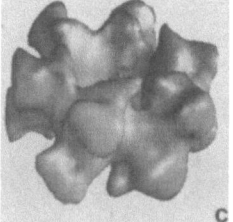

Figure 5. Three-dimensional structure of hydrogenase from *T. roseopersicina*. a) - view along 3-fold axis; b) - side view; c) - view from am oblique angle.

The layers are slightly off-set relative to each other, such that the
lower layer is rotated by approximately 20° about the 3-fold axis. The
tentative monomers are dumb-bell shaped with two distinct domains; one
in the upper and one in the lower ring. To the present resolution
the two domains of the monomer are very similar.

Acknowledgements

The authors wish to thank Dr. N. Zorin, Inst. of Soil Science and
Photosynthesis, Pushchino, for purification of the enzyme and Dr. E.
Smirnova, Inst. of Crystallography, Moscow, for growing crystals, Drs.
Hans Hebert and Urban Kaveus, Dept. of Medical Physics, The Karolinska
Institute, Stockholm, are thanked for the help in displaying the 3D
model with their Raster Technology ONE/380 graphics system. We are
grateful also to Profs. B.K. Vainstein and N.A. Kiselev for arranging
the collaborative research.

References

1. Hausinger,R.P. (1987), 'Nickel utilization by microorganisms',
   Microbiological Reviews, 51, 22-42.
2. Voordouw,G., Menon,N.K., LeGall,J., Choi,E.-S., Peck,H.D. and
   Przybyla,A.E. (1989), 'Analysis and comparison of nucleotide
   sequence encoding genes for [NiFe] and [NiFeSe] hydrogenases from
   Desulfovibrio gigas and Desulfovibrio baculatis, J.Bacteriol, 171,
   2894-2899.
3. Kovacs,L.K., Seefeldt,L.C., Tigyi,G., Doyle,C.M., Mortenson,L.E.
   and Arp,D.J. (1989), 'Immunological relationship among hydrogenases'
   J.Bacteriol, 171, 430-435.
4. Wackett,L.P., Hertwieg,E.A., King,J.A., Orme-Johnson,W.H. and Walsh
   C.T., (1987), 'Electron microscopy of nickel-containing methano-
   genic enzymes. Methyl reductase and F420-reducing hydrogenase',
   J.Bacteriol., 169, 718-727.
5. Higuchi,Y., Yasuoka,N., Kakudo,M., Katsube,Y., Yagi,T., and
   Inokuchi,H. (1987), 'Single crystals of hydrogenase from Desulfovibrio vulgaris', J.Biol. Chem., 262, 2823-2825.
6. Niviere,V., Hatchikian,C., Cambillau,C., and Fery,M. (1987), 'Crystallization, preliminary X-ray study and crystal activity of the
   hydrogenase from Desulfovibrio gigas', J. molec.Biol., 195, 969-
   971.
7. Sherman,M.B., Orlova,E.V., Smirnova,E.A., Zorin,N.A., Tagunova,I.V,
   Kuranova,I.P. and Gogotov,I.N. (1987), 'Structure of microcrystals
   of hydrogenase from Thiocapsa roseopersicina', Doklady Akademii
   Nauk SSSR (Rus.), 295, 509-512.
8. Henderson,R. and Unwin,P.N.T. (1975), 'Three-dimensional model of
   purple membrane obtained by electron microscopy', Nature (London)
   257, 26-32.
9. Sjorgen,A., Hovmoller,S., Farrants,G., Ranta,H., Haapasalo,M.,
   Ranta,K., Lounatmaa (1985), 'Structures of two different surface
   layers found in six bacteroids strains', J. Bacteriol, 164, 1278-
   1282.

10. Shaw,P.J. and Mills,G.J. (1981), 'Tilted specimen in the electron microscope : A simple specimen holder and the calculation of tilt angles for crystalline specimens', Micron, **12**, 279-282.

11. Sherman,M.B., Orlova,E.V., Hovmöller,S., Smirnova,E.A., Zorin,N.A., and Gogotov,I.N. (1989), '3-D structure of of hydrogenase from *Thiocapsa roseopersicina*', Doklady Akademii Nauk SSSR (Rus.), **308**, 1489-1493.

12. Sherman, M.B., Orlova, E.V., Smirnova, E.A., Hovmöller, S. and Zorin,N.A. (1990), 'The three-dimensional sructure of nickel-containing hydrogenase from *Thiocapsa roseopersicina*', J. Bacteriol. in press.

CRYOELECTRON MICROSCOPY OF PROTEIN CRYSTALS. SOME REMARKS ON THE
METHODOLOGY

F. ZEMLIN
Fritz-Haber-Institut der Max-Planck-Gesellschaft
Faradayweg 4-6
D-1000 Berlin 33

ABSTRACT. Cryoelectron microscopy is now proven to be an advantageous
tool for structure research of protein crystals. Using a helium-cooled
superconducting electron microscope, the following protein crystals
have been imaged with high resolution in one projection: crotoxin complex 3.5 Å, purple membrane 2.8 Å, matrix porin OmpF 3.5 Å, surface
protein of sulfolobus spec. B12 10 Å. Fine-structure details within
the unit cell were resolved, although even at 4.5 K specimen temperature the crystal still suffers considerable radiation damage. While
the tolerable dose is only 20 e/Å², for atomic resolution a dose at
least 100 times higher is needed. The ladder to these successes had
two steps: First, the image was taken in a special way with a minimum
of pre-exposure. Second, the low-dose images were evaluated by a sophisticated image-processing procedure, the main part of which is the
cross-correlation averaging. The high resolution structure is in all
cases revealed by averaging thousands of low-dose-imaged unit cells.

For a long time the electron microscopical imaging of protein crystals
with high resolution seemed a hopeless goal. The radiation damage was
too great; the crystals were destroyed before an image with reasonable
signal-to-noise ratio could be recorded. This could easily be observed
and measured by the fading of electron-optical diffraction patterns at
increasing electron dose. These measurements, however, only displayed
half the truth, namely the fading of the intensities (amplitudes) of
the Fourier coefficients. The randomization of the phases increases
with increasing dose--the phases are not recognizable in these diffraction patterns. The first decisive step to overcome these major difficulties was made by Unwin and Henderson in 1975 [1]; they recorded low-
dose images of purple membrane and retrieved the information, i.e. the
structure, by an ensuing image processing. In this image processing
the optical aberrations of the microscope were taken into account, but
the main task was the averaging of thousands of noisy imaged unit cells.
The next step in electron microscopical structure research of protein

305

J. R. Fryer and D. L. Dorset (eds.), Electron Crystallography of Organic Molecules, 305–308.
© 1990 Kluwer Academic Publishers.

crystals was the 3D map of purple membrane achieved by Henderson and Unwin [2]. They evaluated a series of low-dose images by tilting the crystal relative to the optical axis. The resolution of this 3D map was about 7 Å.

Although this was a real breakthrough in electron microscopical structure research of protein crystals, the dream of a high-resolution 3D map in which the side chains of the amino acids can be located was not yet fulfilled. One promising idea was to cool the specimen in order to increase the stability of the crystal, which would then permit higher doses and hence higher resolution. Of course, the primary effects, especially the ionization, could not be reduced by cooling, but the decay of the crystal should be slowed down, and if cooling would do that, the temperature should be as low as possible. For this reason, a helium-cooled superconducting cryomicroscope was used with a specimen temperature of about 4.5 K [3]. Its cryo-objective lens was designed by Isolde Dietrich and co-workers at Siemens, Munich. The main advantages of this cryo-objective area: The magnetic field produced by the superconducting coils and formed by the superconducting shielding is extremely constant, like that of a permanent magnet. The optical aberrations are very small (e.g. C_s = 1.35 mm) although the pole-piece gap is S = 7.5 mm. The large pole-piece gap allows tilting of the specimen. The vacuum in the environment of the specimen is better than 10^{-8} Torr and free of water and hydrocarbons, since the specimen lies in the center of a powerful helium-cooled cryopump. There is no measurable specimen contamination by ice or hydrocarbons. The drift of the specimen is less than 0.005 nm/sec. But the most important advantage is the reduction of radiation damage. This effect is, however, not as strong as expected; even at 4.5 K specimen temperature there is considerable radiation damage. This was measured by the fading of the electron-optical diffraction pattern of purple membrane [4]. The reflex intensities fade exponentially:

$$I = I_o \cdot e^{-D/D_e}. \qquad (1)$$

For 9 Å spacing the critical dose is D_e = (13 ± 4) e/Å². At this dose the reflex intensities are reduced to 37% of the original value, i.e. the structure is indeed not totally destroyed. Now the question arises of which dose to choose for obtaining the highest signal-to-noise ratio. In a theoretical investigation Hayward and Glaeser [5] found that an exposure 2.5 times the critical dose yielded images with highest signal-to-noise ratio. Our experimental findings agree with this theoretical estimation; we found 20 e/Å² to be the optimum dose for imaging purple membrane. But for images with atomic resolution and reasonable signal-to-noise ratio one requires about 2000 e/Å². The images with 20 e/Å² are "low-dose" images. They contain high-resolution information which, however, cannot be seen directly--it is hidden in noise. It can be retrieved by computer image processing, especially by averaging. The computer image processing, described in detail elsewhere [6,7], is a sophisticated iteratively running approach. In the following, some essentials of the procedure are listed:

(a) Correction of phase shift due to spherical aberration, defocus
    and astigmatism (taking account of the contrast transfer function).
(b) Correction of phase shift due to beam tilt (axial coma).
(c) Correction of the reflex splitting due to specimen tilt.
(d) Correction of image distortions due to bending of crystals and
    optical distortion.
(e) Substitution of the amplitudes from electron diffraction.
(f) Cross-correlation averaging of thousands of low-dose-imaged
    unit cells.
(g) Merging of all data of the tilt series to generate a 3D model.

   The value of the cryoelectron microscopic imaging procedure combined with the image processing has been demonstrated. For purple membrane in untilted specimens atomic resolution, 2.8 Å, has been
achieved [8], and even in 3D structure research of purple membrane
almost atomic resolution has been obtained. Data from 72 images from
both tilted and untilted specimens were analyzed to produce the phases
of 2700 independent Fourier coefficients. The amplitudes of the Fourier
coefficients were measured from 150 electron-optical diffraction patterns. Together these data represent about half of the full three-
dimensional transform to 3.5 Å. The map has a resolution of 3.5 Å in a
direction parallel to the membrane but lower than this in perpendicular
direction. It shows many features that are resolved from the main density of the seven alpha helices. We interpret these features as the
bulky aromatic side-chains of phenylalanine, tyrosine and tryptophan res-
-idues. There is also a very dense feature which is the beta-ionone
ring of the retinal chromophore [7].
   The utility of cryomicroscopy for protein structure research might
be questionable if only data from one particular protein were available.
In fact, already several years ago it was possible to image protein
crystals of the crotoxin complex in one projection with 3.5 Å resolution [9]. Matrix porin OmpF, as well, could be imaged in one projection
with 3.5 Å resolution [10]. Recently, investigations at the surface of
protein of sulfolobus Spec. B12 yielded the relatively high resolution
of 10 Å [11]. These results could be obtained although the cryomicroscope used is a prototype with rather outdated electronics and without
computer control. A modern cryomicroscope should have the option of
computer-controlled small-spot scanning with simultaneous defocus compensation [12]. This small-spot scanning has proven to reduce the
movement in the specimen due to electron exposure [13]. Moreover, it
is recommendable to equip such a cryomicroscope with a field emission
gun providing high coherence, which would then result in improved phase
contrast. It is hoped that in the near future data sets of protein
crystals can be routinely provided by using a modern cryomicroscope.

Acknowledgements. The author is deeply indebted to the continuous
assistance of E. Beckmann and K. Heinrich.

308

References

[1] P.N.T. Unwin and R. Henderson, J. Mol. Biol. 94 (1975) 425–440.
[2] R. Henderson and P.N.T. Unwin, Nature 2557 (1975) 28–32.
[3] I. Dietrich, F. Fox, E. Knapek, G. Lefranc, K. Nachtrieb, R. Weyl and H. Zerbst, Ultramicrosocpy 2 (1977) 241–249.
[4] F. Zemlin, E. Reuber, E. Beckmann and D. Dorset, in: Proceedings 44th Annual Meeting EMSA, 1986, pp. 10–13.
[5] S.B. Hayward and R.M. Glaeser, Ultramicroscopy 4 (1979) 201–210.
[6] R. Henderson, J.M. Baldwin, K.H. Downing, J. Lepault and F. Zemlin, Ultramicrosocpy 19 (1986) 147–178.
[7] R. Henderson, J.M. Baldwin, T.A. Ceska, F. Zemlin, E. Beckmann and K.H. Downing, J. Mol. Biol. (1989, in press).
[8] J.M. Baldwin, R. Henderson, E. Beckmann and F. Zemlin, J. Mol. Biol. 202 (1988) 585–591.
[9] T.W. Jeng, W. Chiu, F. Zemlin and E. Zeitler, J. Mol. Biol. 175 (1984) 93–97.
[10] H.-J. Sass, G. Büldt, E. Beckmann, F. Zemlin, M. van Heel, E. Zeitler, J.P. Rosenbusch, D.L. Dorset and A. Massalski, J. Mol. Biol. 209 (1989) 171–175.
[11] G. Lembcke and F. Zemlin, in: Proceedings 12th International Congress on Electron Microscopy, Seattle, 1990 (in press).
[12] F. Zemlin, J. Electron Microscopy Techniques 11 (1989) 251–257.
[13] R. Henderson and R.M. Glaeser, Ultramicroscopy 16 (1985) 139–150.

ELECTRON CRYSTALLOGRAPHIC ANALYSIS OF RECONSTITUTED PhoE PORIN

Bing K. Jap, Peter J. Walian and Kalle Gehring
Lawrence Berkeley Laboratory
University of California
Berkeley, CA 94720

ABSTRACT. High resolution images of PhoE porin have been obtained at tilt angles ranging ±60 degrees from the membrane plane. Selected good quality images of tilted PhoE porin were processed to correct for the effects of lattice distortions and lens aberrations. Such corrections improved the signal-to-noise ratio of the diffraction spots in the Fourier transform of the images. The 3-dimensional map has been reconstructed at a resolution of 6.5Å, and the map shows that porin consists of a trimer of elliptical ß-sheet cylinders. We propose that these elliptical ß-sheet cylinders are the basic structural motif for porin monomers in Gram negative bacteria. Within each elliptical cylinder, there is an internal structure which forms a part of the channel structure itself. The internal structure is proposed to play an important role in determining the channel size of the various porins and may be directly involved in the closing and opening of the channel that has been observed in *in vitro* experiments.

1. Introduction

The porins are a family of pore forming proteins that are found in the outer membrane of Gram-negative bacteria such as *Escherichia coli* and *Salmonella typhimurium* (see reviews, Lugtenberg and Van Alphen, 1983; Nikaido and Vaara, 1985; Nakae, 1986; Jap and Walian, 1990). The function of porins is to serve as the pathway for molecules, such as nutrients and waste products, of less than about 600 daltons to diffuse freely across the outer membrane. The outer membrane itself would be impermeable to most hydrophilic solutes were it not for the presence of porins. In *E. coli*, for instance, there are, among others, OmpF, OmpC and PhoE porin. These three porins have no binding sites for substrates to be transported across the membrane. However, PhoE porin has been demonstrated to have some transport selectivity for anionic and phosphate-containing compounds. Porins also serve as receptors for a variety of bacteriophages.

The amino acid sequences of OmpF, OmpC and PhoE, among many others, have been deduced from their DNA sequences. Sequence analysis has shown that OmpF, OmpC and PhoE porins of *E. coli* are highly homologous and contain a high density of charged residues distributed uniformly along their sequences. The secondary structures of porins are predominantly in ß-sheet conformation which is clearly different from that

J. R. Fryer and D. L. Dorset (eds.), Electron Crystallography of Organic Molecules, 309–315.
© 1990 *Kluwer Academic Publishers.*

of other membrane proteins which are often found to have their transmembrane segments in an α-helical conformation.

PhoE porin has been reconstituted with phospholipid to form 2-dimensional crystalline arrays. The 3-dimensional (3-D) electron diffraction data which extends to a resolution of better than 3 Å (Walian and Jap, 1990) shows two major populations in the orientation of the ß-strands: one with ß-strands oriented perpendicular to the membrane plane and the other tilted 35 degrees from the membrane normal. The projection map of PhoE porin at 3.5Å resolution has also been obtained. The map shows a trimeric ring-like structure, and each ring-like structure consists of "beads" that were interpreted as the projection of ß-strands along their axes (Jap *et al*. 1990).

We report here a preliminary structure of PhoE porin determined to a resolution of about 6.5Å using electron crystallography techniques.

2. Materials and Methods

The crystallization of PhoE porin forming crystalline membrane patches that diffract to better than 3 Å has been described previously (Jap, 1988). Basically, the crystallization involves reconstitution of purified porin with dimyristoyl phosphatidylcholine (DMPC) at a protein-to-lipid ratio of 4:1 (W/W) with the use of sodium dodecylsulphate (SDS) as the detergent. The protein/lipid mixture was dialyzed against a buffer containing 10 mM TRIS, 100 mM NaCl, 3 mM NaN$_3$ at pH 7.5. The reconstitution was performed by slowly dialyzing the mixture of purified protein and DMPC that was also solubilized in the same detergent. To reduce the rate of dialysis, SDS was added initially to the dialyzing solution at concentration of about 0.3 % (W/W). The dialysis was performed by replacing only half of the dialyzing solution every 4 hours with the same buffer but without SDS. When visible "precipitation" occured, the samples were dialyzed twice with 5mM TRIS buffer at pH 7.5 to reduce the concentration of salts. The membrane patches were further treated with phospholipase in order to increase the percentage of highly ordered crystalline patches.

Electron microscopy and electron diffraction were performed as described previously (Jap, *et al*., 1990; Walian and Jap, 1990)). Briefly, 1 μl of the reconstituted crystalline patches in suspension (2 mg/ml) and 2 μl of 1 % trehalose were applied to the copper grid side of the carbon-coated electron microscope grid after it has been glow discharged for 2 minutes under 200 mTorr of water vapor. The sample was allowed to stand on the grid for about 1 minute and was then blotted to a thin layer which was air-dried. The specimen was examined in a JEOL 100B electron microscope which is equipped with a field emission gun and a cold stage that was operated at about -120° C. The microscope was also equipped with a computer control for the low-dose and spot-scan imaging (Downing and Glaeser, 1986). Images were recorded using both flood-beam illumination and the spot-scan technique at a magnification of about 55,000.

Images of specimens tilted at 30, 45 and 60 degrees were recorded using both conventional flood beam illumination and spot-scan techniques. High resolution images of untilted samples have been obtained previously (Jap *et al*., 1990). Images were evaluated on the basis of the quality of their optical diffraction patterns. Images with sharp, high resolution reflections were selected for computer processing. Processing of tilted images followed the procedure as described by Henderson *et al*. (1990), using computer software from the MRC (Medical Research Council, Laboratory of Molecular Biology, Cambridge, England) without any modification. After lens aberration and lattice distortions had been corrected, images were merged to a common phase origin. The structure amplitudes of the images were replaced by the corresponding values derived from our 3-D diffraction data set (Walian and Jap, 1990). Fifteen tilted and eight untilted images were used in our preliminary 3-D reconstruction.

3. Results and Discussion

Correction of crystal distortions is needed in order to obtain reliable high resolution phase information. Although we have selected images which show strong, sharp optical diffraction spots that extend to high resolution for computer analysis, many weak diffraction spots in the Fourier transform of the images are not easily detected. The signal-to-noise ratio can be enhanced by correction of lattice distortion and/or by the use of a large number of unit cells. Since the unit cell dimensions of PhoE porin crystals are rather large, compared to those of bacteriorhodopsin, a significantly larger crystal is needed in order to achieve the same level of signal-to-noise ratio. The size of the patches and of electron micrograph film has often put an upper limit to the size of images. Therefore it has been essential to enhance the signal-to-noise ratio of diffraction spots. Correction of lattice distortions and the use of the spot-scan imaging technique have been shown to increase the high resolution image contrast.

We have used both conventional flood beam illumination and the spot-scan technique in collecting our high resolution images. We have found that the percentage of good images obtained using the spot-scan technique is significantly greater than that obtained by conventional flood beam illumination. We have noted that the number of diffraction spots in the Fourier transform of an image obtained using the spot-scan technique is usually larger than that obtained by the flood beam illumination method although the resolution that can be obtained using these two techniques is comparable. This is consistent with the notion that the spot-scan technique reduces beam-induced motion that degrades high resolution image contrast.

Lattice and image distortions are corrected as described by Henderson *et al.* (1990). Figure 1 shows typical distortion and cross-correlation peak maps of a tilted specimen in which the effects of defocus were compensated for. The bottom parts of the respective maps have an area with either large distortion vectors or small cross-correlation peaks which correspond to the in-focus or zero contrast region of the image. For images at a large tilt angle, an in-focus band is often found within the processed image. On opposite sides of this in-focus band, the image is in opposite contrast and is not well correlated. In that case, performing the correlation analysis for lattice distortion without also correcting for the defocus effects would produce an improper lattice distortion map of the image. The use of such a map to correct for image distortion could introduce additional phase error.

Figure 1: (a) Distortion vectors (amplitudes enlarged 10 times) which are displacements of the centers of gravity of the correlation peaks from the expected perfect lattice points and (b) cross-correlation peak heights. Cross-correlation analysis was performed between a reference area (a relatively small region of the filtered image) and the whole filtered image.

The quality of diffraction spots in the Fourier transform of an image of a 31°-tilted specimen, before and after image distortion corrections have been applied is shown in Figure 2. It is clear that the quality of the diffraction spots as judged from the number of diffraction spots with a high signal-to-noise ratio has been greatly improved after lattice distortions have been corrected. The spots normal to the tilt axis appear to be limited to a lower resolution than those along the tilt axis. This is due in part to the lack of specimen flatness, but also due to the nature of the diffraction spots of the specimen. Electron diffraction patterns tilted at 30 degrees from the membrane normal show that there is a region of low intensity diffraction normal to the tilt axis, but at a resolution zone centered at about 5Å, there is a clustering of intense reflections. This pattern of strong spots has been analyzed and interpreted to indicate the existence of a major fraction of ß-sheet strands which are tilted by 35 degrees from the membrane normal (Walian and Jap, 1990).

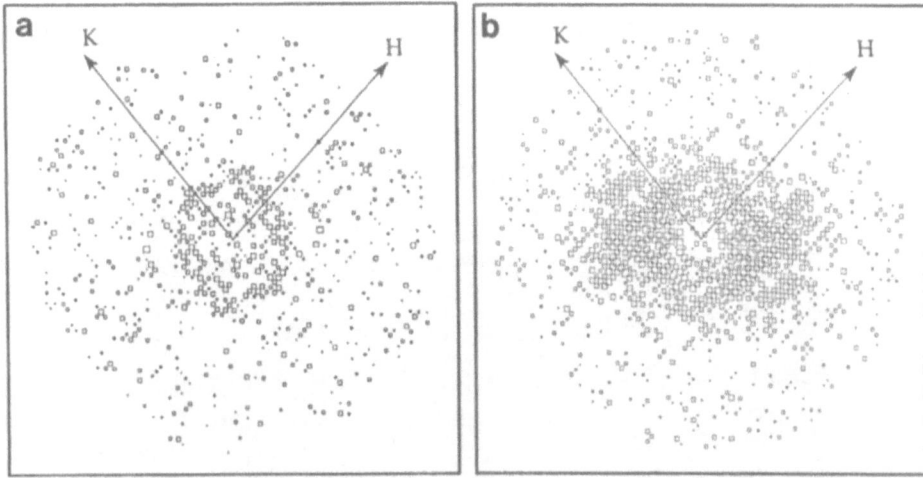

Figure 2: Diffraction spots in the Fourier transform of the image of a 31°-tilted PhoE sample after the effects of defocus and lens aberration have been corrected. The size of the square symbol is proportional to the amplitude of the spots: (a) lattice distortion not corrected and (b) lattice distortion corrected. The edge of the plot corresponds to a resolution of 3.5 Å.

The preliminary 3-D structure of PhoE porin to a resolution of about 6.5Å has been reconstructed by combining the structure factor amplitudes that can be obtained more accurately from our 3-D electron diffraction data and the phase values from images. The 3-D map shows that PhoE porin consists of a trimer of elliptical ß-sheet cylinders with a major axis of 38Å and a minor axis of 28 Å as measured from the center of the cylinder wall. The axis of the elliptical cylinder structure appears to be almost normal to the membrane plane. The cylinder wall has been previously interpreted to be in a ß-sheet conformation with many of the ß-strands oriented normal to the membrane plane and tilted about 35 degrees from the membrane normal (Walian and Jap, 1990). Figure 3 shows the central section of our 3-D map. We propose that this elliptical cylinder of ß-sheet is the basic structural motif of porins in Gram-negative bacteria. Within each elliptical cylinder there exist substructures which form a lining that reduce the pore size of the cylinder. We propose that this internal substructure plays an important role in determining the channel size of the various porins and may be directly involved in the closing and opening of porin channel which has been observed in *in vitro* experiments.

314

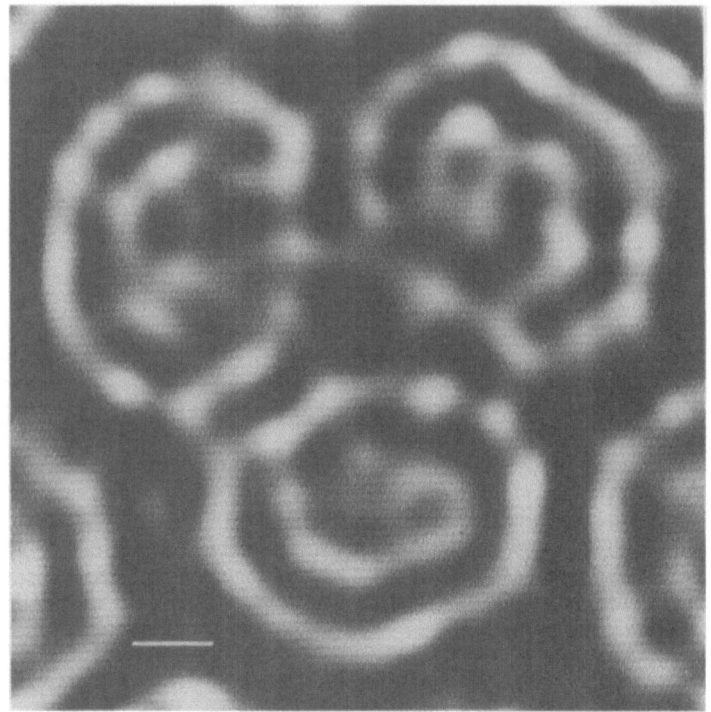

Figure 3: Grey scale plot of the mid-membrane section of the 3-D map at 6.5 Å resolution. The elliptical ring-like structure represents a section of an elliptical β-sheet cylinder which is proposed to be the basic structural motif of porins. Within each ring-like structure, there is internal substructure which is believed to form a lining that reduces the pore size of the cylinder. High density, corresponding to protein, is white, and the scale bar represents 10 Å.

4. Acknowledgements

We would to thank Drs. Kenneth Downing, Robert Glaeser and Richard Henderson for helpful discussions. This work is supported by the office of Health and Environmental Research, US Department of Energy, under Contract DE-AC03-76SF00098, and National Institutes of Health Program Project Grant GM36884.

5. References

Downing, K.H. and Glaeser, R.M. (1986) 'Improvement in High Resolution Image Quality of Radiation-sensitive Specimens Achieved with Reduced Spot Size of the Electron Beam', Ultramicroscopy **20**, 269-278.

Henderson, R., Baldwin, J.M., Ceska, T.A., Zemlin, F., Beckmann, E. and Downing, K.H. (1990) 'A Model for the Structure of Bacteriorhodopsin Based on High Resolution Electron Cryo-Microscopy', J. Mol. Biol. in press.

Jap, B. K. (1988) 'High-Resolution Electron Diffraction of Reconstituted PhoE Porin', J. Mol. Biol. **199**, 229-231.

Jap, B. K. (1989) 'Molecular Design of PhoE Porin and Its Functional Consequences', J. Mol. Biol. **205**, 407 - 419.

Jap, B.K., Downing, K.H. and Walian, J.P. (1990) 'Structure of PhoE Porin in Projection at 3.5Å Resolution', J. Struct. Biol. in press.

Jap, B.K. and Walian, P.J. (1990) 'Biophysics of the Structure and Function of Porins', in press.

Lugtenberg, B. and Van Alphen, L. (1983) 'Molecular Architecture and Functioning of the Outer Membrane of *Escherichia coli* and other Gram-Negative Bacteria', Biochim. Biophys. Acta **737**, 51-115.

Nakae, T. (1986) 'Outer-membrane Permeability of Bacteria', CRC Crit. Rev. Microbiol. **13**, 1-62.

Nikaido, H. and Vaara M. (1985) 'Molecular Basis of Bacterial Outer Membrane Permeability', Microbiological Reviews **49**, 1-32.

Walian, P.J. and Jap, B.K. (1990) 'Three-Dimensional Electron Diffraction of PhoE Porin to 2.8Å Resolution', submitted for publication.

CONCERNING THE SURFACE LATTICE OF MICROTUBULES

R.H.Wade, D.Chrétien, E.Pantos*
Laboratoire de Biologie Structurale (CEA & CNRS URA 1333)
Fédération des Laboratoires de Biologie,
CENG, 85X-38041 GRENOBLE Cedex, FRANCE

*Daresbury Laboratory, Warrington, England

ABSTRACT. Microtubules are important components of the cytoskeleton in eukaryotic cells. *In-vitro* assembled microtubules have been observed by cryo-electron microscopy of frozen-hydrated samples. The individual microtubules can be easily characterised in terms of the number of their constituent protofilaments by direct examination of their image contrast. The contrast variations along the microtubule lengths show that the protofilaments are skewed except for the 13 protofilament case. In the light of these observations we propose that it is the skew which allows the surface lattice to adapt to different protofilament numbers and that *in-vitro* assembled microtubules are likely to have a mixed-lattice organisation.

1. Introduction

We are attempting to obtain information on the structure of a highly dynamic biological polymer, namely the microtubule. At present we are far from a reasonable goal of around 1 nm resolution at which indications of secondary structure should begin to appear. Fortunately an enormous amount of significant information can be obtained at much lower resolution.

Microtubules are filamentary structures found in the cytoplasm of eukaryotic cells where, together with actin filaments and intermediate filaments, they form the components of the so-called cytoskeleton. They have a wide range of functions and can show various levels of structural complexity as witnessed by the singlet, doublet and triplet structures involved in the architecture of the cytoskeleton, of centrioles, of basal bodies, of cilia and of flagella.

The low resolution structure of microtubules was first revealed by electron microscopy of thin sectioned and of negatively stained material. Present knowledge still stems from these two methods combined with information available from X-ray fibre diffraction and from three dimensional reconstructions based on electron micrographs of axoneme microtubules and of various sheet forms of tubulin [1,2].

On the basis of these investigations the commonly accepted microtubule model consists of a hollow tube some 25nm in diameter. The wall of this tube is built from 13 paraxial protofilaments each of which is a string of tubulin hetero-dimers aligned head to tail [3]. In this model adjacent protofilaments have a slight longitudinal shift with respect to each other with the result that the α and β subunits of the tubulin dimers follow a three-start helix with a longitudinal repeat of 12 nm. The dimer packing is considered to correspond to a lattice in which the α and β subunits alternate along each 12 nm pitch helix ; this is called the A lattice. There is some evidence that in the case of *in-vitro* assembled microtubules there may be an alternative packing in which the α

317

J. R. Fryer and D. L. Dorset (eds.), Electron Crystallography of Organic Molecules, 317–325.
© 1990 *Kluwer Academic Publishers.*

318

subunits make up one helix, the β subunits the next, and so forth ; this is the B lattice. Some authors have proposed that a mixed lattice may occur.

Whatever the details of the subunit packing it is now clear that the number of protofilaments built into the microtubule wall can vary significantly since protofilament numbers in the range of 12 to 16 have been observed both in cellular material and in *in-vitro* assembled microtubules,[4,5]. Consequently the surface lattice must have the capacity to adapt itself to these different numbers of protofilaments. There are many reasons why it is important to understand the details of this surface lattice. Microtubule stability may well depend on its organisation and this stability is itself of central importance in cell morphogenesis and motility. The organisation of the lattice may also play a role in the strictly defined intracellular movements along microtubules. There are many other unanswered questions.

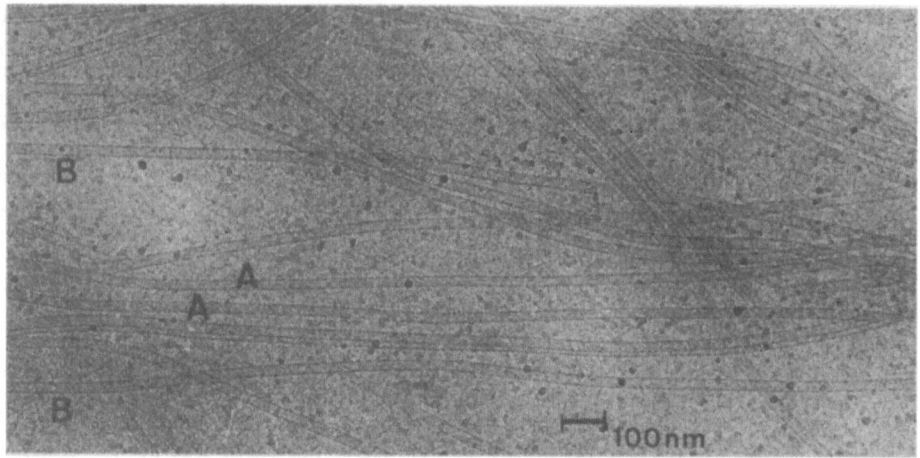

Figure 1. Electron micrograph of a frozen-hydrated sample of microtubules assembled from pure tubulin *in-vitro*. Their highly characteristic contrast allows the microtubule images to be classified into two types, marked A and B, and which are typical respectively of 13 and of 14 protofilament microtubules.

Recently we have been using cryo-electron microscopy to investigate the structural aspects of microtubule assembly. We find that useful information concerning the microtubule protofilament numbers can be obtained from the details of the image contrast [6]. This is extremely attractive since quick freezing followed by cryo-electron microscopy can allow rapid and accurate surveys of microtubule populations in unstained and unfixed specimens prepared *in-vitro* and perhaps *in-vivo*. Our observations and interpretation differ in some significant aspects from the conclusions drawn from previous work on frozen-hydrated microtubules by Mandelkow et al [7]. We have therefore re-examined in some detail the question of image interpretation and have compared our cryo-electron microscope results with thin sections of similar specimens and with images of frozen-hydrated axoneme outer doublets. We can establish unambiguous rules relating the image contrast to the number of protofilaments in the corresponding microtubule. We can also show how the characteristic contrast variations along the microtubule images are related to a skew mechanism by which the surface lattice accomodates extra protofilaments.

In the results section we will give a brief summary of experimental observations which have already been described in detail elsewhere [6]. We will then discuss how the microtubule surface lattice adapts to a variable number of protofilaments and, to close, we will briefly point out some consequences of the proposed surface lattice accommodation.

2. Materials and Methods

Cold labile microtubule protein from beef brain was isolated by 3 cycles of assembly and disassembly in MME buffer, see Wade et al [8] for details and source references. Pure tubulin was isolated from microtubule protein by phosphocellulose column chromatography. The protein was exchanged by a centrifugation-filtration method on a Biogel P6 column into PME buffer. It was adjusted to final concentrations of 7.5 mg/ml tubulin, 10 mM Mg Cl_2, 10 mM acetyl phosphate and 5×10^{-4} mg/ml acetate kinase. Assembly was initiated by warming samples to 37°C, sometimes in the presence of 10^{-4} M GTP. Assembly could be monitored in a spectrophotometer equiped with a constant temperature chamber and 4μl samples were extracted to prepare frozen-hydrated specimens. These were observed in a ZEISS 10C electron microscope and micrographs recorded at a magnification of 20,000, some 1 to 2μm underfocus. Specimen preparation and observation conditions are described in detail in references [6,8].

3. Results

3.1. RESUMÉ OF ELECTRON MICROSCOPY OBSERVATIONS

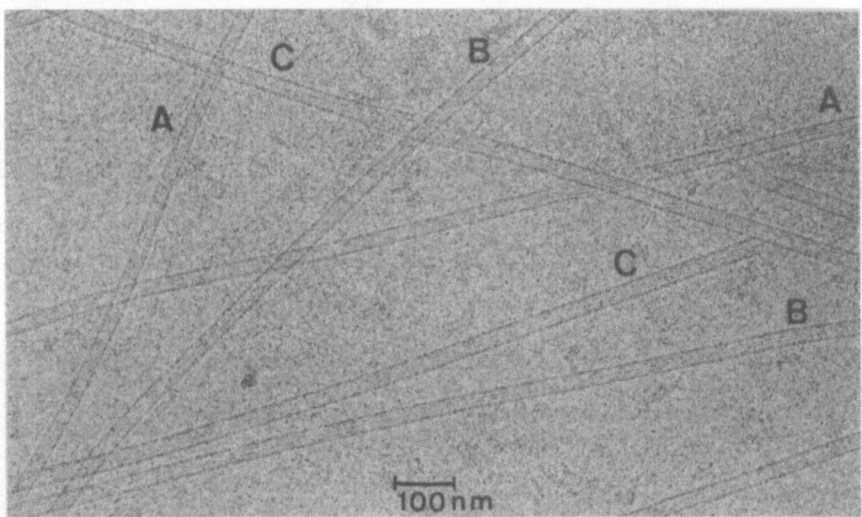

Figure 2. Frozen-hydrated microtubule specimen prepared from "microtubule protein". The micrograph shows type A, B and C contrast images. The type C contrast is typical of 15 protofilament microtubules.

Figure 1 shows images of microtubules obtained using PC tubulin under oscillating assembly conditions [8]. As far as the microtubule image contrast is concerned we obtain identical results in the cases of oscillating or monotonic assembly and with either so-called microtubule protein or pure tubulin [6]. Note that the microtubule images marked A and B have distinctly different contrasts.

The type A images have two inner fringes offset towards one or other of the edges. The edge away from the offset is wider than that on the near side. The fringes run continuously over considerable distances. The fringes can switch their offset from one side of the microtubule image to the other ; this occurs usually after a long blurred stretch. This type of image has an average width of 24.7 nm.

The type B images show a regular repeat of three dark fringes, blur, two dark fringes. The repeat distance is around 410 nm ; the average width of the images is 26.9 nm ; the fringes are centrosymmetric.

Figure 2 shows an image obtained using monotonic assembly of microtubule protein. We again observe the type A and B contrasts and an additional type C contrast consisting of the repeat of three offset fringes, blur, three fringes with the opposite offset. The repeat distance is about 220 nm and the average image width is 28.9 nm.

3.2. IMAGE CHARACTERISATION

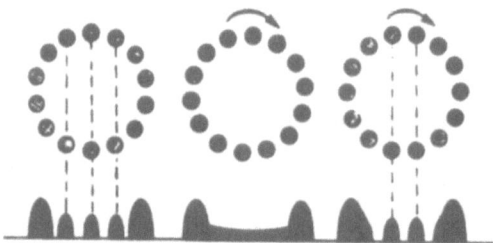

Figure 3. Schematic cross-sections and projected structures of a 14 protofilament microtubule indicating how the projected structure varies with the cross-section orientation. The projections are centrosymmetric, this is always the case for even N, where N is the number of protofilaments. For odd N the projection is asymmetric when the protofilaments project in register.

As we have shown previously the images give a direct indication of the number of protofilaments constituting the microtubules. The basic idea in the interpretation is that the fringe contrast is due to the superposition in projection of protofilaments from the upper and lower surfaces of the microtubule ; this is shown schematically in figure 3. There are four essential elements in the image characterisation :
  (i) the number, symmetry and longitudinal behaviour of the fringes,
  (ii) the image widths,
  (iii) the amplitude,
  (iv) and the phase distribution along the equator of the Fourier transform of a centred portion of the microtubule image.

These four elements together with images from thin sections of pelleted material and with images of frozen-hydrated fragments of axoneme outer-doublets, which are known to have a thirteen protofilament component, allow us to propose the following image characterisation.

type A images <——> 13 pf microtubules,

type B images <——> 14 pf microtubules,
type C images <——> 15 pf microtubules.

4. Discussion

4.1. HOW DOES THE SURFACE LATTICE ACCOMMODATE EXTRA PROTOFILAMENTS ?

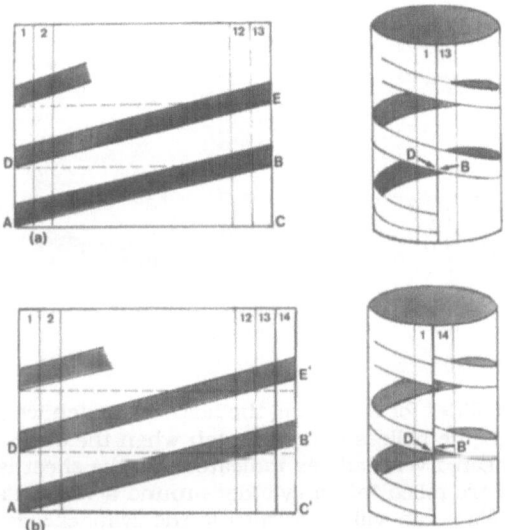

Figure 4. (a) Simplified representation of the 13 protofilament microtubule surface lattice. The vertical columns represent protofilaments. The diagonal strips AB, DE represent the positions of one of the three-start helices. When the sheet is rolled into a cylinder the helix matches at the "seam" between protofilaments 1 and 13.

(b) The addition of an extra protofilament implies that one edge of the helix strip has a pitch AD and the other a pitch B'C'. Consequently the rolled up surface lattice shows a mismatch of the helix at the "seam", giving the jog B'D.

The organization of the tubulin dimer packing in the wall of the microtubule is best visualised in terms of the radial projection [9] opened out as a flat sheet. This is shown in figure 4 (a) in the "ideal" case of a model thirteen protofilament microtubule. The intersection of the three start helix line AB with the right hand edge of the flat lattice is such that BC = AD = helix pitch. In a microtubule AD = 12.0 nm, corresponding to the separation of three subunits along a protofilament.

A problem arises if we add an extra protofilament to the right hand edge, figure 4 (b), since B'C' = AD and B' does not coincide with D when the sheet is rolled up to form a cylinder. There are two fairly obvious ways of obtaining a fit at the "seam".

(1) Shear the structure parallel to the length of the protofilaments, figure 5 (a) so as to slide B' down to the level of D. The structure can now be successfully rolled into a cylinder with a geometrical match of the three-start helix at the "seam". The rise between each adjacent element along the 3 start helix is slightly reduced by the shear.

322

In the 14 protofilament case this amounts to a downward shift of each protofilament by less than 0.1 nm with respect to its neighbour. The protofilaments remain parallel to the cylinder axis.

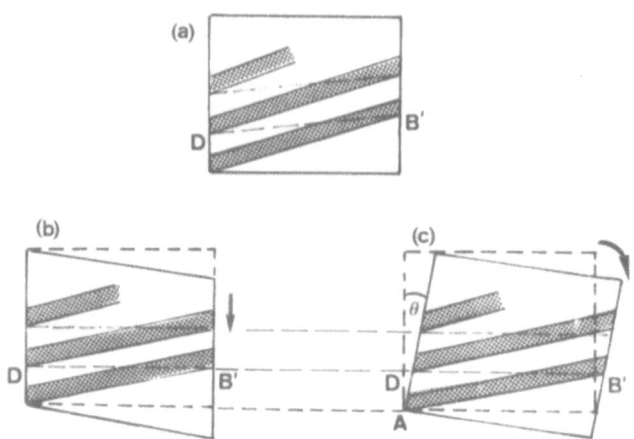

Figure 5. Showing two possibilities for correcting the helix mismatch when N > 13.
(a) Since B' is higher than D the helices cannot match when the sheet is rolled into a cylinder. (b) The sheet is sheared vertically as indicated. (c) The sheet is rotated bodily around A. When the sheets are rolled into a cylinder around a vertical axis the helices now match. In (b) the protofilaments will be parallel to the cylinder axis and in (c) they will be skewed. The shear and rotation operations are exaggerated in the drawings.

(2) Maintain the structure exactly in its original configuration and make a rigid body rotation around A figure 5 (b), until B' is lowered to the height of D. The required rotation can be calculated quite easily and on the basis of a three-start, thirteen protofilament structure with a pitch of y_0 and protofilament separation x_0 the rotation angle is given by $\tan\theta = y_0 (13\text{-}N)/13 N x_0$, where N is the number of protofilaments.
The next step is to roll the sheet into a cylinder around the same vertical axis as previously. An extremely small adjustment in the diameter is necessary to allow the surface lattice to fit but it should be noted that no adjustment of the bonding between protofilaments has been made. The protofilaments are now skewed with respect to the cylinder axis.
The experimental results indicate that it is this second solution which is adopted since the longitudinal variations of the fringe contrast can now be interpreted in terms of the superposition in projection of protofilaments which rotate from point to point along the microtubule as indicated in figure 3.
In the case of a simple model in which the tubulin subunits are represented by spheres aligned along the protofilaments, the surface lattice is shown in figure 6 for the 12,14,15 and 16 protofilament cases, the protofilament skew can be clearly seen. The mass projections in the 12 to 15 protofilament cases are shown in figure 7.
Note that the fringe structures resulting from the mass projections are very similar to those observed, particularly as concerns the image symmetries and the longitudinal variations.

Figure 6. Showing computer model structures of 12, 14, 15 and 16 protofilament microtubules with B type surface lattices. The α and β tubulin subunits are represented by spheres. The last protofilament is paler and shows the skew and the packing mismatch for even N.

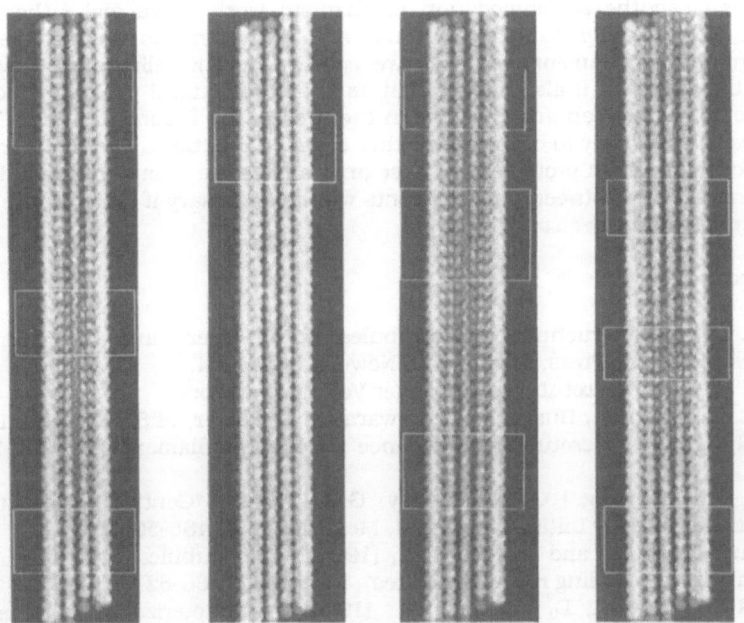

Figure 7. Showing computed mass projections of the model structures of microtubules having from 12 to 15 protofilaments. The fringe contrast is constant for N=13 and in all other cases has a characteristic longitudinal repeat, this can be compared to the experimental images.

4.2. SOME CONSEQUENCES OF THE SURFACE LATTICE ACCOMMODATION MODEL

[1] Once one has learned to "read" the images the structures can be recognised at a glance. At the correct defocus and ice thickness all the microtubules in a field of view can be directly characterised in terms of the number of constituent protofilaments. We have data showing that the microtubule populations of different systems evolve in the course of time even when macroscopically, assembly appears to be complete.

[2] Predictions can be made about probable surface lattices for various protofilament numbers. For example, our first observations concerned the A and B contrasts, i.e.13 and 14 protofilament structures. We then observed the C type contrast, arising from 15 protofilaments.

[3] We have since looked for the other predicted contrasts and now believe that we can recognise images from 12 to 17 pfs. The 16 and 17 pf cases are interesting since it is likely that these will accommodate to a four-start helix structure.

[4] The protofilament skew occurs because the lateral, interprotofilament contacts between tubulin subunits must be constrained, by highly specific intermolecular interactions, to maintain a fixed geometry.

[5] The skew mechanism produces a geometrical accommodation of the surface lattice. It cannot solve the problem of the molecular packing discontinuity which occurs in the case, for example, both for the A lattice and B lattice packing in 14 protofilament microtubules.

[6] How then can the accommodation mechanism work ? We make the following proposition. The α an β tubulin subunits are equivalent as far as the lateral, protofilament to protofilament, contacts are concerned. This eliminates the packing problem at the seam but it also implies that as far as the lateral packing is concerned adjacent protofilaments can arbitrarily be in the A or in the B configuration. Overall, a mixed-lattice is more likely to occur than either the A or B lattices. The same arguement must hold for the thirteen protofilament case and implies that some means of imposing a given lateral packing between protofilaments will be necessary if either of the A or the B lattices are important in-vivo.

5. References

[1] Amos, L.A. (1979) "Structure of microtubules", in K. Roberts and J.S. Hyams (eds.), Microtubules, Academic Press, London and New-York, p. 1-64.
[2] Dustin, P. (1984), "Microtubules", Springer Verlag, New-York.
[3] Tilney, L.G., Bryan, J., Bush, D.J., Fujiwara, K., Moseker, M.S., Murphy, D.B. and Snyder, D.H., (1973), "Microtubules : evidence for 13 protofilaments", J. Cell Biol. 59, 267-275.
[4] Scheele, R.B., Bergen, L.G. and Borisy, G.G., (1982), "Control of the structural fidelity of microtubules by initiation sites", J. Mol. Biol. 154, 485-500.
[5] Eichenlaub-Ritter, U. and Tucker, J.B., (1984), "Microtubules with more than 13 protofilaments in the dividing nuclei of ciliates", Nature 307, 60-62.
[6] Wade, R.H., Chrétien, D. and Job, D., (1990), "Characterization of microtubule protofilament members", J. Mol. Biol. (in press).
[7] Mandelkow, E.-M. and Mandelkow, E., (1985), "Unstained microtubules studied by cryo-electron microscopy : subtructure, supertwist and disassembly", J. Mol. Biol. 181, 123-135.

[8] Wade, R.H., Pirollet, F., Margolis, R.L., Garel, J.-R. and Job, D., (1989), "Monotonic versus oscillating microtubule assembly : a cryo-electron microscope study", Biol. Cell., 65, 37-144.
[9] Klug, A., Crick, F.H.C., Wyckoff, H.W., (1958), "Diffraction by helical structures", Acta Cryst. 11, 199-213.

CORRELATION AVERAGING AND FILTERING OF ELECTRON MICROGRAPHS.

E. V. Orlova
Institute of Crystallography USSR Acad. of Sci.
59 Leninsky pr.
117333 Moscow
USSR

ABSTRACT. Structure of imperfect crystals of creatine kinase and individual particles as well as ATP synthase complex has been studied by electron microscopy of negatively stained specimens and correlation averaging.
Lattice imperfections of high resolution images of GaAs/InAs multilayered heterosystems has been studied by special Fourier filtration technique.

Introduction

Transmission electron microscopy allows one to investigate the objects structure from their images. However complicated dependence between intensity distribution of image and electron density of real object, low signal-to-noise ratio due to specimen preparation and radiation damages preclude interpretation of electron microscopy data. Image enhancement should help extracting reliable object details while suppressing noisy features of different copies of similar objects. It requires the use of statistical methods of image processing, powerful computers and appropriate software.

Methods

A program package for image processing by use of Fourier filtering (up to 512*512 pixels), correlation analysis of images of individual particles and three-dimensional reconstruction of objects was developed at the Institute of Crystallography. All the programs were coded in FORTRAN-77 and implemented on a NORD-100 computer and later adapted for IBM PC.
Electron microscopy investigations were carried out at the Institute of Crystallography in a Philips EM 400 (biological specimens) and in a EM 430 ST operating at 300 kV (material science). Electron micrographs were digitized with a Perkin Elmer PDS 1010A flat bed microdensitometer. The images of individual particles were selected interactively with graytone display system and then circular mask was apllied. The variance of densities inside the mask was normalized.

J. R. Fryer and D. L. Dorset (eds.), Electron Crystallography of Organic Molecules, 327–332.
© 1990 Kluwer Academic Publishers.

Creatine Kinase

The structure of mitochondrial creatine kinase associates [1] was studied by images of imperfect two-dimensional crystals (Fig. 1) with the use
of a correlation averaging. Sampling distance was equal to 0.2 nm. A
well ordered small part of an image (3 * 3 unit cells) was enhanced by
band filtering, 4-fold symmetrized and used as a reference for the calculation of the correlation map (Fig. 1b). Parts of two-dimensional
crystal image having correlation peak heights larger than the threshold
were selected and averaged (Fig. 1c).

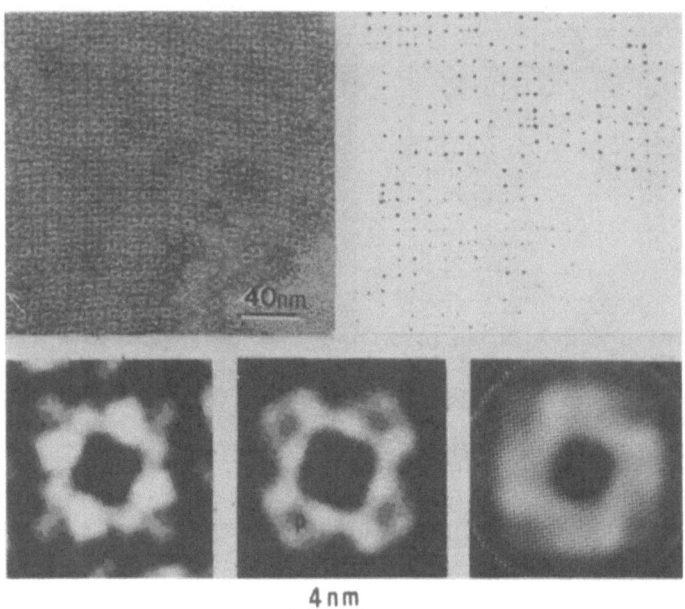

Figure 1. Associates of creatine kinase from beaf heart mitochondria
negatively stained with uranyl acetate; a) micrograph of two-dimensional
imperfect crystal; b) cross-correlation function between 'a)' and small
reference patch; c) averaged image of an associate in crystal; d) 4-fold
symmetrized image of individual associate; e) average image of individual creatine kinase associates.

Correlation averaging was also performed for the images of individual creatine kinase associates. The translational alignment was done by
using cross-correlation function. Orientational pre-alignment was performed in polar coordinates using amplitude part of Fourier transforms
only. The final refinement was done using both real and imaginary parts
of spectra in the vicinity of an angle found at the first stage of the

procedure (and that plus $180°$ was tested also). It was found that cross-correlation coefficients of different particles during orientational alignment were of the same order as correlation coefficient of a particle with itself rotated by $90°$. Then it was reasonable to perform 4-fold symmetrization of averages (Fig. 1d,e). It was 25 images included in averaging of 75 scanned ones. The resolution attained for average images of individual particles was worse than that in case of mono-layers. But these two averages were quite similar. Size of associates was approximately 11 by 11 nm^2. Thus it was possible to assume that creatine kinase associates of P4 symmetry consist of 8 subunits. But neither averaged crystal images nor images of individual particles showed mirror symmetry. This can be explained by different staining of the upper and lower sides of particles.

ATP Synthase

The structure of membrane-bound ATP synthase from beef heart mitochondria has been studied by electron microscopy (Fig. 2a) and image processing. Selected areas of micrographs were scanned at 0.5 nm intervals. ATP synthase is formed by F_1-ATPase and memrane-embedded F_0-sector. The averaging was done in two stages. At first stage all images were filtered with band-pass filter and summed together indepedently of their cross-correlation coefficients. The obtained image was used as the reference for a second stage. At this stage only those images were taken for averaging which had correlation coefficients higher than given threshold. The result image was formed using approximately one half of all scanned images. The average showed three morphologically distinct regions of protein distribution: one extrinsic to the membrane - F_1-ATPase, the stalk and the embedded in membrane F_0-sector. The F_1-ATPase was 8 by 10 nm, F_0-sector - 7 by 8 nm and stalk was 4 - 4.5 nm long and about 3.5 nm wide [2]. It was possible to estimate average distance between the comlexes on the membrane, which was egual $\cong 12nm$ (Fig. 2b).

Unfortunately it was impossible to determine the relative arrangement of subunits of F1-ATPase in this image. It may be explained by slight inclination of long axis of complexes to support film, because the ATP synthase view perpendicular to its long axis was seen only on the fold of membrane. At the same time complexes may rotate along their long axis, therefore particles could orient arbitrarily relative to each other on the surface of membrane. It is possible that more detailed information about the arrangement of subunits in F_0-sector and F_1-ATPase may be obtained by use of multivariate statistics in analysing the images of ATP synthase.

330

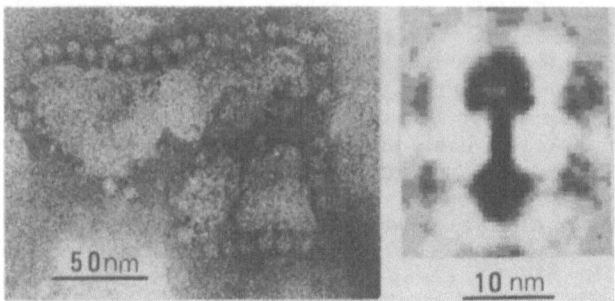

Figure 2. a) micrograph of negatively stained by uranyl acetate submito-
chondrial particles; b) average image of membrane-bound ATP synthase
complex.

Inorganic Multilayered Heterosystems

In case of processing of inorganic crystal images it is often the detection of various crystal lattice defects required. Then one should apply
special filters that single out pairs of reflections including some
background around them. This kind of filters differs from ones used in
biological studies with a regular set of pinholes providing the study of
the averaged unit cell.

   Filtering when a Friedel pair of reflections is allowed to pass
could be described by the equation [3]:

$$\mathcal{F}^{-1}\{ \ \mathcal{F}\{f(\vec{r})\} \ [W(\vec{R}) \ \otimes \ [\delta(\vec{R}-\vec{A})+\delta(\vec{R}+\vec{A})]]\} =$$

$$= f(\vec{r}) \ \otimes \ [w(\vec{r})Cos(2\pi\vec{r}\vec{A})] \tag{1}$$

where $f(\vec{r})$ is the two-dimensional intensity distribution for an image,
$\mathcal{F}$ denotes Fourier-transform operator, $W(\vec{R})$ is the filter window function
$(W(\vec{R}) \equiv 1$ if $|\vec{R}| \leq B$; $W(\vec{R}) \equiv 0$ if $|\vec{R}| > B$), $w(\vec{r})=\mathcal{F}\{W(\vec{R})\}$, $(\vec{a}\vec{A})=1$, $\vec{a}$ is a
lattice vector. The right part of the above equation is the convolution
of $f(\vec{r})$ with the weighting function of $Cos(2\pi\vec{r}\vec{A})$ enveloped by $w(\vec{r})$.
Obviously that the weight is equal to unity if $\vec{r} = n\cdot\vec{a}$, and to zero if $\vec{r}$
perpendicular to $\vec{a}$. Half-width (B) of windows $W(\vec{R})$ determines that of
the main maximum of $w(\vec{r})$ (b). If $B \Rightarrow 0$, then $b \Rightarrow \infty$ and integration in
the equation should be performed over the whole image. The result is in
fact a fully averaged image showing the average periodicity in $\vec{a}$ direc-
tion. If $B \Rightarrow \infty$, then $w(\vec{r}) \Rightarrow \delta(\vec{r})$ and the filtered image is equivalent
to the initial one. Therefore in order to detect the lattice imperfec-
tions of the order of $|\vec{a}|$ in size , one should choose the window with
$B \cong 0.5|\vec{A}|$. Then image is averaged over the region of aproximately $|\vec{a}|$
width.

The described method permits one to discriminate between image regions matching the periodicity of the filter function and the regions containing some defects of the periodicity.

Investigations of real crystal structure using the filtering required use of different approach to the process of treatment. In a practical work visual comparison of original and processed images was required. That was rather difficult for unexperienced person. It was useful then to add a part of slow changing original background to processed image for reliable visual identification of defects in processed images.

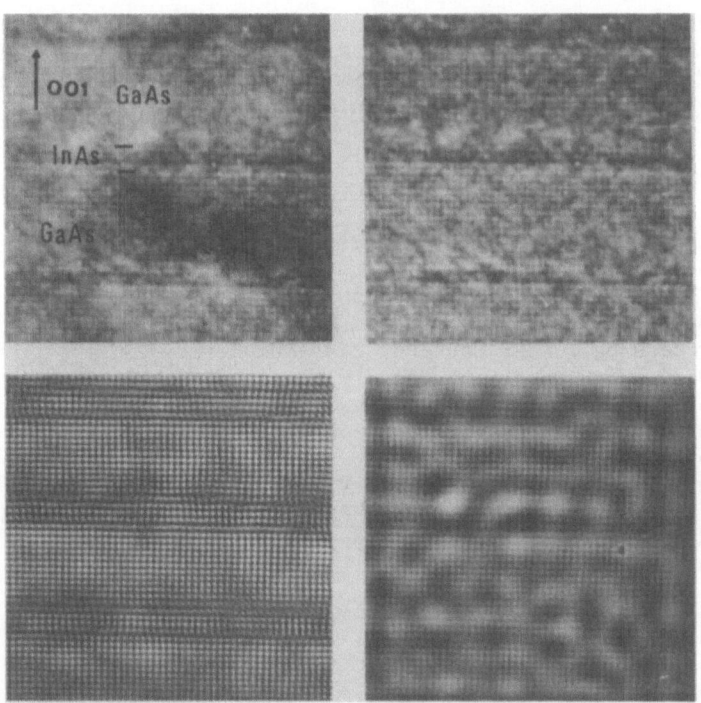

Figure 3. a) <100> cross-section image of an InAs/(001)GaAs multilayered structure; b) high-pass filtered 'a)' image; c) reconstructed 'a)' image with 200 and 020 reflections allowed to pass; d) 'c)' image with a partially added background

This technique simplify the recognition of InAs and GaAs layers in multilayered heterosystems and discrimination between the two lattices and was illustrated by processing of images of InAs/GaAs multilayered structure [4]. Image treatment was carried out for the structure projection along the <110> and <100> zone axes. InAs layer thickness was equal approximately to 1 nm and that of GaAs to 7 nm. Micrographs were digitized at 0.07 nm intervals. Results of processing are presented in

Fig. 3. An original image of <100> cross-section of InAs/GaAs (Fig. 3a) was high-pass filtered (R = 1 nm^{-1}) to eliminate a slow-changing background (Fig. 3b). Then the filtering was done using the 200 and 400 reflexes including their vicinities. In this case size of window function was equal to approximately 1/3 |$\vec{a}$| (Fig. 3c). After all a fraction of background was added to the result to facilitate pattern recognition. This method allowed one to enhance the periodicity in selected directions.

Acknowledgements

Author would like to thank for the electron microscopy data provided by V.Ya.Stel'mashchuk, V.L.Tsuprun (biology) and by V.Yu.Karasev (inorganics) (Institute of Crystallography, Moscow) and profs. B.K.Vainstein and N.A.Kiselev for their helpful discussions.

References

1. Fedosov,S.N. and Belousova,L.V., (1988). Biochimia (Rus.), **53**, Num 4, 551-565.
2. Tsuprun,V.L., Orlova,E.V., and Mesyanzhinova,I.V. (1989) 'Structure of the ATP-synthase studied by electron microscopy and image processing', FEBS Letters, **244**, Num 2, 279-2823.
3. Zheng,Y. (1987) 'Structure fine des dislocations dans les feldspaths alcalins - étude par microscopie électronigue à haute résolution (METHR) et modélisation', These de doctorat de l'universite Paris YI.
4. Karasev,V.Yu., Kiselev,N.A., Orlova,E.V., Gutakovski,A.K., Pintus,S.M. and Rubanov,S.V. (1989) 'HREM of InAs and GaAs-based multilayered heterosystems', Inst. Phys. Conf. Ser. No 100: Section I. 33-38.

Electron Crystallography in an Industrial Laboratory

G R Duckett, D White and A H Grace
BP Research, Sunbury Research Centre
Sunbury-on-Thames, Middlesex. TW16 7LN. UK.

J L Hutchison and H D Cochrane
Department of Metallurgy and Science of Materials
University of Oxford, Parks Road, Oxford. OX1 3PH.

Abstract.

The functions of an industrial microscopy laboratory involve the
combination of scientific capabilities and commercial requirements.
The various operational roles are exemplified. These include quality
control, troubleshooting and analytical support work on semiconducting
materials. Structure determination of a new material and atomic level,
dynamic studies of catalyst surfaces demonstrate research applications
and include aspects of microscopy technique development.

1. Introduction

The benefits of using electron beam techniques for the identification
and characterisation of materials, at high spatial resolution, are well
established (1). In a modern electron microscope a wealth of
structural detail is obtainable which can be correlated with
macroscopic properties, thereby enhancing our understanding of
materials and their behaviour. Such detail is often difficult or
impossible to obtain (especially at the nm scale) using other
techniques. This capability is becoming increasingly important in the
development and commercial application of high performance materials.
In a large international company with diverse products, the provision
of resources to investigate microstructural detail is extremely
important. This applies to research and development of new products
and processes and to technical support for existing interests.

It is worth noting however, that the information sought from a
microstructural study relates to the commercial need as well as the
scientific capabilities. Therefore, an industrial microscopy group
will generally have roles in research, analytical support and
troubleshooting, each affecting the level of crystallographic
characterisation required. Often the boundary between these roles is
not clearly defined and a "routine" support activity or troubleshooting

333

J. R. Fryer and D. L. Dorset (eds.), Electron Crystallography of Organic Molecules, 333–341.
© 1990 *Kluwer Academic Publishers.*

334

task may necessitate a more detailed approach. This paper is aimed at demonstrating how the above considerations influence the direction taken in specific work examples.

2 Analytical Support/ Troubleshooting

The microscopy group provides a resource base of specialist skills and techniques to support the company's interests. In typical operational problems, these capabilities are often required to provide scientifically detailed answers, but on a short timescale.

Example 1 : Troubleshooting

A batch of single crystal silicon for semiconductor fabrication was found to be defective in quality control testing. X-ray topography revealed crystallographic defects in the material and prompted an electron microscope study to determine their precise nature.

Investigation of suitably prepared specimens (see, for example, (2)) by TEM revealed "strings" of small inclusions within the single crystal silicon matrix - figure 1. Individual inclusions had projected diameters of around 20nm (inset) approximating to hexagonal in outline. Such low magnification images suggested an orientation relationship between the inclusions and the silicon host. EDX in an analytical STEM showed peaks for copper and silicon.

The precise orientational relationship between the inclusions and the silicon crystal was examined by lattice imaging with the silicon crystal tilted into the <110> zone. Lattice fringes in the inclusion were consistent with the spacing of {111} planes in copper and these lay parallel to the {111} planes of silicon (figure 2). The information obtained enabled a faulty processing step to be identified.

Example 2: Analytical Support

The determination of fine-scale crystallography was employed in an investigation of Beta-SiC grown on <100>Si for high temperature semiconductor applications (3).

Planar thin sections were used in initial selected area electron diffraction studies (figure 3). Inspection of the patterns and comparison with simple diffraction nets suggested that the likely orientation relationship was <100> zones parallel. This was confirmed using cross-sectional specimens; figure 4 shows a lattice image of a boundary region after tilting into <110> orientation. This approach also enables the nature and density of defects at the interface to be determined (4).

3 Research

This role involves programmed work which is longer term and more fundamental in nature, concerned with new materials/processes, often including development of the analytical techniques themselves.

Example 3: Zeolite structure determination

The success of some synthetic zeolites as shape selective catalysts in the petroleum industry has maintained interest in the search for others. A new type of high silica synthetic zeolite, Theta-1, has been discovered and its crystal structure determined (5).

It proved difficult to grow large crystals of Theta-1 for conventional, single crystal, X-ray diffraction however, selected area electron diffraction patterns from sub-micron single crystals provided unit cell parameters and information on the likely space group symmetry (6). This information, together with solid-state NMR and sorption data, provided a model used to interpret the powder X-ray diffraction pattern and hence a solution of the structure. An essential aspect is a channel of medium pore size (0.6nm diameter) running through the structure, parallel to one of the crystal axes. The arrangement of such channels was confirmed by direct imaging in the HRTEM, as shown in figure 5, together with the projected model structure and appropriate simulated image (6).

Example 4: Surface studies by on-line, Atomic Resolution TEM

Electron microscopy has had limited success in the correlation of microstructural detail with catalytic behaviour – the precise characteristics of the "active site", such as electronic and geometric structure on an atomic scale, are not easily ascertained. Nevertheless, improvements in the resolving power of the ARTEM, together with the development of surface-profile imaging (7), has renewed interest in the investigation of surface atomic detail. Furthermore, the emergence of fast, easy to use, digital image processing systems has enabled dynamic phenomena to be considered. The integration of this approach with the on-going development of controlled environment TEM (8) will be invaluable.

Figure 6 shows the surface profile image of a ceria (CeO_2) catalyst support material in <110> projection. Work reported elsewhere (9) has shown that the image features can be understood in terms of microscope operating conditions and specimen thickness. Dark contrast "dots" correspond to projected cation columns and surface features such as mono-atomic steps (arrowed) can be seen; the nature of the anion lattice and the absolute surface cleanliness are uncertain.

The addition of a highly dispersed, second (metal) phase produces a typical supported metal catalyst – figure 7 shows a small (~1nm) platinum particle on a (111) surface of a ceria crystal. Such systems can then be used to study both metal and support and their interaction as well as catalytic behaviour in microreactor studies (10).

The capabilities for computer aided studies of dynamic phenomena have been demonstrated by the observation of electron beam induced rearrangement of the surface atomic configurations of ceria (11) and by changes in the wetting behaviour of supported platinum and gold particles (10,12). Figure 8 shows the particle in figure 7 having undergone a change in morphology and metal/support interaction. Thus, catalytically important information may be obtained.

Conclusions

This paper has attempted to show, by example, the various roles required of an industrial EM unit and the interplay between the scientific content and the commercial relevance of the work carried out.

Acknowledgements

The authors would like to thank Dr H Reehal for providing the semiconductor samples and Dr S A I Barri for the zeolite specimen. BP plc is acknowledged for permission to publish and NATO for the invitation to present this work.

References

1. "High Resolution Transmission Electron Microscopy and Associated Techniques", Eds: Buseck P R, Cowley J M and Eyring L. Oxford University Press (1988).

2. Bravman J C and Sinclair R. "The preparation of Cross-section Specimens for Transmission Electron Microscopy", J. Electron Micros. Tech., 1 pp 53-61, (1984).

3. Powell A J, Matus L G and Kuczmarski M A. "Growth and Characterisation of Cubic Single-Crystal Films on Si", J. Electrochem. Soc.: Solid-State Science and Technology. 134, pp 1558-1565. (1987).

4. Carter C H, Davis R F and Nutt S R. "Transmission Electron Microscopy of process-induced defects in beta-SiC thin films", J Mater. Res. 1(6), pp 811-819, (1987).

5. Barri S A I, Smith G W, White D and Young D. "Structure of Theta-1, the first unidimensional medium-pore high-silica zeolite", Nature, 312, pp 533-534, (1984).

6. White D, Ramdas S and Millward G R. "Zeolite theta-1 - structural studies by TEM", Inst. Phys. Conf. Ser. 78. EMAG (1985).

7. Sinclair R, Yamashita T and Ponce F A. "Atomic Motion on the surface of a cadmium telluride single crystal", Nature 290 pp 386 (1981).

8. Parkinson G M. "High resolution, in-situ controlled atmosphere transmission electron microscopy (CATEM) of heterogeneous catalysts", Cat. Letters, **2**, pp 303-308, (1989).

9. Cochrane H D, Hutchison J L and White D. "Surface Studies of Catalytic Ceria using Atomic-Resolution TEM", Ultramicroscopy **31**, pp 138-142, (1989).

10. Cochrane H D, Hutchison J L, White D, Parkinson G M, Dupas C and Scott A J. "High Resolution Electron Microscopy of Ceria Supported Catalysts" - submitted to Ultramicroscopy.

11. White D, Hutchison J L and Ramdas S. "Surface Atom Studies of Ceria and Thoria", Inst. Phys. Conf. Ser. **90**, pp 249-252, EMAG (1987).

12. Cochrane H D, Hutchison J L and White D. "The structure of small metal crystallites on cerium dioxide", submitted to ICEM XII, Seattle, Washington, (1990).

338

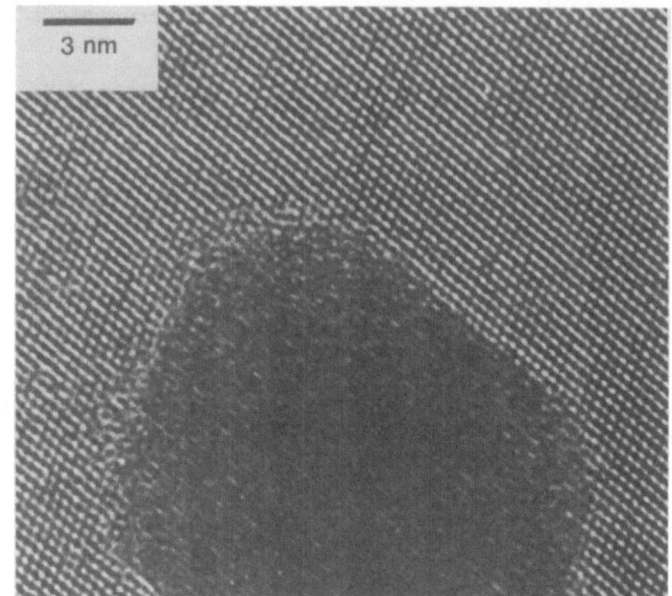

Figure 1: Inclusions in single-crystal Si.

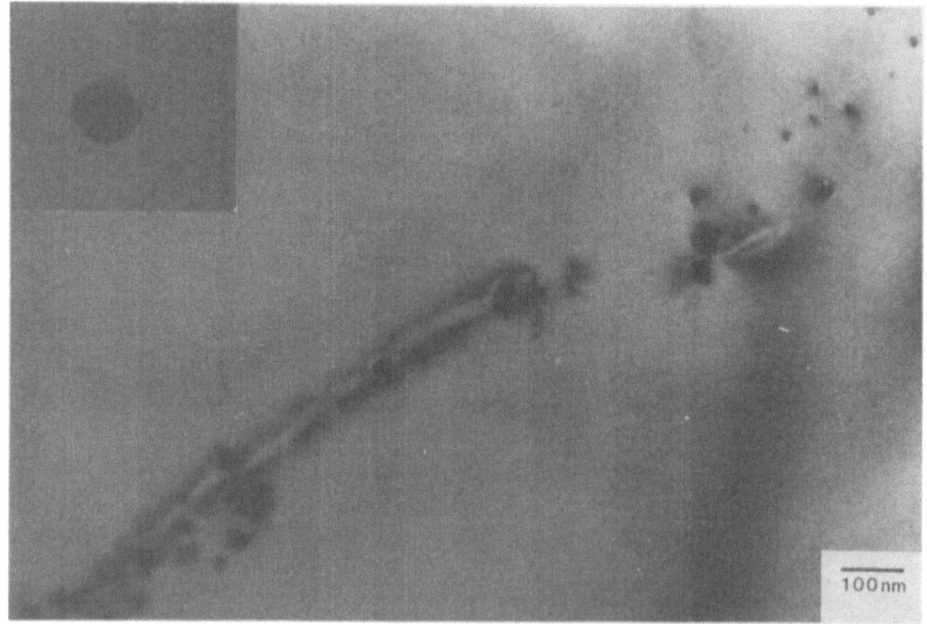

Figure 2: Lattice image showing epitaxy.

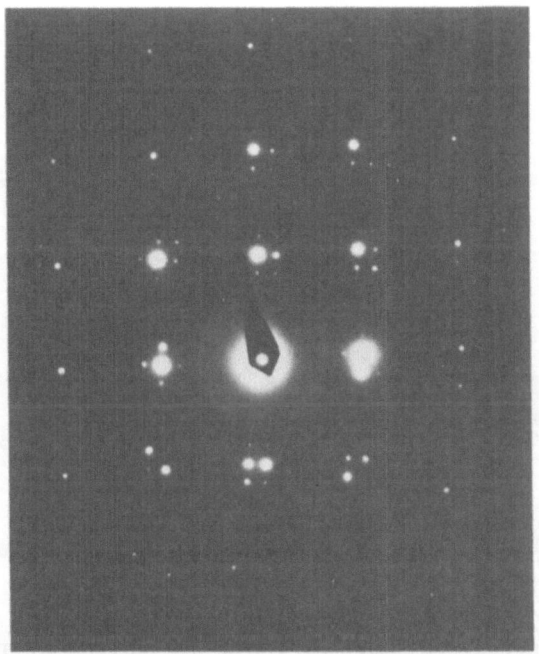

Figure 3: Diffraction pattern from planar SiC on Si.

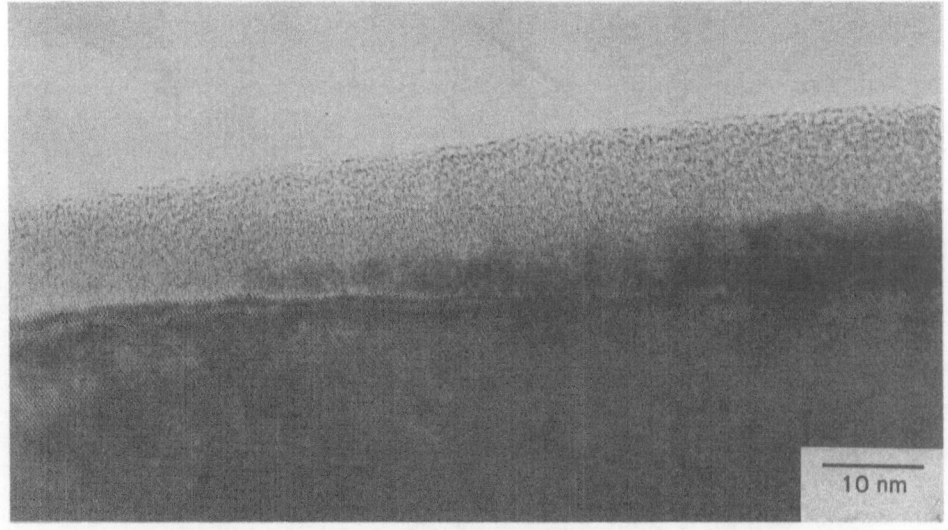

10 nm

Figure 4: Lattice image of SiC/Si cross-section.

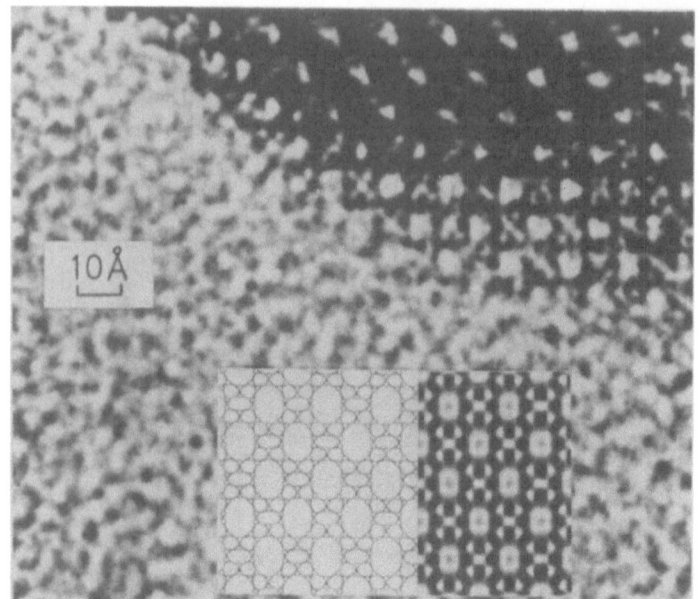

Figure 5: Structure image of [001] projection of Theta-1.

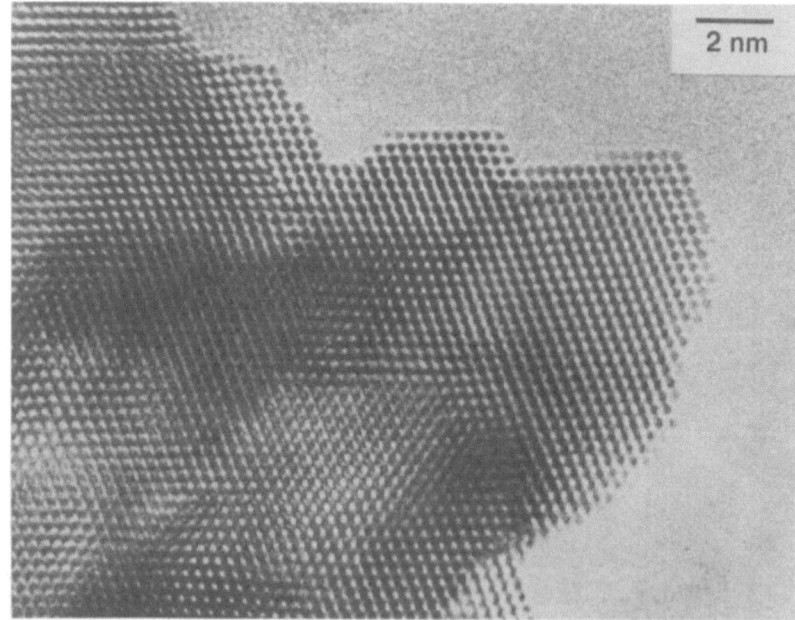

Figure 6: Surface profile image of <110> ceria.

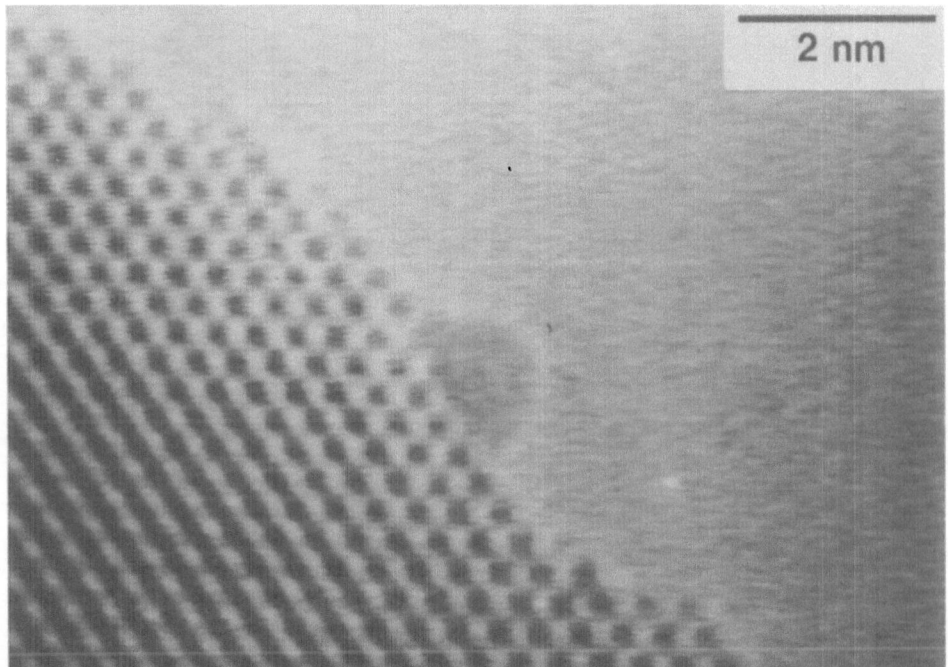

Figure 7: Platinum particle on ceria surface.

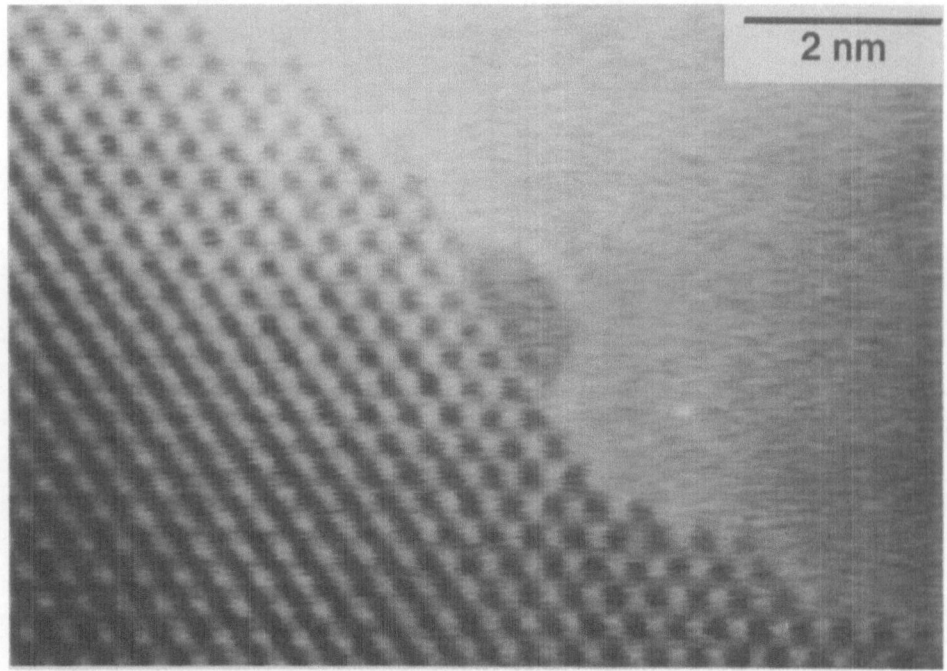

Figure 8: Particle from figure 7 after morphological change.

THE STRUCTURE ANALYSIS OF AN ORGANO-AZO-CALCIUM SALT BY HIGH RESOLUTION ELECTRON MICROSCOPY AND IMAGE PROCESSING.

G. Miller, J.R. Fryer. Electron Microscope Centre, Chemistry Dept., University of Glasgow, Glasgow G12 8QQ.

W. Kunath*, K. Weiss. Fritz-Haber-Institut der Max-Planck-Gesellschaft, Faradayweg 4-6, D-1000 Berlin 33.

* Wolfgang Kunath died January 1990.

Abstract.

Image Processing of High Resolution Electron Microscope images has been used to reveal the molecular arrangement within the crystal unit cell for an Organo-Azo-Calcium salt. Images with a dosage of 10 el/A 2 were used. Those showing strong calculated diffractions were kept for processing. It was found that the recorded crystal images showed variations throughout, which could be explained by crystal bending and beam damage. Results from averaging over the whole 512 * 512 pixel image in reciprocal space revealed no information below 0.8nm. Improvements in the image detail were obtained by using only areas within the image which showed coherency and had a power spectrum showing the presence of structural detail to bettter than 0.4nm. In this manner the molecular arrangement becomes apparent for the surface layer. A molecular modelling package was then used to show 3-D information.

Introduction.

The molecular arrangement found in the crystal unit cell is of importance in understanding the crystal properties of the compound. Various single crystal x-ray analyses of other organo- azo compounds have been carried out [1-6]. These have shown that molecules of this type, which can exist in a tauto. meric (figures 1 and 2) form, are found predominantly as the hydrazone tautomer (figure 2). In this form the molecule is also essentially planar [1-6]. There are a number of these compounds, especially the metal salt derivatives, which are unsuitable for single crystal x-ray analysis due to the small crystallite size. The aim of this work was to try and obtain crystal structure information from such crystals by using High Resolution Electron Microscopy (HREM). Microscopy has been used with success in determining the crystal structures in a variety of inorganic materials, [7-10] and with increasing success with beam sensitive organic, polymeric and biological crystals [11-16]. The reason for this is the impovements in the image collection systems [17- 18], and the low dose techniques being used which help reduce the amount of radiation damage and thus preserve the internal crystal structure [19-22]. Another important consideration in obtaining high resolution images is the crystal thickness [23], with the

343

J. R. Fryer and D. L. Dorset (eds.), Electron Crystallography of Organic Molecules, 343–353.
© 1990 Kluwer Academic Publishers.

image. To allow for peaks not fitting the reciprocal lattice vectors exactly a crossed sinc profile value was calculated for a 4*4 pixel region giving a better value for the lattice component. This result in the recovery of one unit cell. It is then a relatively easy task to display any number of these unit cells to represent the surface of the crystal.

Materials and Experimental.

The material used was obtained as a fine powder or in a granular form. The material is ground then subjected to ultrasonics to give dispersion in a water media. A drop was then added to a larger water vessel with a little detergent solution to reduce surface tension. A carbon coated copper grid was then drawn through the fine crystal suspension at the surface. In this way the heavier crystals tended to sink, leaving crystals of a thin nature and preferentially orientated along the h0l plane on the grids. The ultrasonics tended to break up the crystal aggregates rather than reduce the crystal size in any dramatic fashion.

The microscope used was the DEEKO 100 [20], with a Cs = 0.5mm, Cc = 0.7mm wavelengh of 0.00370nm a stability of specimen drift of <0.001 nm/sec. and a resolution of 0.27nm with Scherzer defocus.

Low Dose Microscopy

Since radiation damage is dependent upon the number of incident electrons striking the crystal (24) the following method was used to limit the damage by the electron beam.
The following was carried out using the on line T.V. system.
1. Single crystals were searched for in the diffraction mode with a limiting field aperture of 300nm .
2. The electron diffraction was inspected. If it showed point like reflections down to 0.2nm and the intensities were symetrical, it was stored.
3. The electron beam was then moved off the axis by 1.5 micro.m in a manner described by Fryer[22]. In the off axis image mode the astigmatism was corrected and focus adjusted to 1-2 Scherzer units.
4. The electron beam was then moved on axis and a series of four 10 electron/A^2 images were collected with a recording time of 50ms.
5. The diffractogram of the first image was calculated and displayed. If this showed second order reflections in two directions or better, the sequence of images were stored for further investigation.
6. A high dose image was collected to give information about astigmatism and true focus.

Results by this method showed that about ten percent of crystals which satisfied the step two condition could fulfill the condition in step 5. It was also found that crystals greater than 20nm gave poor contrast. A number of micrographs at 100K and 200K were taken. These gave optimum

maximum resolution obtainable from a crystal with a thickness R using a
microscope with an electron wavelength lamda being given as,

$$R = (d^2/(2*pi*lamda))$$ equation 1.

where d = resolution

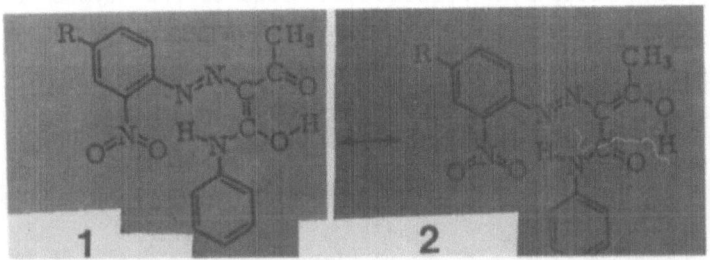

Figures 1 and 2. Tautomeric forms of the molecule

The disadvantage to low dose images is the correspondingly low signal to
noise ratio. To improve such images computer [24,31] and optical
filtration techniques [25] can be adopted. This paper is primarily
concerned with the collection of images via a television camera system,
where the images are displayed on a monitor and can be stored onto a
host computer. A number of systems of this type have been described in
the literature [32]. The system in question is described by Kunath
[26].

Image processing.

In both Berlin and Glasgow the Semper and Micro-Semper   processing
systems are in use respectively. This arrangement allows easy transfer
of data from one location to the other. The low signal to noise images
are improved by digital processing by histogram equalisation and methods
based on Fourier space manipulation where motifs shrouded in noise can
be recognised and extracted. A circular mask was used around the image
to reduce horizontal and vertical artifacts being introduced into the
fourier transform [27].   For display purposes the power spectrum was
calculated and displayed as the natural logarithm. This allows the
extent of the periodicity to be seen and reciprocal base vectors to be
chosen (U,V).   These base vectors were improved by a least squares
method given a list of peak positions in the transform. This results in
more accurate base vectors (U2,V2). The Fourier transform could then be
filtered so as to allow only fourier components close to the defined
lattice sites to be in the reconstructed image.
  The   unit cell was recovered from the image by lattice averaging using
a transform peak profile fitting method(FLC), which fits   fourier
components to the isolated peaks found in the transform of the periodic

diffractions similar to the calculated diffractions.
Images were then stored on a vax 11/780.

Results

From the microscopy and electron diffraction results two types of
crystal were found.

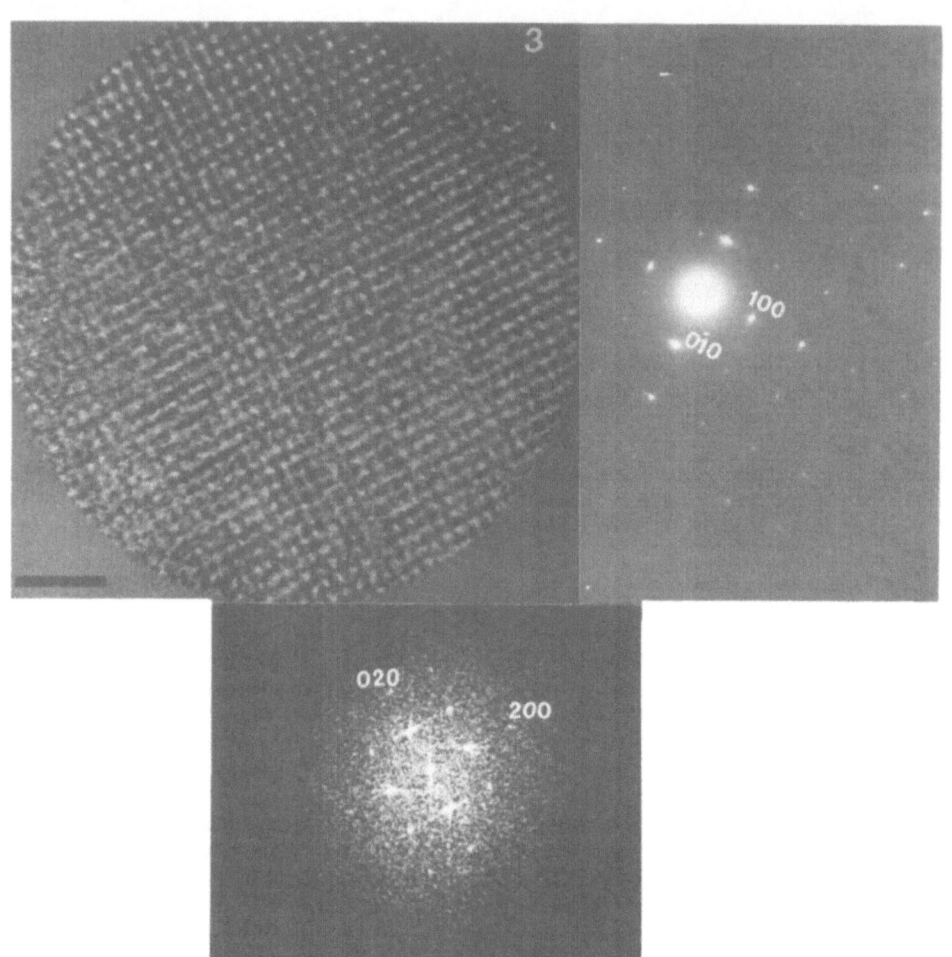

Figure 3. Low dose image showing crossed lattices. Scale bar is 5nm.
Figure 3.1   Calculated Power Spectrum of image in figure 3. This
              shows second order reflections in two directions.
Figure 4. Electron diffraction pattern from a type 1 crystal.

Type 1 were thin crystals in the range of 5 micro.m diameter and in the range of 10nm to 20nm thick. These (figure 3) showed crossed lattices with calculated power spectra (figure 3.1) resembling the characteristic diffraction pattern (figure 4).

Type 2 crystals were thicker crystals having a thickness of a 100nm or more. Diffraction pattterns and calculated diffractograms (figure 5) both show the C* plane down to the 9th and 5th order respectively. The power spectra also show a smeared band at a distance of 1/0.8 nm perpendicular to the c* axis. This is indicative of disorder within the crystal [9]. It was not possible to index any single reflection within this side band. Lattice images (figure 6) of the corresponding power spectrum show the 3.2 nm lattices but no lower resolution detail. This is probably due to the disorder and also the thickness of the crystal. Using equation 1, for crystals thicker than 100nm we will not see detail bellow 1.5nm, with lamda =.0037nm.

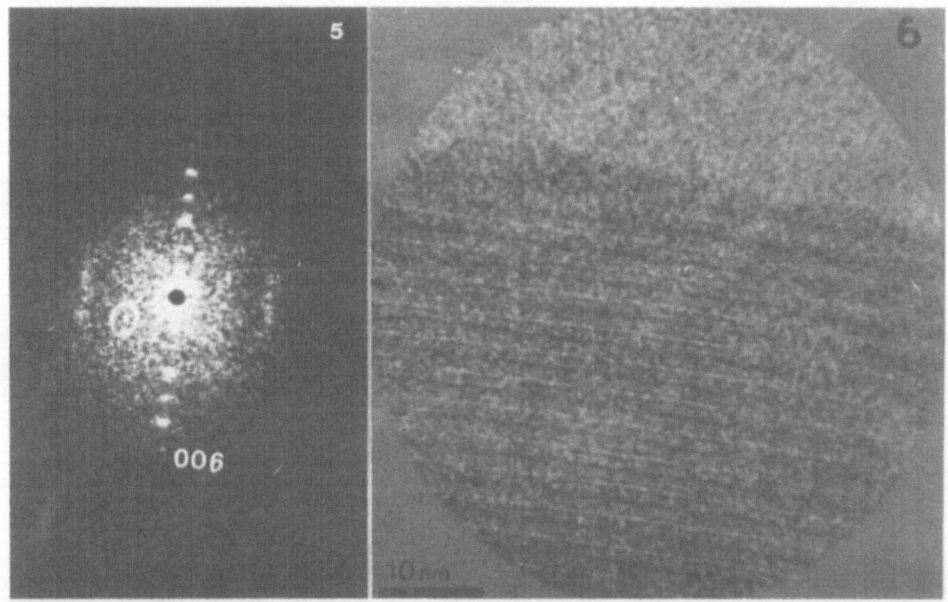

Figure 5. Calculated Power Spectrum showing reflection to the 6th order and side bands corresponding to 0.8nm
Figure 6. Type 2 crystal with a 3.2nm lattice spacing. Scale bar =10nm

With these results the unit cell was initally thought of as being tetragonal. When looking at the amplitudes of the respective diffraction patterns and transforms it became apparent that there was no four fold symmetry and only two fold symmetry so the unit cell was revised to being orthorhombic with a = 0.78nm, b = 0.78nm and c = 3.2nm.

Unit cell recovery.

Unit cell averaging over the whole crystal image by the FLC method gave results such as shown in Fig.7. Although the major periodicities are clear,there was no finer molecular detail apparant. The problem in averaging over such a large crystal area becomes apparent when one looks at the filtered image (figure 8).   The crystal suffers from irregularities throughout due to bending and beam damage [30]. To overcome this a coherent area which shows structural information present from each pair of reflections is extracted for unit cell averaging. The result of this (figure 9), is that structural information present to better than 0.4nm is apparent. It was then calculated that given the stated cell dimensions there would be four molecules per unit cell.

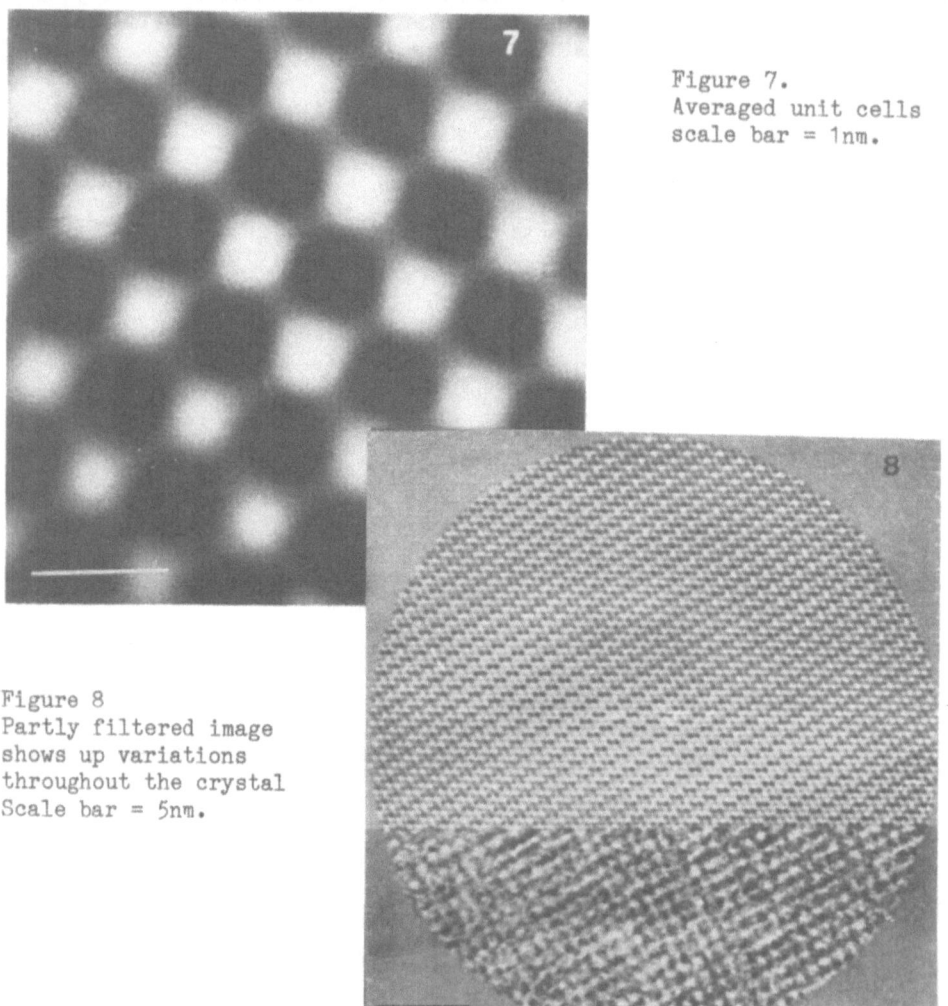

Figure 7.
Averaged unit cells
scale bar = 1nm.

Figure 8
Partly filtered image
shows up variations
throughout the crystal
Scale bar = 5nm.

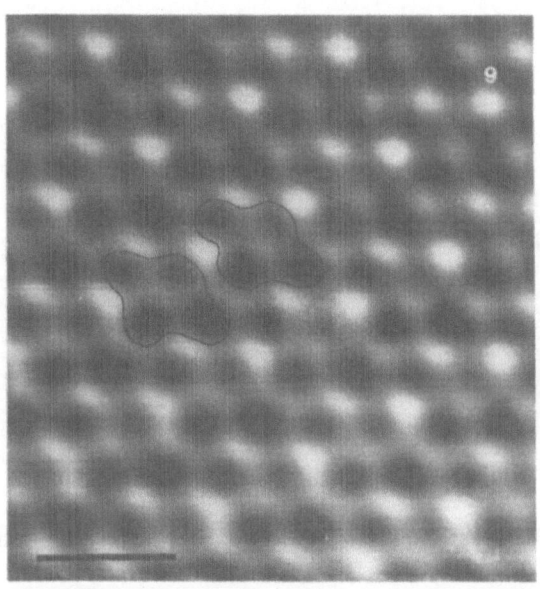

Figure 9. Averaged unit cell showing detail to better than 0.4nm
           Scale bar = 1nm

Molecular Modelling.

Molecular mechanics have proved useful for the investigation of a
diversity of topics associated with organic chemistry including
molecular conformations, molecular thermodynamic properties and the
interaction of molecules with other molecules. Using a system developed
at Glasgow by D. White [28 -29] a molecular reconstruction of the
organo-salt was built to the same general shape as described
earlier.(figure 10) This structure was energy minimised by a two
stage Newton-Raphson procedure [28]. The dimer was then built using the
result from the minimisation and was again energy minimised. (figure 11
) By rotating the molecule about the x-axis we see how the minimised
molecule retained its planarity (figure 12). The task now was to arrange
the molecules so as to conform with the unit cell dimensions and the
molecular images found by the image processing.The arranged molecules
were then minimised (Figure 13). The 3.2 nm lattice spacing is easily
explained by these layers stacking on top of each other.figure 14

350

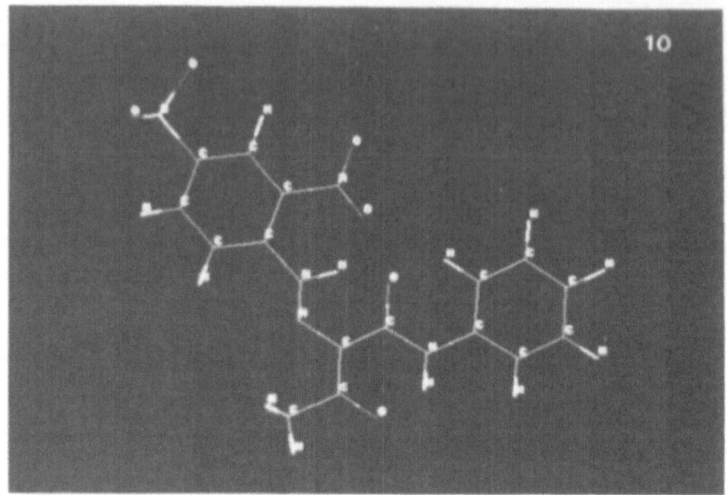

Figure 10. Molecule reconstructed to show the molecular shape

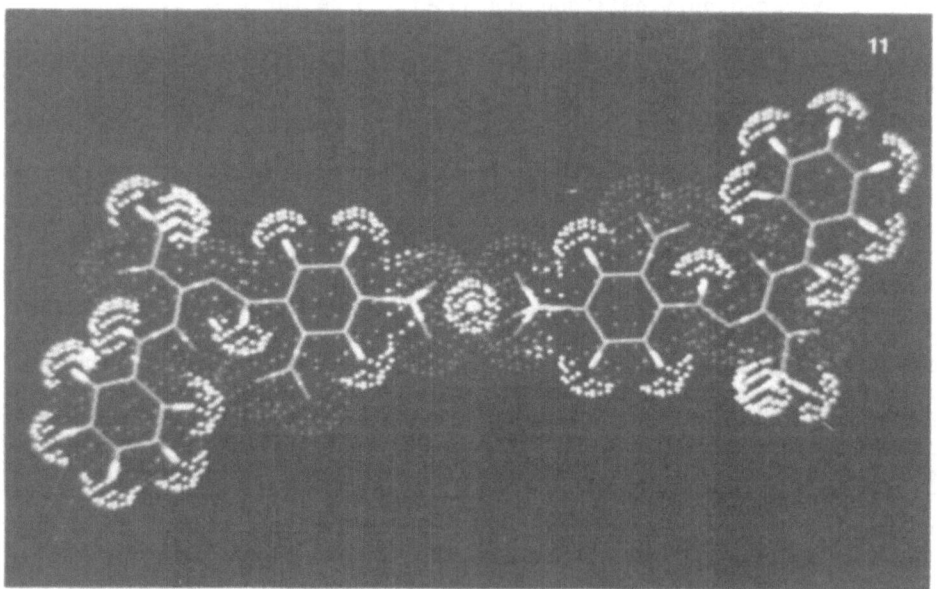

Figure 11. Two molecules showing how the dimer may be formed in the unit cell.

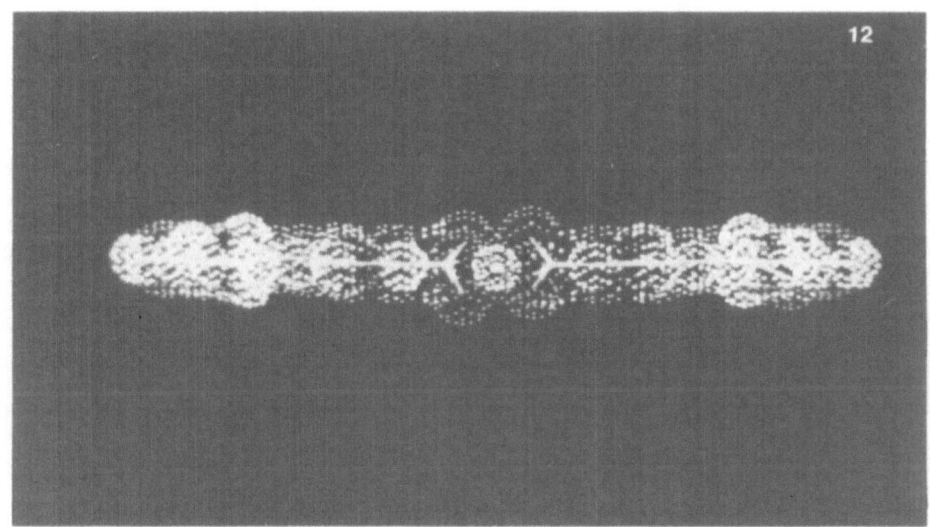

Figure 12. Dimer as in figure 11, rotated by 90 deg. This shows the planarity which is found in this type of molecule.

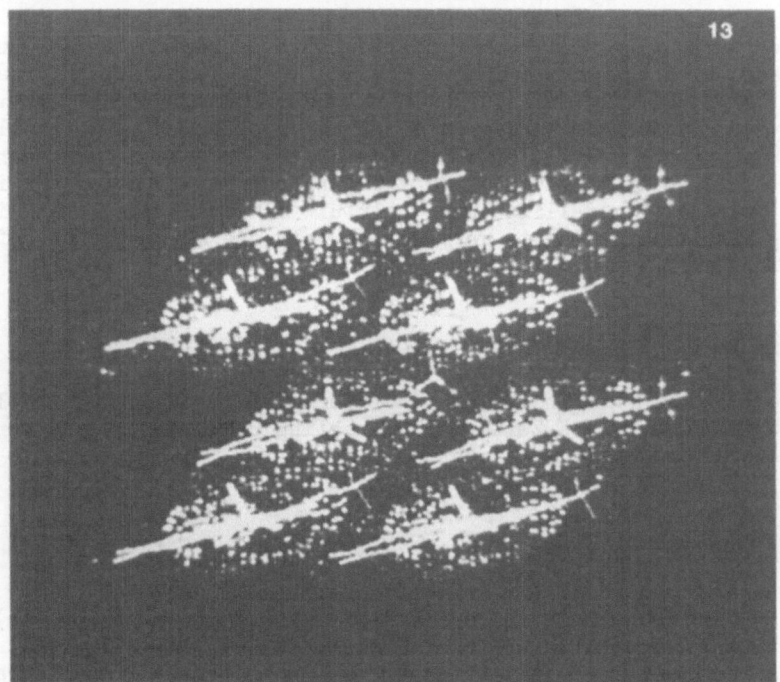

Figure 13. Eight dimers arranged to represent the image processing results for the hk0 plane.

352

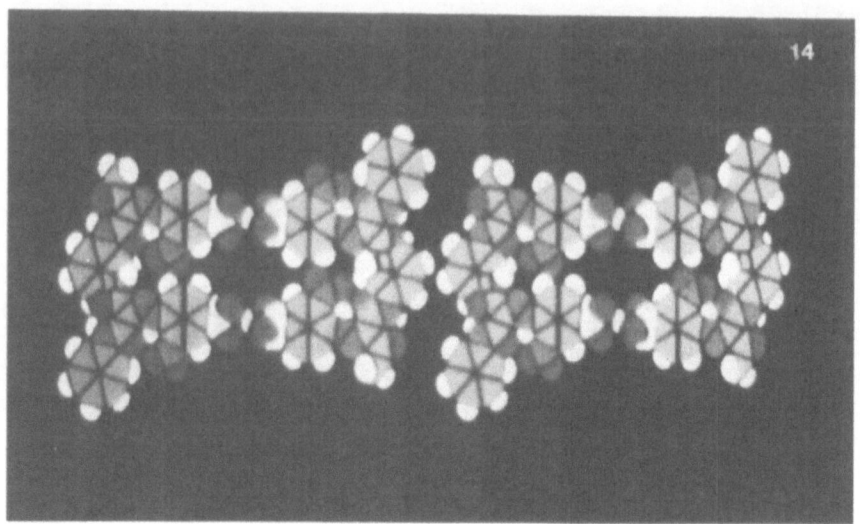

Figure 14. Four dimers in an arrangement that explains the 3.2nm lattice
spacing.

Conclusion.

Crystals of this type will remain difficult to characterise as long as
we only have high resolution images giving two dimensional information
about the crystal. The limitation in this work proved to be the thick
type 2 crystals. Another advantage of thin crystals is that it will be
acting as a weak phase object and intensity and phase data can be
collected for use in maximun entropy experiments to improve the image
quality. Molecular modelling was helpful in confirming the image
processing results and in its very nature lets one view the predefined
atomic arrangements.

Acknowlegments

Thanks go to David R. White for the use of his molecular modelling
package (CHEMMOD) and to E.Zeitler for enabling G.M. to stay and work at
the Fritz-Haber-Institut.

Refrences
1. A. Whitaker, Zeitschaft fur Kristallographie 163, 19-30 (1983)
2. A. Whitaker, Zeitschaft fur Kristallographie 163, 139-149 (1983)
3. A. Whitaker, Zeitschaft fur Kristallographie 166, 177-188 (1984)
4. A. Whitaker, Zeitschaft fur Kristallographie 167, 225-233 (1984)
5. E.F.Paulus, W.Rieper and D.Wagner,Zeitschaft fur Kristallographie

6. E.F. Paulus, Zeitschaft fur Kristallographie 167, 65-72 (1984)

7. S.Hovmoller, X.Zou, D.N.Wang, J.M.Gonzalez-Calbet and M.Vallet -Regi, J. Solid State Chem. 77,316-321 (1988)

8. D.X. Li and S. Hovmoller, J. Solid State Chem. 73,5-10 (1988)

9. C. Rosique-Perez, J. Gonzalez-Calbet, M. Vallet-Regi and M.A. Alario-Franco,J. Solid State Chem. 76,313-318 (1988)

10.J.M. Gonzalez-Calbet and M. Vallet-Regi, J. Solid State Chem. -Regi, J. Solid State Chem. 68,266-272 (1987)

11. L. Kihlborg, M. Sundberg and O. Savborg, Ultramicroscopy 18 (1985) 191-196

12. P. Bullough and R. Henderson, Ultramicroscopy 21 (1987) 223- 230

13. M. Tsuji, S. Isoda, M. Ohara, A. Kawaguchi and K. Katayama Polymer, 23 (1982) 1568-1574

14. P.N.T. Unwin and R. Henderson, J. Mol. Biol. (1975) 94,425- 440

15. D.L. Stokes and N.M. Green, Biophys. J. 57 (1990) 1-14

16. E.L. Thomas and D.G. Ast, Polymer 1974, vol 15, 37-41

17. P.T.E. Roberts, J.N. Chapman and A.M. MacLeod, Ultramicroscopy 8 (1982) 385-396

18. W. Krakow, Ultramicroscopy 18 (1985) 197-210

19. J.P. Martinez, D. Locatelli, J.L. Balladore and J. Trinquier, Ultramicroscopy 8 (1982) 437-440

20. A. Boudet and L.P. Kubin, Ultramicroscopy 8 (1982) 409-416

21. J.R. Fryer and K. Holland, Proc. R. Soc. Lond. A. 393, 353- 369 (1984)

22. J.R. Fryer, Ultramicroscopy 23 (1987) 321-328

23. W.O. Saxton, Topics in Current Physics, Vol. 13, Computer Processing of Electron Microscope Images, Ed. P.W. Hawkes, Springr-Verlag Berlin Heidlberg New York 1980

24. P. Padere and E.L. Thomas, submitted to Ultramicroscopy, March 1989

25. R.W. Horne and R. Markman, Application of Optical Diffraction and Umage Reconstruction Techneques to Electron Micrographs, in Practical Methods in Electron Microscopy, Ed. A.M. Glaubert (North Holland,, 1972)

26. W. Kunath, F. Zemlin and K. Weiss, Optic 76 No. 6 (1987) 122- 131

27. M. Tomita, H. Hashimoto, T. Ikuta, H. Endon and Y. Yokota, Ultramicroscopy 16 (1985) 9-18

28. D.N.J. White and M.J. Bovill, J. Chem. Soc. Perkin II, 1610 (1977)

29. D.N.J. White, J.N. Ruddock and P.R. Edgington, Molecular Simulation, 1989, Vol. 3, 71-100

30. D.L. Dorset and F. Zemlin, Ultramicroscopy 21 (1987) 263-270

31. Image Analysis. 30 papers on computer aided microscopy in medicine. Applied Optics, 26(16) (1987) 3199-3416

32. W.O. Saxton, Ultramicroscopy 4 (1979) 343-354

RADIATION DAMAGE IN ELECTRON CRYSTALLOGRAPHY

E. Zeitler
Fritz-Haber-Institut der Max-Planck-Gesellschaft
Faradayweg 4-6
D-1000 Berlin 33

ABSTRACT

The manifestation of radiation damage in the diffraction patterns of
small crystals is discussed. The Wilson method, well known in X-ray
crystallography, is also well suited for the problems in electron
crystallography. Association of inner-shell excitations caused by
electron energy losses with data from desorption experiments on
surfaces leads to new insights (Auger cascades). Dynamical features
must be considered for a better understanding of radiation damage.

1. INTRODUCTION

"At some time every practicing microscopist becomes painfully aware of
the damaging potential of the beam when he observes subtle or even
drastic changes such as shrinkage, fragmentation, melting, sputtering
or bubbling taking place in his specimen. On the other hand, the beam-
induced fixation effect is an accepted feature of the microscope and in
some cases is looked upon as beneficial to specimen examination because
it grossly stabilizes the object and thus prevents object change. Conformation also appears to play a role in determining which constituent
groups are affected by radiation, e.g., residues of specific amino
acids suffering the most damage vary from protein to protein. For this
reason, unless structural analogies can be made one generally cannot
predict which of the constituent amino acids of a given protein will
suffer damage."

This quotation, taken from a classical paper by Stenn and Bahr [1]
written twenty years ago, still is valid and may be seen as an explanation for the fact that every meeting on electron microscopy specialties
dedicates at least one session to the topic of this paper.

On account of their more intense interactions with matter, electrons can be used to investigate minute and thin crystals which are not
available in sizes suitable for examination with the more gentle X-rays.
But the price for this stronger interaction is the heightened chance
for destruction of the sample. This is the permanent dilemma of electron microscopy. For the newcomer to the field we must mention the work

355

J. R. Fryer and D. L. Dorset (eds.), Electron Crystallography of Organic Molecules, 355–360.
© 1990 *Kluwer Academic Publishers.*

of K. Kobayashi, whose hope that he might succeed in understanding the
damage in the chemically uniform polymers [2] led him to pioneer the
development of high voltage electron microscopes for polymer studies.
Of course, ionic crystals and metals also undergo changes in the electron beam. The most dangerous radiation damage in metals is the one
caused in the retaining metal shields of nuclear reactors; it can be
simulated and studied in the high voltage electron microscope. Very
good reviews on all those aspects, which have not aged, are available
[3-5].

The main concern of electron crystallography is the unravelling of
the structure units which occur periodically in a three-dimensional arrangement. In some cases it is a gift of nature to provide such crystals;
in other cases it takes the wit, skill and endurance of a dedicated re-
searcher to come up with an "artistic" arrangement of a periodically
recurring motif of interest. The obvious advantage for the structural
investigation of, say, a macromolecule is the simultaneous representation of a multitude of individual molecules, each and every one in the
same orientation and environment. Every result derived from such an
arrangement has the character of an average statement with high statistical significance and validity. But the interaction with radiation
will also change averaged results during the exposure of the object.
Radiation will affect the crystal in two aspects. It will either cause
disorder of the regular arrangement of the network or injure the individual building blocks whose structure is the very object of the research. In order to arrive at statements about the virgin structure,
the course of the decay should be known so that it can be traced back
in time.

This contribution should help as a guide for the extrapolation
from the damaged structure to that at the onset of irradiation, using
established facts instead of fantasy. We all know that diffraction
patterns, discrete spots or Debye-Scherrer rings manifest the crystallinity of the specimen. We also know that those diffraction patterns
fade during electron exposure. For a long time study of these fading
curves represented a major part of radiation damage in electron microscopy. Rather late in the game, Clark et al. [6] came up with a theory
to model fading curves. This theory was modified and enlarged by
Reimer and Spruth [7] to a multi-hit model. Experiments yield values
for the number of hits necessary to damage a sensitive unit, the cross
section of a single hit and an activation energy of the temperature-
dependent processes. Van Dyck and Wilkens [8] maintained that the fading of the diffraction pattern is more than just an indicator for radiation sensitivity. It provides (unused) detailed information about the
damage itself. The extraction of this information does not necessarily
require the knowledge of the perfect crystal structure.

2. TWO ARCHETYPES OF RADIATION-INDUCED STRUCTURE CHANGES

In the ideal case, when the atoms in the many subunits (macromolecules)
occupy identical places, the diffractive behavior of each unit cell can
be described by one and the same structure factor F_O. The intensity in
the Bragg spots of the lattice is proportional to F_O^2. Two extreme

types of damage can be considered:

1. The structure factors of the individual unit cells are altered by the radiation, but the lattice remains without any strain. The Bragg spots stay at their location but the intensity is lower because the structure factor F_O now has to be averaged over the damaged unit cells and reduced to $\bar{F}$. The individual deviations of the structure factors from the average lead to a diffuse background intensity around and between the Bragg spots. The previous theories consider the average structure factor as derived from a binary mixture of undamaged and of (permanently) damaged unit cells. The radiation dose D merely changes the "mol" fraction n(D) of the two components:

$$\bar{F} = n(D) \ F_O + (1 - n(D)) \ F_D.$$

The difference between the two theories lies in the assumed models for n(D). This simple two-unit model renders the general trend of the fading curves, but it cannot account for all experimental details. Van Dyck and Wilkens put the radiation damage within the unit cell and assume at this level a division into two groups of atoms, one fraction not hit by electrons and thus remaining at its atomic location and the other hit and moved with an assumed likelihood Q(r) over a distance r. Hence there exists a probability for the damage of a unit cell which can be written as

$$P(r,D) = n(D) \ \delta(r) + (1 - n(D)) \ Q(r).$$

The models differ in the various Q's. For example, like in Brownian motion the mean square displacement could be assumed to be proportional to a diffusion coefficient and the time of irradiation. Then Q becomes a normalized Gaussian $\exp(-r^2/\bar{r}^2)$ whose Fourier transform is again a Gaussian reminiscent of a Debye-Waller factor. It is the Debye-Waller factor which accounts for the dissimilarity of the various unit cells caused by the incoherent thermal movement of the atoms forming the unit cell. The assumption that radiation damage has a similar effect, at least to first order, seems reasonable and justified.

The remaining question--namely, how one can determine the temperature factor, in our case the "displacement factor", of a yet unknown structure--was answered nearly fifty years ago by Wilson [9] addressing the equivalent problem in X-ray structure determination. He observed that in the ideal case, i.e. without displacement, the total intensity in the Bragg spots of a certain region in reciprocal space is, to a close approximation, equal to the intensities which the participating single atoms would diffract into the same region. As the scattering amplitudes of the single atoms are known, the ratio of the experimentally determined accumulated Bragg intensities (in optics this quantity is called encircled intensity) to the intensity scattered by the atoms can be determined. It will be independent of the scattering angle only if the atoms stay in place. If, however, their displacement is similar to a thermal motion, the experimental ratio should decay exponentially with the square of the scattering angle. The general case does not follow such a simple law. This approach, described in detail by

Van Dyck and Wilkens, has the advantage that it avoids the scientist's fixation on the fading curves of a single Bragg reflex and, instead, utilizes the entire diffracted information, inherently more reliable. Jeng and Chiu [10], in their study of the crotoxin complex, applied this Wilson method. The average intensities were calculated by dividing reciprocal space into concentric zones assigned with a suitably averaged scattering angle, each containing about 40 Bragg spots. Their result, obtained for different exposures and at different temperatures, confirms that the crystalline disordering by radiation can be described by a Gaussian factor equivalent to the Debye-Waller temperature factor, thus justifying the name "disordering factor". The root mean square of thermal displacement is replaced by displacement distances proportional to the square root of the electron exposure which, in the exponent of the Gaussian, effects the well established decay of fading curves with dose.

2. For the second extreme to occur, individual atoms must be dislodged either into interstitial places or totally removed from the crystal. This will lead to strain and other lattice defects, dislocation loops and eventually mass loss. The strain fields can be visualized by the so-called weak-beam method in which the specimen is oriented such that only one of the weakly excited diffracted beams is utilized for the image formation. This method, pioneered by Cockayne [11], has found successful application in the crystallography of metals but not yet in the crystallography of organic or biological materials. The mass loss, however, of organic and biological specimens in the electron beam is a very well known fact that has been studied extensively [12]. The question of whether the mass loss is accompanied by or even the reason for the lattice defects still remains to be answered.

3. NEW INSIGHTS INTO THE DAMAGE MECHANISM

As long as we do not know the mechanisms involved, descriptive cataloging of radiation damage is necessary. Every experience with a new specimen will be entered in this taxonomy of radiation damage for the benefit of other researchers. Fortunately, from very separate scientific activities--like TV recordings of small metal clusters, measuring the desorption of ions from surfaces and pondering about the happening (possible occurrence) of Coulomb explosions--comes a fresh outlook to this stale field. Reporting about these findings seems to be a fitting final section.

In his thesis, M. Isaacson [13] noticed that the effective damage cross sections for the nucleic acid bases are in magnitude very much the same as the cross section for K-shell ionization. He went on to propose that K-shell ionization might be the key process in radiation damage. The difference from the direct excitation of the valence of the bonding valence electrons is that an inner-shell ionization requires a high energy of several hundred electron volts (depending on the atomic number of the absorbing atoms), and this energy is concentrated and stored in a single atom as the epicenter of an ensuing quake (de-excitation). The localized concentration of energy is more disruptive than the same energy imparted in smaller amounts to the outer electrons from whence it can be dissipated less harmfully. Crewe [14] was the first to capitalize on the possibility of this mechanism. The implications are obvious. If the K-shell ionization is the cause, electrons whose energy does not permit such ionization should not damage.

Surface scientists have studied the desorption of particles (ions or neutrals) as a consequence of their primary electronic excitation by impinging electrons or photons. We might look at these phenomena as a special type of radiation damage. For the interested electron microscopist, D. Menzel surveys the detailed knowledge available about this damage [15]. The primary energies involved are a few kV. The most remarkable fact is that the primary electronic excitation responsible for desorption is a K-shell ionization. In the low-Z material relevant to biology (also investigated by Menzel), the ensuing rapid electronic rearrangement (10^{-15} - 10^{-16} sec) amounts to an internal conversion in which the inner hole is filled by an outer electron, and the extra energy is used to expel one further electron, creating at least two "outer holes" without the occurrence of a photon (Auger effect). For the damage to occur, a transfer of the electronic motion to the nucleus must take place. And again it is the competition between the local action of the primary excitation and its delocalization which determines the efficacy for damage [16].

The photon-induced desorption yields of various ions from carbon monoxide adsorbed on a rubidium surface (001) as a function of the photon energy parallel the curves for Auger yield. That means a desorption is preceded by an Auger effect, i.e. a K-shell ionization. The detailed analysis shows that this K-shell ionization may occur in either the O atom or the C atom of the adsorbed CO.

The fact that photons were used for this investigation is irrelevant for the result. Electrons would have produced the same Auger effects. But with analytical electron microscopy advancing, other surface scientists might use their EELS machines to confirm that fact and/or investigate similar systems. Other low-energy investigations by Howie et al. [17] confirmed Isaacson's supposition that K losses determine radiation damage. Significant damage in p-terphenyl is observed only when the primary electron energy is in excess of 1.5 keV, which is about six times the K energy loss of carbon. In a later paper [18], they include a number of other organic crystals; again they observe a damage threshold at about 1 keV and refute, on account of this finding, the standard picture that damage occurs as a result of low-energy valence excitations, maintaining that the carbon K-shell ionization initiates the damage. It should be emphasized that the dose rates as well as the doses applied fall into the range of regular electron microscopy.

The present state of the art permits one to record in real time the fate of an electron microscopic specimen during exposure, although with dose rates close to those in scanning transmission electron microscopy (ca. 10^5 e/Å² s). The "actors" of such video clippings are small clusters of gold, 20-1000 Å in diameter. One's outlook on radiation damage will change drastically after having seen such a movie [19]. Let me quote from a "film" critique which stems from A. Howie [20], in which he connects the aforementioned Auger cascades and the inner-shell

360

ionization with the "abrupt kaleidoscopic image change." The resulting Coulomb explosions are the propellant for the "breakdance" of these particles: "A certain particle can be seen to jump intermittently between several states--single crystal, decahedron, icosahedron and complex multiple twin. Sometimes these states seem to be quasi-stable, persisting for a few tenths of a second and exploding (e.g. local details of surface facets that minimize the energy). At other times the particle is clearly highly excited and appears to be molten or spinning on the support.

"Although these abrupt changes of overall structure seem to be suppressed in larger gold particles, 100 Å, equally startling and sudden convulsions of surface structure have been observed. . . . Their images show what appear to be atomic clouds of gold atoms outside the particle that interact with individual atomic columns viewed in projection at the particle surface."

This account by Howie recaptures the puzzlement that those movies create in the beholder. I hope that this description acts as an effective advertisement for these movies, or that it at least directs the microscopist's present ideas about radiation damage away from the static to a more dynamic point of view.

REFERENCES

[1] K. Stenn and G.F. Bahr, J. Ultrastr. Res. 31 (1970) 526.
[2] K. Kobayashi and K. Sakaoku, in: Quantitative Electron Microscopy, Eds. G. Bahr and E. Zeitler (Williams & Wilkins, Baltimore, 1965), p. 359.
[3] M. Isaacson, in: Principles and Techniques of Electron Microscopy, Vol. 7, Biological Applications, Ed. M.A. Hayat (Van Nostrand, 1975), p. 1.
[4] Cryomicroscopy and Radiation Damage, Ultramicroscopy 10 (1982) 1; Ultramicroscopy 14 (1984) 1.
[5] L.W. Hobbs, in: Quantitative Electron Microscopy, Eds. J.N. Chapman and A.J. Craven (Edinburgh University Press, 1984), p. 437.
[6] W.R.K. Clark, J.N. Chapman, A.M. MacLeod and R.P. Ferrier, Ultramicroscopy 5 (1980) 195.
[7] L. Reimer and J. Spruth, Ultramicroscopy 10 (1982) 199.
[8] D. Van Dyck and M. Wilkens, Ultramicroscopy 14 (1984) 237.
[9] A.J.C. Wilson, Nature 150 (1942; Acta Cryst. 2 (1949) 318.
[10] W. Jeng and W. Chiu, J. Microscopy 136 (1984) 35.
[11] E.J.H. Cockayne, I.L.P. Ray and M.J. Whelan, Phil. Mag. 20 (1969) 1356.
[12] J. Berriman and K. Leonard, Ultramicroscopy 19 (1986) 349.
[13] M. Isaacson, J. Chem. Phys. 56 (1972) 1813.
[14] A. Crewe, Ultramicrosocpy 1 (1976) 267.
[15] D. Menzel, Ultramicrosocpy 14 (1984) 175.
[16] P.J. Feiblman and M.L. Knotek, Phys. Rev. B18 (1978) 6531.
[17] A. Howie, J.F. Rocca and U. Valdre, Phil. Mag. B52 (1985) 751.
[18] A. Howie, M.N. Mohd Muhid, J.F. Rocca and U. Valdre, Inst. Phys. Conf. Ser. 90 (1987) 155.
[19] J.O. Bovin et al., Nature 317 (1985) 47; S. Iijima and T. Ichihashi, Phys. Rev. Lett. 56 (1986) 616; J. Hutchison, private communication.
[20] A. Howie, Nature 320 (1986) 684.

ELECTRON BEAM DAMAGE OF POLYETHYLENE (PE) SINGLE CRYSTALS.

D. VESELY
Brunel University
Department of Materials
Uxbridge
Middlesex UB8 3PH, UK

K. H. DOWNING
Donner Laboratory
Lawrence Berkeley Lab.
Berkeley
California 94720, USA

ABSTRACT. An attempt is made to contribute to the understanding of the mechanism of crystallinity loss. It is
shown that the crystallinity decay curve follows an expone-
tial law and that the increase in diffracted intensity of
some reflections is due to changes in the specimen flatness.

1. EXPERIMENTAL.

Specimens of polyethylene (HDPE; BP 006-60) were prepared
at 64°C by self-seeding from a solution in tetrachlorethyl-
ene. After deposition on a glass slide, the crystals were
vapour coated with a 20 nm layer of carbon, using a heat
protection technique. The crystals were visualised opti-
cally, using DIC and PC microscopy. They were about 10 um
long and free from parasitic epitaxial multiple layers.
Their thickness, as stated in the literature for the above
crystallisation conditions, is 10.8 nm. Selected areas
with a suitable density of isolated crystals were mounted
on 200 mesh copper supporting grids for electron microscopy. A JEOL 100B microscope, with field emission gun, was
equipped with special computer control of the beam position and the lens currents. This enabled the spot scanning
technique to be utilised for HR images. Most of the work
was performed at temperatures around 170 K. Conventional
diffraction patterns as well as multiple dark field images
were also recorded. The HR images were digitised on a
Perkin Elmer densitometre and computer processed using
modified programmes from the MRC suite.

2. HIGH RESOLUTION IMAGES.

The reduction in image contrast can be measured from the
decay of the electron diffraction pattern. Our measure-

J. R. Fryer and D. L. Dorset (eds.), Electron Crystallography of Organic Molecules, 361–364.

ments show that reduction of the temperature by 120 K will increase the lifetime about 2.5 times. At a low temperature, an exposure of 4 electrons/A^2 (64 C/m^2) corresponds to a reduction of the diffracted intensity to 1/e (critical dose). Lattice images were recorded with a wide range of exposures. However, the best images, recorded on Agfa Scientia films, were made with the relatively low exposure of about 2.5 el/A^2. At this exposure we could record up to three HR images from the same area. An example is shown in Figure 1. The lattice image however changes significantly from the first to the second exposure and after the third exposure the lattice has suffered a large amount of degradation. This corresponds to changes seen in the electron diffraction patterns and multiple dark field images. It is important to know to what extent the first image is representative of the undamaged structure. The type of decay function will influence the critical exposure. As for the purely exponential curve, no induction time is present. No swelling, shrinkage or phase transformation from orthorhombic to hexagonal (as for paraffins) or to monoclinic (as on deformation) has been observed.

3.DIFFRACTION.

The diffraction patterns (Figure 2) show some distortion of the lattice, which is manifested by a small broadening of the diffraction spots at high exposures. This broadening might give the impression that the overall intensity of the diffraction pattern is improving slightly with irradiation, before it starts to fade away. The optical image transform (Figure 3) and the Fourier transform of digitised images show a similar effect. There are however some experimental results, which support the idea that the crystallinity is decaying purely exponentially.
To overcome the problem with the saturation of the emulsion, the diffraction spots were defocussed, producing multiple dark field images, so that the whole crystal could be seen in each spot (Figure 4). One of the series of multiple dark fields was digitised on a densitometre and the density distribution measured for different reflections and exposures. The bend contours can be seen to move significantly under the beam, indicating some rocking of the crystal. This motion is apparently a result of stress induced by the collapse of the crystals onto the substrate. The changes in bend contours can produce large anomalous effects in the intensities of a diffraction pattern. However, when the intensities are followed carefully within the multiple dark field images, it can be seen that the peak intensities (corresponding to the maxima on the bend contours) fade smoothly and exponentially. Figure 5 shows the intensity maxima averaged for different reflections.

Within the scatter of the data, all reflections show unambiguous exponential decay. This behaviour is also supported by the gradual increase in the intensity of amorphous rings, while the 'subjective' intensity of the diffraction spots is increasing. This behaviour is direct-ly related to the chemical changes. The FTIR spectroscopy shows purely exponential changes in peaks corresponding to carbon-hydrogen stretching, bending and deformation. The most important changes are the loss of hydrogen and the formation of double bonds.

4.CONCLUSIONS.

Our present understanding of the crystallinity loss can be summarised as follows: The incident electrons are transfer-ring their energy to the polymer chain randomly. The excited molecules reduce their energy by emitting an elec-tron or radiation. The loss of electrons results in ioni-sation and chemical changes, mainly loss of hydrogen atoms and formation of double bonds. The damaged part of the chain no longer satisfies the thermodynamical requirements for crystallinity and an amorphous region is formed. The size of this region is related to the stiffness of the molecular chain, which is temperature dependent. These small regions gradually fill up the volume of the crystal, leaving smaller and smaller domains of undamaged lattice, which provide diffraction, and it is increasingly difficult for the eye to recognise the symmetry of the image. The growth of the amorphous volume (or loss of the crystalline volume) is an exponential function of exposure and follows the chemical changes in the polymer.

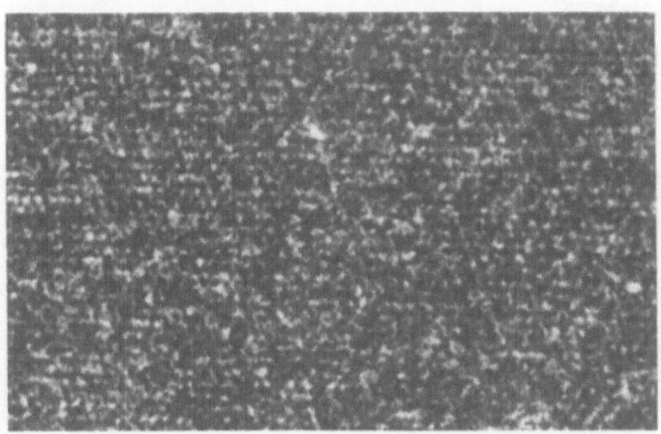

|——10 nm ——|

Figure 1. High resolution image of PE single crystal

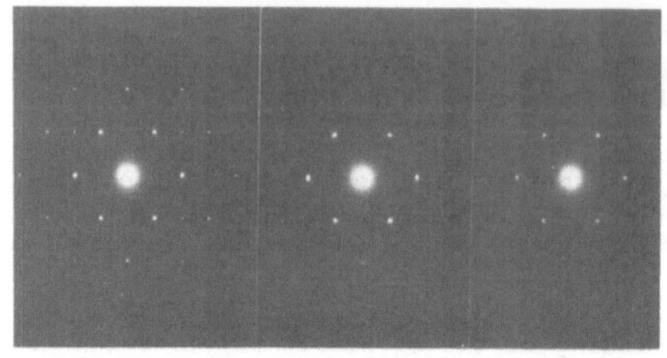

Figure 2.
Diffraction
0.7, 3 and 8
el/A^2

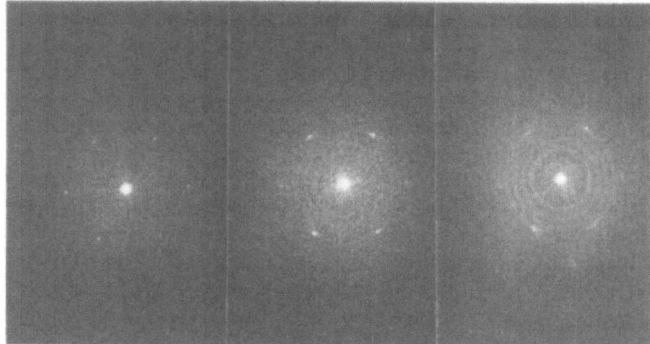

Figure 3.
optical
transform
2.5, 5 and
7.5 el/A^2

Figure 4.
Multiple
dark field
image

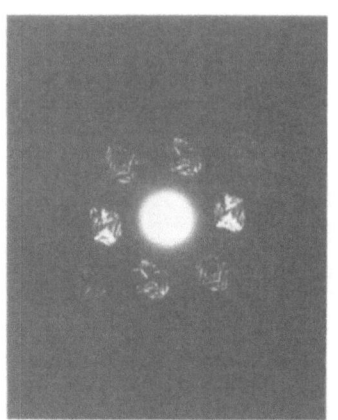

RELATIVE INTENSITY Vs. EXPOSURE
RELATIVE INTENSITY
100
90
80
70
60
50
40
30
20
10
0
0
1
2
3
4
5
EXPOSURE (el/Å²)

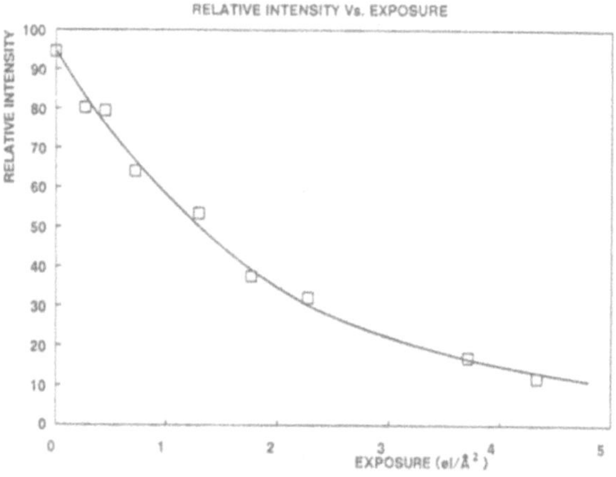

Figure 5. Maximum image contrast
measured on digitised multiple
dark field images and averaged
for four pairs of diffraction spots

AN ANALYSIS OF THE BROADENING INDUCED BY BEAM DAMAGE IN TRANSMISSION ELECTRON DIFFRACTION SPOTS FROM AN ORIENTED ALIPHATIC MONOLAYER

R.STEITZ and I.R.PETERSON
Institut für physikalische Chemie
Johannes Gutenberg Universität
Jakob Welder Weg, 11
D6500 Mainz FRG

ABSTRACT. We have analysed the progressive changes in diffraction spot shape during prolonged transmission electron diffraction observation of a soap monolayer supported on a thin polymer film. The material used to form the monolayer was cadmium eicosanoate (arachidate). The observed changes cannot be explained at all in terms of the chemical crosslinking which is known to occur as a result of beam damage, nor completely in terms of the strain fields caused by unbound dislocation defects of the crystalline lattice. The most plausible explanation involves the formation of linear dislocation aggregates which resemble grain boundaries but yet which are not linked into a continuous network. The evolution of the spot profiles can be explained in terms of three processes: firstly the mobilisation of existing dislocations; secondly, the creation of new dislocations; and finally the growth of dislocation aggregates. A mechanism for the creation of dislocations is proposed.

1. Introduction

Electron beam, or radiation, damage is one of the most important problems in the electron microscopy of organic materials at atomic resolution [1], and perhaps the most beam sensitive of these are the synthetic polymers [2]. Beam damage results in a significant loss of resolution, amounting at times to an order of magnitude degradation in the minimum observable feature size. In materials displaying diffraction peaks, this loss of resolution is more than can be accounted for by the accompanying reduction in scattering intensity of higher order spots [3].

There are a number of techniques which have been used experimentally to reduce the severity of the problem. An obvious precaution is to minimise the sample exposure using low-dose techniques. Cooling to cryogenic temperatures is also effective in reducing the resolution loss, although the degree of protection conferred varies widely from experiment to experiment [4-7]. A third alternative is to incorporate in the sample certain chemical moieties such as covalently-bound halogens [8] and naphthacene [9]. To explain the resulting reduction in damage, it has been proposed [9] that the primary damage event is the formation of charge carriers or excitons which

J. R. Fryer and D. L. Dorset (eds.), Electron Crystallography of Organic Molecules, 365–375.

are relatively harmlessly trapped by the protective moiety before they can propagate to the final site of damage. Alternatively the propagating species may be a free radical or ion [8,10]. Fourthly, the action of layers of carbon, metals [11] and silicon monoxide [12] has been ascribed to the 'cage' effect, ie the physical retention of the volatile products of radical decay. To complicate the comparison of experimental results, the amount of damage may vary with sample thickness [13].

Since the extensive work of Patel and Keller [14] it has been known that a very important effect of ionising radiation on organic materials, specifically polyethylene and its oligomers, is the formation of chemical cross-links between the chains. Since the cross-links clearly distort the crystalline lattice in their vicinity, they must be responsible for at least part of the loss of resolution.

In the investigation of the effects of beam damage, it is common to use thin lamellae consisting largely of long, hexagonally close packed aliphatic chains. These lamellae may be conveniently prepared by crystallisation from solution [3,13], amphiphilic self-assembly [15] or from a monolayer on a water surface [16]. They provide a convenient model system of reproducible structure analogous to a range of polymers and biological membranes. A related way of producing a lamella of close-packed aliphatic chains with controllable thickness and lateral structure is Langmuir-Blodgett (LB) deposition onto a solid substrate [17,18]. Although to our knowledge beam damage has not yet been investigated in films produced by this technique, transmission electron diffraction (TED) has been used to investigate the lateral packing of their molecules [19,20]. TED patterns from these films often display a considerable number of diffuse peaks, indicating the presence of long range order of a crystalline type in the films. In most previous studies, only the parameters of an average unit cell have been extracted from the positions of the diffraction peak centres, and the cause of the diffuseness has not been investigated, although Garoff et al did suggest that it originates in specific lattice defects [21].

This suggestion has been pursued in a recent work [22] which considered several different types of lattice defect. It was assumed that the effect of each defect is not purely local, but is spread by mechanical contact between molecules according to Hooke's law. It was shown that the effect of paracrystal-type defects (i.e. defects which do not interrupt the lattice rows) was to convert the initial δ-function profile of the peaks into a power law, with the scattered intensities at the positions of the reciprocal lattice points still infinite. The signature of a density of loosely-bound dislocations in thermodynamic equilibrium was found to be a Gaussian profile with tangential and radial widths in the ratio $\sqrt{3}:1$. The signature of a density of dislocations linked together into grain boundaries is conventional Debye-Scherrer rings.

TED profiles for two aliphatic monolayers displaying hexagonal symmetry were measured and compared to the theoretical predictions. The observed diffuse profiles could be explained by assuming the simultaneous existence in the lattice of dislocations and paracrystalline defects, but were inconsistent with the presence of grain boundaries.

In a recent report of beam damage in lamellae of a long-chain alkane [23], the initially circular TED spots were found to evolve to an elongated shape. The ratio of the tangential to radial spot widths was not determined, but can be judged from the figures to be not far from $\sqrt{3}$. It therefore seemed of interest to pursue the suggestion of Van Dyck and Wilkens [25], and to perform a detailed analysis of the progressive changes in line profile induced by electron irradiation. Because of the familiarity of the authors with the LB technique, the experiment was performed on a monolayer of cadmium eicosanoate supported on a thin Formvar substrate. It is expected that the results will also be applicable to other aliphatic lamellar systems.

2. Experimental

The equipment used has been described previously [22]. The eicosanoic (arachidic) acid was obtained from Fluka (purity > 99%) and used without further purification. It was dissolved in p.a. grade chloroform at a concentration of approx. 1 g/l to form the spreading solvent. The subphase was adjusted to a pH of 6.8 with $NaHCO_3$ and contained $CdCl_2$ at a concentration of either 1.4×10^{-3} M (in experiments 1 and 3) or 5×10^{-4} M (in experiment 2). Transmission microscope grids were trapped between a thin Formvar film and a glass microscope slide using the technique of Reimer [24,26]. The Formvar thickness was determined using an optical interference technique [27] to be typically 20 nm thick. Monolayers of the soap were deposited onto these sandwich structures using a Lauda FW1 film balance and FL1 dipping assembly at a temperature of $20^{0}C$ and surface pressure of 20 mN/m (experiments 1 and 3) to 30 mN/m (experiment 2) and at a withdrawal speed of 83 μm/s.

The grids were removed without deformation and observed by TED in a Philips EM300 microscope at an accelerating voltage of 80 kV. The patterns were recorded on Ilford PANF film and developed using Kodak D19 developer for 5 minutes. The resulting negatives were back-illuminated by a diffused light source with less than 0.5% brightness variation and scanned in two dimensions with a Hamamatsu C2400-07 Plumbicon camera whose output was digitised in real time by the A/D converter of an Eltec APAL-1.2/68K image processing system and averaged over 32 frames. The resulting array of 512×512 8-bit brightness values was stored on diskette. In the previous study [22] it was found unnecessary to compensate for film nonlinearity.

The measured scattering intensities for each individual (100) scattering peak were least-squares fitted to the law $A \cdot \{1+0.5948 \cdot (r^2/\alpha^2+t^2/\beta^2)\}^{-5}$, where r and t are radial and tangential coordinates relative to a spot centre which was allowed to vary as part of the fit, α and β are radial and tangential full width half maxima (FWHM), respectively, and A is a peak amplitude. This formula has no theoretical justification but was found in the previous study [22] to be a better fit than the 1D Lorentzian, 2D Lorentzian or Gaussian profiles. The six radial FWHM for each TED negative were combined to give an average and standard deviation, as were the six tangential values.

3. Results

The monolayers proved to be comparatively resistant to beam damage. While each TED negative received an adequate exposure after 4 s from a total diffraction area of diameter 100 μm, distinct (100) diffraction peaks could still be observed after 200 s irradiation. Unfortunately the instrument has as yet no facility for beam current calibration, so that absolute dose figures cannot be quoted. Figure 1a and b show negatives corresponding to 12 s and 132 s total exposure, respectively. The features observed in previous work of broadening and reduction in intensity are apparent. The peaks of Fig. 1a are somewhat extended tangentially, and it is difficult to judge by eye whether the aspect ratio has changed in Fig. 1b.

Fig 2a, b and c show the raw data for the average spot amplitude and widths together with their standard deviations as a function of exposure time. It can be seen that the amplitude remains almost constant for an initial induction period before decreasing monotonically. This behaviour is as

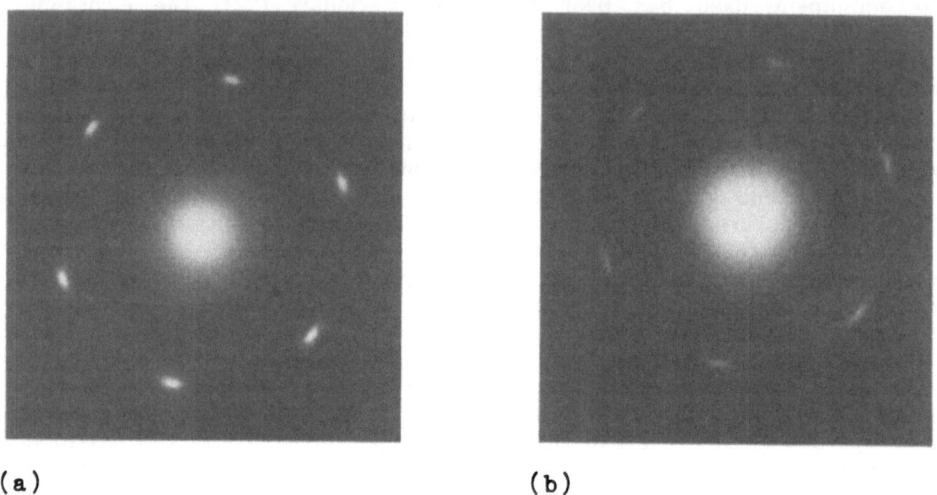

(a)   (b)

Fig 1. TED patterns recorded from a monolayer after (a) 12 s exposure (b) 132 s exposure

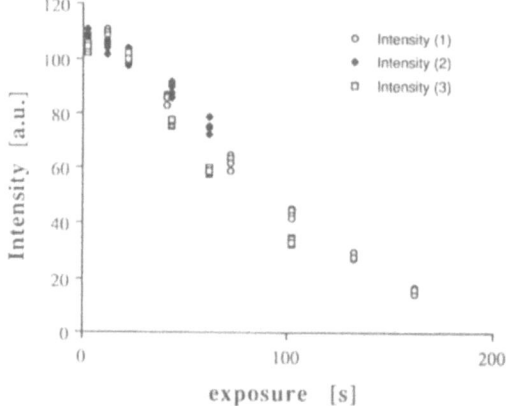

Fig. 2a. Best-fit amplitude for each peak as a function of exposure time

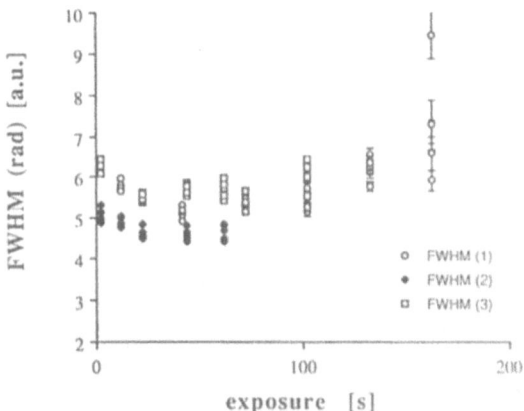

Fig. 2b. Best-fit radial full-width half maximum (FWHM) for each peak as a function of exposure time

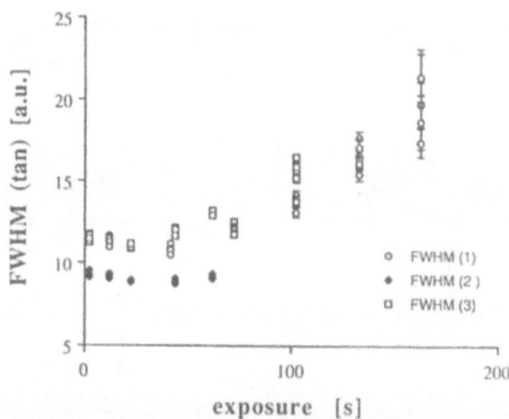

Fig. 2c. Best-fit tangential FWHM for each peak as a function of exposure time

expected. However both the radial and tangential width decrease slightly during the induction period before subsequently increasing monotonically.

Figure 3 shows the evolution of logarithm of spot amplitude versus time. As also observed by Ohno [13] and Hui [16], the behaviour can be fitted by linear regression lines, although there are quantitative differerences.

Figure 4 shows the time dependence of the ratio of the two widths. This appears to increase linearly from an initial value just above $\sqrt{3}$ to a significantly higher value.

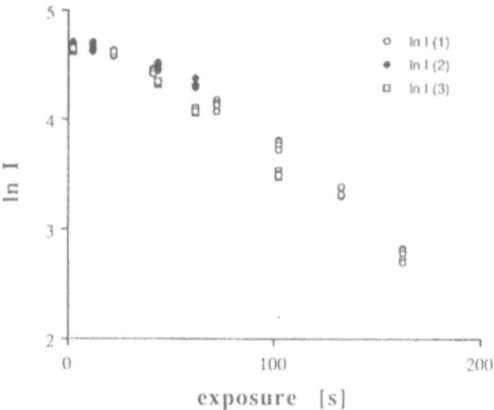

Fig. 3. Time dependence of the logarithm of the spot amplitude

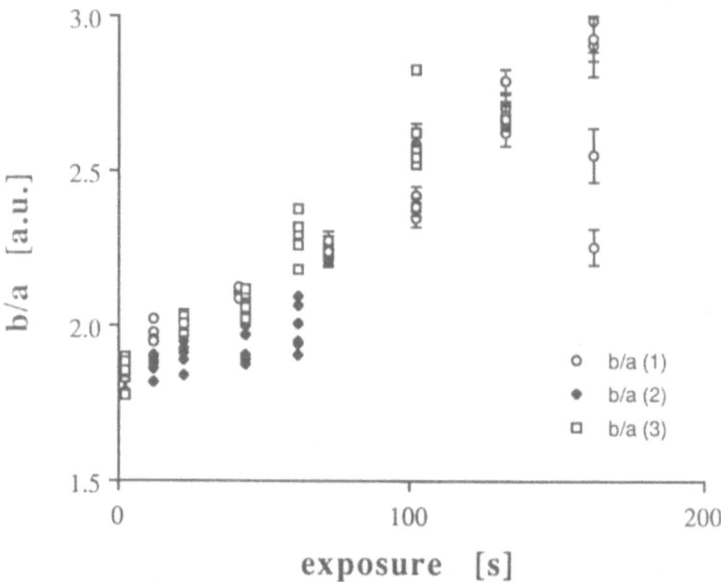

Fig. 4. Time dependence of the ratio of tangential to radial spot width.

4. Discussion

Shiomi [9] has proposed that beam damage is mediated by hot charge carriers or excitons, and ascribed the relative robustness of metals to 'electron delocalisation'. Fryer et al [8] comment that the critical dose for TTF-TCNQ is comparable to that of nonconductive materials of similar structure, and argue that therefore charge carriers and excitons cannot be the full explanation. They prefer a mechanism whereby free radicals are the entities which propagate between the initial electron impact site and the final damage site.

However it is clear that on the hot quasiparticle model the effectiveness of a protective group will depend on how rapidly it can dissipate the energy of the quasiparticle to below the threshold of several eV necessary to break bonds. Since the actual breaking of a bond corresponds to the excitation of a molecule to a multiple optical phonon state, a robust group will preferentially generate acoustic phonons, which in addition dissipate at the speed of sound. Rapid de-excitation by this route requires the existence of a ladder of electronic states separated by less than the maximum phonon energy over the whole energy range. In graphite, metals and silicon monoxide, the conduction band forms a continuum of states many eV wide and fulfils this criterion. The maximum acoustic phonon energy is typically 0.1 eV.

Although the expression 'electron delocalisation' appears to describe the spatial extent of molecular orbitals in a molecule, polyethylene is not normally considered to display it, even though it has a band structure with conventional Bloch orbitals of infinite extent [29] and high carrier mobilities [30]. The normal operational meaning of the term is that the gap between successive electronic states of an isolated molecule is small, typically a few eV. Shiomi's hypothesis does seem to accord well with the observed relative beam sensitivities of aliphatic and aromatic materials [31]. Electron delocalisation should not be confused with bulk conductivity. Although the former is a necessary condition for the latter, many TTF and TCNQ salts are nevertheless insulators. Bulk conductivity depends critically on the energy level structure of the condensed state within 0.1 eV or so of the Fermi level. The de-excitation of potentially damaging quasiparticles depends instead on its energy level structure up to perhaps 10 eV away.

Compared to metals, organic conductors typically have a very narrow conduction band [32], not much larger than kT, with acoustic phonon energies less than 0.01 eV. Hence with a suitable interpretation of the somewhat vague expression 'electron delocalisation', Shiomi's hypothesis is by no means ruled out by the known facts.

If two molecules in the monolayer lattice become linked by a covalent bond, their separation will decrease, and this change will be communicated by elastic forces to the surrounding molecules. However the lattice will remain 'good crystal' in Burgers' sense if it was initially: no dislocations are necessarily involved. If the major effect of beam damage to the molecular organisation in the film were merely the pairwise linkage of molecules, then the mathematical analysis detailed in reference [22] shows that the radial and tangential FWHM of the diffraction peaks would be essentially unaffected. The profile would change somewhat, but essentially only in the wings. However experimentally the peak widths do change significantly.

Crosslinking is also inadequate by itself to explain the observed difference between the reductions in primary and secondary diffraction spot intensities [3]. The strain fields due to crosslinking decay with distance as r^{-2}, so that the strain at a point is dominated by local defects. Only dislocations give rise to the strains correlated over long distances which can explain this anomaly. Hence in addition to crosslinking of molecules, radiation damage must also involve the reorganisation and probably the creation of dislocation defects in the monolayer lattice.

If the dislocations of the monolayer after irradiation remained largely unbound and independent of one another, then although the radial and tangential peak widths might change, their aspect ratio would remain constant and equal to $\sqrt{3}$. However experimentally the ratio of the two widths increases steadily with increasing irradiation. Hence the assumptions of the high-temperature Debye-Hückel limit for the hexatic phase of the two-dimensional lattice become progressively worse.

There is another limiting case of the Nelson and Halperin model of a two-dimensional crystal in which the peak aspect ratio can be predicted. This is the low temperature limit, below the temperature of the transition to the hexatic phase. In this case the phase is polycrystalline, and dislocations cannot exist in isolation, but link up into connected strings forming the boundaries of regions of constant lattice orientation. The diffraction pattern given by such a phase consists of Debye–Scherrer rings, for which the peak aspect ratio is large, essentially infinite.

There are experimental grounds for believing that a Langmuir–Blodgett film deposited on a solid substrate does not represent the thermodynamic ground state of the system. The molecular tilt elevation in the case of 22-tricosenoic acid multilayers has been observed to remain constant at the value prevailing immediately after deposition for periods of months, although it can be varied over a range equal to 40% of the mean value by changing deposition conditions [33]. On heating and recooling a monolayer of cadmium eicosanoate, there is a permanent change in the diffraction pattern [34]. This behaviour is typical of nonergodic systems, of which the most common example is glass. Nonergodic behaviour can be observed even in metals by the expedient of sufficiently rapid cooling of the melt [35]. In the deposition of an LB film, the transition of the monolayer from the water surface to the solid substrate most probably involves a phase change [36] which is completed within 1 μs.

It is well known that the fluctuations in a glass can be assigned a fictitious temperature considerably higher than the prevailing one and roughly equal to its glass transition temperature [37]. Hence it is plausible that the statistics of the molecular organisation in the deposited LB film correspond to a temperature considerably higher than the physical film temperature. Increased mobility of film defects will result in an annealing process, with film parameters evolving towards those appropriate for the true temperature. In particular the diffraction spot aspect ratio will increase. This is the behaviour actually observed on irradiation. It can be concluded that irradiation allows existing dislocation defects in the film to become mobile. There are two possible outcomes: either recombination with a dislocation of opposite Burgers vector, or trapping by a dislocation of the same Burgers vector. The end result of the second process is the building up of a linear array of dislocations, which may be considered to be a detached grain boundary. Eventually these will link up with other such arrays to form a conventional polycrystalline texture. This process of dislocation mobilisation, recombination and link-up can explain the initial decrease in the radial FWHM of the diffraction spot profiles.

Experimentally, after the initial induction period, both FWHMs start to increase. This can only be explained by an increase in the density of dislocations. Hence they must be created by some process involving a large disturbance to the crystalline lattice. Dislocations are topological lattice defects, whose presence can be detected at large distances by the Burgers circuit construction. Hence their creation involves the perturbation of the spatial relationships between many molecules. However the processes of radical rearrangement normally considered seem inadequate to provide such a disturbance. The 10^0 bond angle change from sp^3 to sp^2 hybridisation at a C· centre can be readily accommodated by rotation about adjacent C-C bonds, without breaking the lattice rows.

It is well known that, under conditions of high dielectric stress, polyethylene like other polymers undergoes a process of slow breakdown, involving the formation of dendritic regions of increased electrical conductivity [38]. Eventually the polymer is partly converted to graphite and all insulation capability is lost. The initial process involves the injection of charge into the polymer [39]. There is clearly a similarity to the processes of beam damage.

Perfectly crystalline polyethylene possesses a wide electronic bandgap of 8.3 eV [30,40]. However the normal, partly amorphous material possesses in addition to the band tails a small density of both donor and acceptor midgap states, at energies within 0.5 eV of the Fermi level and hence readily accessible [41]. While some of these states are associated with impurities, others have been unequivocally associated with unusual conformations of the polymer chain itself [42]. To conform to the observed chemical inertness of aliphatic chains, it must be assumed that these states are only formed when the chain is forcibly deformed by the surrounding molecules. Once formed, the states can readily ionise, particularly the donor states because of their very small ionisation potential of ~1.5 eV [41]. The behaviour of these states would appear to be very similar to those of the DX centres in GaAs associated with donor impurities [43].

Once one of these donor states is ionised, the mechanical energetics of the situation changes, because it now requires an additional 5.3 eV to promote the hole to the valence band if the chain is straightened out. If a hole is injected into the material, it is plausible that a chain will spontaneously deform to release the excess energy of the order of eV. This process can explain the slow but progressive degradation process during dielectric breakdown in polymers. We propose that this deformation induced by hot holes also explains the generation of dislocations in aliphatic monolayers on exposure to an electron beam.

The participation of directly-generated free radicals as propagating species cannot be ruled out as one of the other processes involved in beam damage. However we would point out that the proposed hot hole process will take place preferentially in regions where the hole is trapped by small fluctuations in the energy of the valence band edge, that is to say at lattice defects, or in their absence, on the lamella edges. Since the process creates disorder it is autocatalytic. It is known that multilayer LB films are more ordered than monolayers [19]. The process can therefore explain Ohno's observed increase

374

of critical dose and induction period with increasing number of docosanoic acid layers on a protective carbon layer [13].

5. Conclusion

We have used a new technique to extract information about the processes of electron beam damage in oriented aliphatic monolayers on a thin polymer substrate. This technique of line profile analysis depends critically on the ability to assign molecular causes to measured changes. Using this technique we have demonstrated that the formation of chemical cross-links between molecules is not by itself responsible for all the different effects occurring during irradiation. Beam damage is also accompanied by the generation, mobilisation and linking up of dislocation defects of the monolayer's crystalline lattice.

We have pointed to similarities between the processes of beam damage and of slow dielectric breakdown in polymers. Of particular interest are the energetics of deep-donor states of deformed aliphatic chains, for which a certain amount of evidence exists and which could account for the observed dislocation formation as well as other aspects of beam damage.

6. Acknowledgements

This work was financed jointly by the Bundesministerium für Forschung und Technologie within the Ultrathin Polymer Film Project No. 03 M 4008 C3, by the Deutsche Forschungsgemeinschaft within the Special Research Initiative SFB 262 on Nonmetallic Glasses, and by the Materials Science Research Center of the University of Mainz. The authors would like to thank Professor E.W.Fischer for useful discussions. Special thanks are due to Professor H. Möhwald for having initiated the project, and for his continued encouragement and support.

8. Bibliography

1. e.g. Glaeser, R.M. (1979) Chapter 16 in *Introduction to Analytical Electron Microscopy*, Eds Hren, J.J., Goldstein, J.I. and Joy, D.C. , Plenum, New York
2. Vesely, D. (1984) *Ultramicroscopy* **14**, 279
3. Henderson, R. and Glaeser, R.M. (1985) *Ultramicroscopy* **16**, 139
4. Wade, R.H. (1984) *Ultramicroscopy* **14**, 265
5. Knapek, E., Formanek, H., Lefranc, G. and Dietrich, I. (1984) *Ultramicroscopy* **14**, 253
6. Dubochet, J., Knapek, E. and Dietrich, I. (1981) *Ultramicroscopy* **6**, 77
7. Grubb, D.T. and Groves, T.W. (1971) *Philos. Mag.* **24**, 815
8. Fryer, J. (1984) *Ultramicroscopy* **14**, 227
9. Shiomi, N. (1966) *J. Phys. Soc. Japan* **21**, 907
10. Fryer, J.R. and Holland, F.M. (1984) *Proc. Roy. Soc. (London)* **A393**, 353
11. Salih, S.M. and Cosslett, V.E. (1974) *Philos. Mag.* **A30** (1225
12. Fryer, J.R. and Holland, F.M. (1983) *Ultramicroscopy* **11**, 67

13. Ohno,T. (1984) *Ultramicroscopy* **15**, 319
14. Patel, G.N. and Keller,A. (1975) *J. Polymer Sci. Polym. Phys. Edn.* **13**, 303, 321,333,339,351,361
15. Giorgio, S. and Kern,R. (1987) *Ultramicroscopy* **21**, 157
16. Hui,S.W. (1980) *Ultramicroscopy* **5**, 505
17. Tredgold, R.H. (1987) *Rep. Prog. Phys.* **50**, 1609
18. Peterson, I.R. (1990) *J. Phys. D* **22**, in press
19. Bonnerot,A., Chollet,P.A., Frisby, H. and Hoclet,M. (1985) *Chem. Phys.* **97**, 365
20. Peterson, I.R. and Russell,G.J. (1985) *Brit. Polym. J.* **17**, 364
21. Garoff,S., Deckman, H.W., Dunsmuir, J.H. and Alvarez,M.S. (1986) *J. Phys. France* **47**, 701
22. Peterson, I.R., Steitz,R., Krug, H. and Voigt-Martin,I. (1990) *J. Phys. France* **51**, 53
23. Downing, K.H. (1988) *Ultramicroscopy* **24**, 387
24. Reimer,L. (1967) *Elektronenmikroskopische Untersuchungs- und Präparationsmethoden*, Springer, Berlin, p322
25. Van Dyck, D. and Wilkens,M. (1984) *Ultramicroscopy* **14**, 237
26. Orth, H. and Fischer,E.W. (1965) *Makromol. Chem.* **88**, 118
27. Laxhuber, L.A., Rothenhäusler,B., Schneider, G. and Möhwald,H. (1986) *App. Phys.* **A39**, 173
28. Albrecht, O., Grüler, H. and Sackmann, E. (1978) *J. Phys. France* **39**, 301
29. Ueno, N., Fukushima, T., Sugita, K., Kiyono,S., Seki, K. and Inokuchi, H. (1980) *J. Phys. Soc. Japan* **48**, 1254
30. Less, K.J. and Wilson,E.G. (1973) *J. Phys. C* **6**, 3110
31. Reimer,L. (1984) *Ultramicroscopy* **14**, 291
32. Gutmann, F. and Lyons,L.E. (1967) *Organic Semiconductors* Wiley, New York
33. Peterson,I.R., Russell,G.J., Earls,J.D. and Girling,I.R. (1988) *Thin Solid Films* **161**, 325
34. Böhm,C., Steitz, R. and Riegler,H. (1989) *Thin Solid Films*, **178**, 511
35. Chen, H.S. and Goldstein,M. (1972) *J. Appl. Phys.* **43**, 1642
36. Peterson, I.R. and Russell, G.J. (1985) *Thin Solid Films* **134**, 143
37. M.H.Cohen and G.S.Grest (1979) *Phys. Rev. B* **20**, 1077
38. Mason, J.H. (1981) *IEE Proc* **128A**, 193
39. Vijh, A.K. and Crine, J.P. (1989) *J. Appl. Phys.* **65**, 398
40. Ueno, N., Gädeke, W., Koch,E.E., Engelhardt,R., Dudde,R., Laxhuber,L. and Möhwald, H. (1985) *J. Mol. Electron.* **1**, 19
41. Tanaka, T. (1973) *J. Appl. Phys.* **44**, 2430
42. Markiewicz A. and Fleming, R.J. (1988) *J. Phys. D* **21**, 349
43. Lang, D.V. (1986),in *Deep Centres in Semiconductors*, Chap. 7, Ed. S.T.Pantelides, Gordon and Breach, New York

DISCUSSION SESSION

The discussion was led by speakers representing different facets of
the field of electron crystallography.Their brief was to speculate on
the direction of future work and what difficulties may be encountered.
The session was tape recorded but the transcription from the tape was
edited.Thus I apologise for any errors or omissions arising from the
                    John Fryer.

Chairman.D.L.Dorset.

THEORETICAL STRUCTURE PREDICTION

A.Gavezzotti
The objective of theoretical crystallographic studies is to predict the
structure of organised matter-not necessarily crystalline-by modelling.
 The difficult step is to generate geometric structures since it is
possible to calculate potentials.To obtain an organised structure
starting with a molecular structure requires a medium sized computer
and a crystallographer capable of making interactive judgements.
 It is possible to achieve an ab initio prediction of a three
dimensional crystal structure for molecules containing 30-40 carbon
atoms.The sucess rate for this size of molecule is approximately
60%.The minimum requirements are adequate computing and a good graduate
student who can carry out the calculation and interpretation of one
space group per day,thus for the five most common space groups which
are adopted by hydrocarbons one could obtain a correct crystal
structure in one week in 90% of the cases..Polar molecules require
better potential values,but the limiting factor is the speed with which
the human interpretation is done.

Discussion:
D.Martin.To comment on the organisation of matter,it maybe of value to
consider other physical systems as analogies for the calculation of
geometry and energetics although taking place on a different
scale.Examples are block co-polymers,polystyrene spheres and silica
droplets.
S.Hovmoller.When you said that you had 60% success in prediction of
structure for a range of molecules do you mean the space group and unit
cell parameters?To what degree of accuracy do you achieve this?
A.Gavezzotti.Success means taking a test molecule and calculating the
crystal structure from the molecular shape.to achieve a unit cell
within 5% of the experimentally determined values.There is insufficient
data for a full assessment of accuracy.

J. R. Fryer and D. L. Dorset (eds.), Electron Crystallography of Organic Molecules, 377–383.
© 1990 *Kluwer Academic Publishers.*

J.Fryer.Success presumably means a minimum energy structure,but is not necessarily the structure found experimentally,which depends on crystallisation environment.An example of this is starch shown earlier.
A.Gavezzotti.I agree.
S.Perez.We cannot solve crystal structures just by knowing the numbers of atoms in the unit cell.We are not God nor are we dumb,so with five or six possible structural answers we must compare these with the experimental information.To take a pragmatic approach we must put constraints on the speculation and get information by other means-NMR and other spectroscopic techniques-to help interpretation.
D.Dorset.Electron diffraction may be of use here.
J.Lando.Particularly electron diffraction can be used as a control against gross error-especially in the future as techniques and instruments improve,even if humans don't°
R.Scaringe.The object of the theoretical side is to do ab initio predictions-this is not possible at present.The theoretical work produces a number of answers,e.g.because of polymorphism.It is difficult to assign probabilities to these answers at present but it should improve.The minimum requirements for an answer are unit cell constants and space group.The achievement of this is close for small molecules,polymers are a little further away and proteins some distance away at present.

SAMPLE PREPARATION and SURFACE INTERACTIONS

B.Lotz
What is the future for the determination of surface structure?The STM has not yet reached atomic resolution on organic specimens,the best being 5A on calcium arachnidate.The difficulty with scanning techniques is that the force between the sensor and the sample must be less than the forces determining the surface structure.TEM uses stereo-techniques 3D reconstruction,decoration,shadowing and staining methods.
 Surfaces are interesting because they may help us to orient the sample.Large lattice spacings can assist in setting specific tilt angles,and an oriented crystal confers a specific tilt angle.Thus it is necessary to adapt the growth substrate to the material to be oriented.We have a good range of substrates suitable for polymers.The objective is to use the substrate to limit the number of orientations involved.The development from this is to modify the structure via surface interactions,to provide substrates which produce specific polymorphs.An example of this is polydiacetylene which can be oriented to optimise non-linear optical properties.

Discussion:
R.Scaringe.Do you have any ideas on technological applications of these growth techniques.
B.Lotz.Jean Wittmann and myself have been interested in the epitaxy of polymers.This effect disappears rapidly with thickness and if there is an inherent tendancy to disorder,one cannot grow more than a very thin oriented film.An application which we have studied is the growth of oriented polyvinylidene fluoride in the ferroelectric modification.

S.Hui.For many materials,as soon as the substrate is removed the material will start to deform.Is there any way to use epitaxy to hold crystals and not change long range interactions?

B.Lotz.We have not addressed this question so far.As I said, if there is a tendancy to disorder,epitaxy is lost after a few molecular layers from the substrate.For example if polyethylene is crystallised on a substrate which produces an unstable modification,then epitaxy is lost after 30-40A.That is within an EM sample thickness.

These techniques can be used for materials which are difficult to crystallise,or to orient,or have a high density of nucleation. It is advantageous if the sample consists of small crystals all in the same orientation.

J.Wittmann.Large crystals of organic acid salts are readily available and have been used to make oriented films of polydiacetylenes.It is also of importanace to note the work on controlled poisoning of crystal faces done by Addadi,Leserowitz and Lehav. This looks very promising.

J.Lando.Thicker oriented samples can be obtained by epitaxial polmerisation.The monomer is ordered and then polymerised up to 10um in thickness whilst retaining orientation. This has been done for diacetylenes and SN_x.

D.Dorset.A suggestion was made by A.MacPherson that protein crystals could be oriented.Can you comment on this?

B.Lotz.We have only experience with fibrous proteins such as silk,and these have been epitaxially crystallised. In general proteins require much larger periodicities of the substrate. The ratio of the interactions between the substrate and the crystal, compared to those between the crystal and the crystal will be quite small, and this will make epitaxy difficult.

J.Fryer.Could a structure etched on a semiconductor make a substrate for protein oriented growth.

B.Lotz.I have been very impressed by recent electron lithographic results which can etch gratings down to 100A separation.This is on the scale of globular proteins.

W.Baumeister.In particular such a substrate might help to bring protein molecules into a desired orientation.

B.Lotz.What size of crystal is required for protein studies?

W.Baumeister.The substrate crystal should be about 0.5mm in diameter and would need a grating of 2-3um.

QUANTITATIVE TECHNIQUES FOR ELECTRON CRYSTALLOGRAPHY

F.Brisse.

The intensities of electron diffraction reflections from small molecules can be obtained,and three dimensional information from tilting experiments or epitaxial growth.The problems such as bending and dynamical effects are largely understood,and therefore not insuperable.Hence direct methods of phase determination can be applied.

Thus for a few crystals it has been possible to obtain structure factors and the structure has been solved by a combination of dierect methods,reconstruction,packing analysis and high resolution imaging. Nothing has been said about structure refinement because of the limited

data available,but such programmes as LALS-which apply some constraints as mentioned by Serge Perez earlier-should be applicable.Also the use of comparative information availble in data banks is a good idea.
For the future we would like to get three dimensional intensities much faster than one plane at a time,and put structure determination on a routine basis with an analogous program to MULTAN using electron diffraction intensities.
There remain some questions unanswered.Why are some structures obtained to near atomic resolution despite the very limited amount of data available?Non-synthetic chiral polymers often show a much higher symmetry in their diffraction patterns than expected,for example 6mm. How is this to be interpreted?Is it twinning?There is at present no answer to this.

Discussion:
D.Dorset.With reference to the routine basis of structure determination I do not think that X-ray crystallography will be be supplanted by electron crystallography.We are concerned with disordered materials and those which can only be crystallised with difficulty.
S.Perez.There are two problems.They are the symmetry of the base plane and the validity of three dimensional intensities.The intensities are suspect as radiation damage does not affect them all equally,and some recent work from Chanzy's laboratory shows that this is particularly true for three dimensional intensities.Therefore it is not possible to extrepolate back to zero time.
D.Dorset.Dynamical scattering will also alter intensity ratios between zones.
F.Brisse.The damage effects can be minimised by rapid digital recording of intensities so that decay can be extrepolated for each reflection pair.
D.Dorset.The use of a new high voltage microscope as described by Professor Uyeda would also be advantageous.

PROSPECTS FOR PHASING

F.H.Li
It is best to combine the techniques of electron crystallography with those of X-ray crystallography.Different approaches I have already described but a disadvantage is the lack of three dimensional information and the limited number of reflections. This disadvantage can be reduced by high resolution imaging which will provide pahse information for some of the reflections.
Dynamical effects can be overcome by the use of very thin crystals.

IMAGING
HIGH RESOLUTION ELECTRON MICROSCOPY OF LINEAR POLYMERS

J-F.Revol
There are two catagories of people who are concerned with high resolution imaging.They are those who work with beam sensitive specimens and those who do not.The latter are mainly concerned with

technological developments in instrumental resolution,whereas the damage people are working to improve the stability of their samples,and methods by which information can be retrieved prior to severe specimen degradation.

There is much in common between those who study small organic molecules and those who study proteins,but the biggest difference is that at 7A resolution there is usually sufficient structural information to build a molecular model of the protein,whereas for small molecules it is necessary to get to 4A before any structural information is obtained and 2A is desirable-although not possible at present.A problem with proteins which is less evident for small molecules,is that the molecules are so large that the high resolution information is masked by the noise level caused by inelastic scattering. Therefore the use of an electron energy loss spectrometer to filter off the inelastically scattered electrons,may prove beneficial for the collection of electron diffraction data.

There are two approaches to the acquisition of better data from radiation sensitive specimens.

1.High voltage microscopy.Up to now the reduction in damage from the higher voltage has been matched by the loss of sensitivity of the photographic recording medium.This should no longer be the case for the electron image plates described by Professor Uyeda, which are independent of the incident electron voltage.

2.Cryoprotection.A difficulty here has been the limited access available to high stability cryomicroscopes.The separate cryostages are not normally of the stability required for high resolution imaging.Thus there is a need for such high stability cooling stages to be made.

Discussion:

D.Dorset.The access to the cryomicroscope in Berlin is not as difficult as you make out.Arrangements to study specimens in that microscope can be easily made.

J.Fryer.I would like to make some observations on the use and choice of high voltage instruments.I have used the Cambridge HRFM and the JEOL4000EX.Both have excellent lens characteristics,but the achievement of the theoretical resolution in these very large machines has proved very time consuming.A major factor is mechanical instability associated with the large column of the microscope.In particular acoustic vibrations can badly affect resolution.It is for this reason that I prefer the 200keV microscope with low spherical aberration,even though the defocus'window' for optimum resolution is not as large.The precautions necessary to obtain high resolution rise exponentially as the resolution is improved.

D.Dorset.The other point you mentioned was that of cryoprotection.The extent to which this of value depends upon whether liquid helium or liquid nitrogen is employed.There is some controversy about the precise values of protection achieved by cryoprotection.I would like to ask Professor Zeitler if he would comment on this.

E.Zeitler.I do not quote values achieved by our cryomicroscope in Berlin.The 'protection factors'are just a number game and it is asinine to think that they have any scientific validity.What is important is the cryoprotection afforded to the particular specimen under investigation. The extent of protection will vary from one specimen to another, and one is soley concerned as to whether the protection is sufficient to obtain the desired result.

DISORDERED MATERIALS

I.Voigt-Martin
We are concerned with liquid crystals of which there is very little known structurally at the molecular level.The textures at the light microscope level are well characterised,but electron microscope studies have been largely done by Steve Hui and ourselves.
  The polymeric liquid crystals are of interest for their good tensile properties,non-lineral optical,ferro-electrical and piezo-electrical characteristics.At Mainz there there is a history of interest in the synthesis and characterisation of these materials,thus it is possible to achieve close collaboration between synthetic work and all aspects of characterisation.
  Our stategy is to compare the differences between the liquid crystal and the perfect crystalline state.The static structural information is obtained by X-ray,electron or neutron scattering,and the dynamic information from light scattering.The perfect crystalline data normally is obtained by X-ray analysis,and this structure is compared to the liquid crystal studied in the electron microscope.Neutron scattering is used to determine chain conformation between smectic layers.
  This cooperation between different areas of expertise is important to our approach.We have found that it is essential that we study the same sample which is used for physical characterisation.It would seem likely that the many subdivisions of liquid crystal which are quoted are not structurally valid.

Discussion:
D.Dorset.In your work on liquid crystals described earlier you presented some excellent line shape analyses.I would like to comment that the continuous diffuse scattering also contains information of value.This aspect is not generally studied but I would recommend Guinier's book on this subject.

PROSPECTS FOR THE FUTURE IN PROTEINS

W.Baumeister
Electron microscopists have been claiming for many years that it was possible to obtain more information from a protein using electrons,than X-rays.This was shown by Unwin and Henderson in 1975.It was not that the individual methods they used were new,but that they had the courage to apply the combined techniques of low dose and image averaging to retrieve structural information from below the noise level.
  Since that time the progress has been embarassingly slow as technological developments and improvements in techniques were made. However,the presentations at this meeting by Fritz Zemlin and Bing Jap

show that all the basic techniques are now available.There remains some lesser problems on specimen preparation.

We need two dimensional crystals and there is a need to invest more effort to study the crystallisation of proteins at this level,and to place the subject on a rational basis.It is necessary to distinguish between membrane and soluble proteins.Membrane proteins are well suited to electron microscopy.It is their purification which is probably the most crucial aspect of their preparation.

Soluble and hydrophilic proteins have a less well defined approach.The most promising appears to be the use of monolayer techniques.This was first suggested by Fromhaus over 20years ago,but is recently undergoing further evaluation.It is a simple method which could be done in various elecgant ways,such as using lipids with appropriate functional ligands to confer an orientation of the water interface.Once in this orientation the surface mobility would help the proteins cone into suitable close contact for crystallisation.It would also be possible to use antibodies or FAB fragments as a matrix for crystallisation.It will also be necessary to develope techniques to monitor the crystallisation these will probably involve light scattering.However,the most important factor remains the purity of the protein.

As far as microscopy is concerned a possible bottleneck is that there is a lot of trial and error with a low success rate.The microscope should be able to combine many imaging modes such as spot scan,dynamic focussing and low dose.This is very tedious and prone to error when done manually.For fast exposure field emission guns are desirable and there should be real time image processing.I would repeat the comments of an earlier speaker in saying that there is real need for cryostages with low drift and vibration.

Image analysis is the most advanced aspect.Image distortions require correction,but if the molecular positions can be located then these distortions can be corrected.
D.Dorset.What aspects do we need to understand for specimen preparation?
W.Baumeister.We do not understand the interactions between the protein and the support film on the grid.Ultimately we will probably all work with ice embedded specimens.
S.Hui.To what extent would environmental stages be of use ?
J.Turner.Such stages are unlikely to be suitable for high resolution imaging because of the scattering in the gas and drift.
G.Duckett.For inorganic specimens it has proved possible to obtain better than 3A under a low gas pressure.
D.Dorset.For hydrated specimens a good test is to use catalase,where the high resolution diffraction information is rapidly lost with dehydration and radiation damage.

END.

INDEX